AF323739

Advances In Environmental Fluid Mechanics

Advances In Environmental Fluid Mechanics

Dragutin T Mihailovic
University of Novi Sad, Serbia

Carlo Gualtieri
University of Napoli Federico II, Italy

Editors

World Scientific

NEW JERSEY · LONDON · SINGAPORE · BEIJING · SHANGHAI · HONG KONG · TAIPEI · CHENNAI

Published by

World Scientific Publishing Co. Pte. Ltd.

5 Toh Tuck Link, Singapore 596224

USA office: 27 Warren Street, Suite 401-402, Hackensack, NJ 07601

UK office: 57 Shelton Street, Covent Garden, London WC2H 9HE

British Library Cataloguing-in-Publication Data
A catalogue record for this book is available from the British Library.

ADVANCES IN ENVIRONMENTAL FLUID MECHANICS

ISBN-13 978-981-4291-99-6
ISBN-10 981-4291-99-4

Printed by FuIsland Offset Printing (S) Pte Ltd, Singapore

To Ljuba Nedeljković, physist from Belgrade.

Dragutin T. Mihailovic

To Susanna and Federico. To my parents.

Carlo Gualtieri

Preface

We live in a changing world. This thought expresses a historical fact entirely present in man's consciousness regardless of whether we consider him as an individual or a social human being. The awareness about the changing world always introduces an amount of anxiety into man's life that can be defined by the following question: whether man can preserve the existing world for the future of his children. To preserve this world for the future of our children, we must all strive for sustainable development - a development that meets the needs of the present generation without compromising options and resources the future generations will use to meet their own needs. It implies environmentally sound development in societies and regions free from threats to life and property. Human security is an essential ingredient of sustainability, which is increasingly threatened by extreme events, both natural and human-induced.

These sentences give insight into the question: Why is there now a focus on environmental problems ? One particular answer can be found in a hierarchy of the main scientific problems for the 21st century as seen by the community mostly consisting of physicists, biologists, chemists, engineers and human science communities. According to them, in the 21st century the scientific community will be occupied by the problems linked to environmental ones that are primarily expressed through problem of climate changes and related issues. A unique characteristic of these problems is their close connection with the questions of the survival of individual human beings on the Earth. This is the first time in the history of science that environmental problems take their place at the front of science – from fundamentals to applications. This science could be defined from the point of view of many single sciences. However, the definition given in Random House Dictionary as *"the branch of science concerned with the physical, chemical, and biological conditions of the environment and their effect on organisms"* can be accepted as the broadest one, because it implicitly includes its main feature – its interdisciplinary nature. The field of environmental science is abundant with various interfaces and is the right place for the application of new fundamental approaches leading towards a better understanding of environmental phenomena. Therefore, we will slightly evolve the above definition, by setting the focus on the concept of the *environmental interface*, defined as an interface between two either abiotic or biotic environments which are in relative motion exchanging energy and substances through physical, biological and chemical processes and fluctuating temporally and spatially. In that sense we define environmental science as *"the branch of science concerned with the interactions in environmental interfaces regarded as the natural complex systems"* that will be exploited in this book. Environmental science encompasses issues such as climate change, biodiversity, water quality, ground water contamination, use of natural resources, waste management, sustainable development, disaster reduction, air pollution, and noise pollution. It is built from many sciences including environmental fluid mechanics as one of the core components.

Environmental fluid mechanics (EFM) is the scientific study of transport, dispersion and transformation processes in natural fluid flows on our planet Earth, from the microscale to the planetary scale. Stratification and turbulence are two essential ingredients of environmental fluid mechanics. Stratification occurs when the density of the fluid varies spatially, as in a sea breeze where masses of warm and cold air lie next to each other or in an estuary where fresh river water flows over saline seawater. Turbulence is the term used to characterize the complex, seemingly random motions that continually result from instabilities in fluid flows. Turbulence is ubiquitous in natural fluid flows because of the large scales that these flows typically occupy. The processes studied by environmental fluid mechanics greatly affect the quality of natural ecosystems and are largely investigated using modeling techniques and software packages.

The objective of this book is to bring together scientists and engineers working in research institutions, universities and academia, who engage in the study of theoretical, modelling, measuring and software aspects in environmental fluid mechanics. It is intended to provide a forum for the participants to obtain up-to-date information, and exchange new ideas and expertise through the presentations of up-to date and recent overall achievements in this field. In this regard, a mixing of professionals from a wide spectrum of different areas is promoted focusing on the EFM theoretical, modeling and experimental approaches, including the issues of software design and measurements.

The book is organized in two broad parts: Theoretical and Modeling Aspects (Part One) and Applicative, Software and Experimental Issues (Part Two). Part One has seven chapters covering various theoretical and modeling aspects of EFM, such as turbulent dispersion, stable atmospheric boundary layers, calculation of albedo on complex geometries, stratified turbulent flows, viscous flow induced by wave propagation over bed ripples, aquatic turbulent flow over rough boundaries and exchange processes in environmental interfaces regarded as biophysical complex systems. Part Two addresses in eight chapters a broad range of EFM applicative issues, such as source identification in atmospheric pollution, fate and transport of mercury in aquatic systems, exchange processes in a river with dead zones, flow resistance in vegetated channels, contaminant intrusion in a water supply network, ecological modeling of coastal waters, integration of spatio-temporal data for fluid modeling in a Geographical Information System (GIS) environment and long-term measurements of energy budget components and trace gas fluxes between the atmosphere and different types of ecosystem.

Part One starts with a chapter by N. Mole, P.C. Chatwin and P.J. Sullivan, that deals with the dispersion of contaminants in turbulent flows at high Péclet number, where both turbulent advection and molecular diffusion were considered to derive exact results for moments and the probability density function (pdf) of concentration for the hypothetical case of zero diffusivity. Also, the results for the moments were modified to account for the presence of slowly acting diffusion. Finally, the authors demonstrated that many of the results obtained with this approach agree well with experimental observations. Following this, the chapter by D.T. Mihailovic and I. Balaz investigates an aspect of dynamics of energy flow based on the energy balance equation. Also, two illustrative issues important for the modelling of interacting environmental interfaces regarded as complex systems were addressed. These were the use of algebra for modelling the autonomous establishment of local hierarchies in biophysical systems and the numerical investigation of coupled maps representing exchange of energy, chemical and other relevant biophysical quantities between biophysical entities in their surrounding environment. The next chapter, by B. Grisogono, is devoted to the atmospheric boundary layer (ABL), which is the lowest part of the atmosphere that is continuously under the influence of the underlying surfaces through mechanical (roughness and shear) and thermal effects (cooling and warming), and the overlying, more free layers. Its proper characterization and modeling is very important in both short- or medium-range weather forecasting and applied micrometeorological studies. The author investigated an inclined strongly stratified stable ABL proposing an improved *z-less* mixing length scale model and a new generalized *z-less* mixing length-scale model. In Chapter 4, V. Fuka, J. Brechler and A. Jirk studied the impact of vertical temperature stratification on a 2D laminar flow structure, modeled via the Boussinesq approximation and by varying the Froude number (Fr). Later, they reviewed several approaches in treating turbulence in modeling studies, with an emphasis on an implicit large-eddy simulation. The results of Taylor–Green vortex computations, performed by using this method, were compared in terms of kinetic energy dissipation rate, probability density functions of turbulent fluctuations, and 3D energy spectra with the results of a direct numerical simulation at moderate Reynolds numbers. The chapter by R.J. Schindler and J.D. Ackerman deals with turbulent flow over rough boundaries, which is a common appearance in nature and the subject of high interest in a range of disciplines. Since recent technological advances have increased our ability to model, visualize and measure flow structure over complex boundaries, the authors reviewed the experimental evidence for rough boundary/flow interactions across different disciplines, pointing out that different approaches have led to the adoption of a variety of parameters that are used to describe boundary roughness, and different criteria were used to evaluate the relative effects of boundary roughness. Furthermore, they stressed that much of the experimental data relates to idealized surfaces or is taken in low Reynolds number flows, and cannot be generally applied to aquatic flows over natural surfaces. In the chapter following this, A.A. Dimas and G.A. Kolokythas present the results of numerical simulations of the free-surface flow, developing by the propagation of nonlinear water waves over a rippled bottom. The flow was assumed to be two-dimensional, incompressible and viscous and the simulations were based on the numerical solution of the Navier-Stokes equations subject to the fully-nonlinear free-surface boundary conditions and appropriate bottom, inflow and outflow boundary conditions. Results pointed out that over the rippled bed, the wave boundary layer thickness increased significantly, in comparison to the one over a flat bed, due to flow separation at the ripple crests, which generates alternating circulation regions. Second, the effect of ripple height and Reynolds number on the distribution of wall shear stress and drag forces was investigated. Part 1 is closed with the chapter by D. Kapor, A. Ćirišan and D.T. Mihailovic that presents a general approach for the calculation of aggregated albedo in complex geometries, where analytic solution is not available. Thus, an efficient numerical procedure, the so-called *ray-tracing Monte Carlo* approach, was developed and tested for known analytical solutions. Later, this method was applied to a two-patch grid cell with a square geometrical distribution and with different heights of its parts. Comparison with the results from conventional approaches pointed out remarkable changes in values of sensible and latent heat fluxes, as well as the corresponding surface temperature.

Part Two begins with the chapter by B. Rajković, M. Vujadinović and Z. Gršić, in which are proposed two methods for locating a possible source of air pollution that combine measurements and inverse modeling based on Bayesian statistics. In both methods a puff model was used to generate the pollutant concentration field and the synthetic observations at predefined measuring points using real meteorological data. In the first approach, the position of the possible source was found with an iterative process, defined as the maximum of probability density function from an ensemble of possible sources. The second approach used a library of records and scenarios, with combinations of values for the meteorological and emission parameters in the problem. The next chapter, by T. Weidinger, L. Horváth and Z. Nagy, presents the status of micrometeorological research activity in Hungary, field measurement programs and instrumentation and flux calculation methodology. Later, the chapter illustrates the development of the new Hungarian basic climatological network for the detection of the possible effects of future climate change, which consists of the standard climate station measurements, soil, radiation, energy budget and CO_2 flux measurements. The chapter by L. Matejicek presents a case study focused on dust dispersion above a surface coal mine which demonstrated the ability of GIS methods to manage, pre-process, post-process and to visualize all the pertinent data. As remote sensing helped to identify and classify the coal mine surface in order to map erosion sites and other surface objects, Global Positioning System (GPS) was used to improve the accuracy of the erosion site boundaries and to locate other point emission sources such as excavators, storage sites, and line emission sources such as conveyors and roads. Wind flow and dust dispersion modeling was successfully integrated in the GIS environment. The chapter by P.A. López-Jiménez, J.J. Mora-Rodríguez, V.S. Fuertes-Miquel and F.J. Martínez-Solano studied pathogen intrusion in water distribution systems by using both experimental and numerical methods. Pathogen intrusion occurs when negative pressure conditions are achieved in the systems, allowing the entrance of water around a leak, causing a problem of water quality. First, laboratory experimental works were carried out. Experimental results were successfully compared with numerical results obtained with Computational Fluid Dynamics (CFD) methods. Issues about the calibration process are also discussed. The chapter by C. Gualtieri deals with the 3D steady-state and time-variable numerical simulations of the flow in a rectangular channel with a lateral square cavity representing a dead zone of a river. This geometry was previously experimentally studied. The exchange coefficient between the main flow and the dead zones was calculated both from the transverse velocity data along the dead zone-main channel interface and from the temporal decay of the concentration of a tracer that was homogeneously injected in the dead zone. Analysis of the flow field demonstrated that the numerical simulations qualitatively reproduced the observed flow patterns but underestimated the exchange rate between the dead zone and the main stream, whereas the concentration data were in better agreement with the experimental data. The chapter by A. Massoudieh, D. Zagar, P.G. Green, C. Cabrera-Toledo, M. Horvat, T.R. Ginn, T. Barkouki, T. Weathers and F.A. Bombardelli covers fate and transport of mercury in aquatic systems. First the chapter identifies the various processes that are potentially important in mercury fate and transport as well as knowledge and uncertainty about these processes. Second, an integrated multi-component reactive transport modeling approach was proposed to capture several of the processes. This integrated modeling framework includes the coupled advective-dispersive transport of mercury species in the water body, both in dissolved phase and as associated to mobile suspended sediments. Some results for the application of the model to the Colusa Basin Drain in California were also presented. The chapter by J.F. Lopes and A.C. Cardoso presents the results of a 3D numerical study of the distribution of temperature and phytoplankton biomass in the near-shore Aveiro coastal zone (Portugal). The study area is located on the western coast of the Iberian Peninsula, characterized by meteorological conditions of strong north/northwest prevailing winds, which favour the upwelling of nutrient enriched waters resulting from the divergences associated with the Ekman transport and, therefore, generating high nutrient availability. The results showed that the model was able to reproduce the horizontal and vertical temperature and chlorophyll-a (*Chl-a*) patterns. Also, they demonstrated the setup of a layer of cold water along the coastal side and the increasing of the declivity of the thermocline and the nutricline toward the coast. The model successfully predicted the values for the maximum *Chl-a* concentration and the depth of the inshore subsurface chlorophyll maximum. The last chapter by S. De Felice, P. Gualtieri and G. Pulci Doria deals with the calibration of a new simplified experimental method to evaluate absolute roughness of vegetated channels. The method is based on boundary layer measurements in a short channel rather than on uniform flow measurements, as usual. The proposed method can be applied to any kind of rough bed, but it is particularly useful in vegetated beds where long channels are difficult to prepare. The results are successfully compared with literature data demonstrating that the proposed method can provide a reliable prediction of absolute roughness in vegetated channels.

The Editors wish to thank all the chapter authors for their continuous and dedicated effort that made possible the realization of this book. The Editors are also grateful to the anonymous reviewers of the project for their thoughtful and detailed suggestions that have improved both the contents and presentation of this book. The Editors finally acknowledge with gratitude the assistance of the Editorial Office of World Scientific and, especially, of Dr.Elena Nash.

Dragutin T. Mihailovic
Carlo Gualtieri

Biographies of Editors and Contributors

Editors

Carlo Gualtieri is currently Assistant Professor in Environmental Hydraulics at the Hydraulic, Geotechnical and Environmental Engineering Department of the University of Napoli *Federico II*, Italy. He received a B.Sc. in Hydraulic Engineering at the University of Napoli *Federico II*, where he received also a M.Sc. in Environmental Engineering. He finally received a Ph.D. in Environmental Engineering at the same University. He has 88 peer-reviewed scientific papers, including 19 publications in scientific journals, 4 book chapters, 44 publications in conference proceedings, and 21 other refereed publications in subjects related to environmental hydraulics and computational environmental fluid mechanics, with over 60 papers, experimental investigations of two-phase flows, water supply networks management and environmental risk. He is also co-author of 2 textbooks on Hydraulics and author of a textbook on Environmental Hydraulics. He is co-editor of the book *Fluid Mechanics of Environmental Interfaces* published by Taylor & Francis in 2008. He is member of the Editorial Board of the journals *Environmental Modelling and Software (Elsevier)* and *Environmental Fluid Mechanics (Springer)*. He contributed as manuscript editor for *Environmental Modelling and Software*, as reviewer in several scientific journals (e.g. *Environmental Fluid Mechanics, Environmental Modeling and Software, Journal of Environmental Engineering ASCE, Journal of Hydraulic Engineering ASCE, Journal of Hydraulic Research IAHR, Advances in Water Resources, Experiments in Fluids, Ecological Modelling, Environmental Science & Technology, Journal of Coastal Research*) and as external examiner for Ph.D thesis in foreign countries. He co-organized sessions dealing with environmental fluid mechanics in international conferences, such as *iEMSs 2004, iEMSs 2006, EMS 2007, iEMSs 2008* and *EGU 2009*. He is also active as expert reviewer for research funding agencies in several countries. He is member of the International Scientific Advisory Board of the EU project *RRP-CMEP*, of the *American Society of Civil Engineers* (ASCE), of the *International Association of Hydraulic Engineering & Research* (IAHR) and of the *International Environmental Modelling & Software Society (iEMSs)*.

Dragutin T. Mihailovic is the Professor of the Meteorology and Biophysics at the Department of the Vegetable and Crops, Faculty of Agriculture, University of Novi Sad, Serbia. He is also the Professor of the Modelling Physical Processes at the Department of Physics, Faculty of Sciences at the same university and the Visiting Professor at the State University of New York at Albany. He teaches various theoretical and numerical meteorology courses to Physics and Agriculture students. He received a B.Sc. in Physics at the University of Belgrade, Serbia, his M.Sc. in Dynamics Meteorology at the University of Belgrade and defended his Ph.D. Thesis in Dynamics Meteorology at the University of Belgrade. He is head of the Center for Meteorology and Environmental Modelling (CMEM) which is the part of the Association of Centers for Multidisciplinary and Interdisciplinary Studies (ACIMSI) of the University of Novi Sad where he has teaching activities. His main research interests are the surface processes and boundary layer meteorology with application to air pollution modelling and agriculture. Recently, he has developed an interest for (i) analysis of occurrence of the deterministic chaos at environmental interfaces and (ii) modelling the complex biophysical systems using the category theory.

Contributors

Josef Ackerman is the Associate Dean of the Faculty of Environmental Sciences at the University of Guelph, Canada, where he administers an interdisciplinary faculty that serves all academic units at the university. He is an Associate Professor in the Department of Integrative Biology where he conducts research on the physical ecology of aquatic plants and animals, as well as environmental issues. Most of this research is focused on small-scale fluid dynamic and ecological processes. He holds adjunct faculty positions in the School of Engineering at Guelph and the Environmental Science & Engineering Programs at the University of Northern British Columbia. Before coming to Guelph, he was a faculty member at the UNBC, where he played a leading role in founding the university's environmental science and environmental engineering programs and held the Canada Research Chair in *Physical Ecology and Aquatic Science*. He is currently an Associate Editor of the journals *Limnology & Oceanography* and *Aquatic Sciences*.

Igor Balaz is the PhD student at the Center for Meteorology and Environmental Modelling (CMEM), Novi Sad, Serbia. He received a B.Sc. in Biology at the

University of Novi Sad and M.Sc. in Microbiology at the same University. His main research interests are (i) modelling of complex biological systems with special emphasize on their evolution and self-organization, and (ii) artificial life.

Tammer Barkouki is currently a graduate student at the Department of Civil and Environmental Engineering at the University of California, Davis, USA. He obtained a Bachelor's degree in Civil and Environmental Engineering from UC Davis in 2008. His work focuses on multicomponent reactive transport applied to uraninite reoxidation, ureolytic-induced calcite precipitation and strontium co-precipitation, and mercury fate and transport in estuaries.

Fabián Bombardelli graduated from the University of Illinois at Urbana-Champaign, USA, with a Ph.D. in Civil and Environmental Engineering, in May 2004. He is now Assistant Professor at the Department of Civil and Environmental Engineering at the University of California, Davis, USA. Before coming to the United States, Dr. Bombardelli obtained in 1999 a Master's degree in *Numerical Simulation and Control* at the University of Buenos Aires, Argentina, being the first student to be awarded this degree. Dr. Bombardelli's research focuses on theoretical and numerical aspects of turbulence in multiphase flow dynamics, including bubble plumes, density currents, sediment transport, hydraulic jumps, and pollutant transport, speciation and fate. The research program involves the development and use of CFD and computational-hydraulics techniques to address problems belonging to the field of environmental fluid mechanics.

Josef Brechler is currently a Associate Professor of Meteorology at the Department of Meteorology and Environment Protection, Faculty of Mathematics and Physics, Charles University in Prague, Czech Republic. He is currently head of this department. He specializes on the problems of air flow in the atmospheric boundary layer and the issue of air quality problems. He is a member of the Czech meteorological society and European meteorological society, he is a member of the international committee of the European Association for the Science of Air Pollution (EURASAP).

Ana C. Cardoso was born in Oporto, Portugal in May 1977. She graduated in Physics – Meteorology and Oceanography in 2003, from University of Aveiro, Portugal, received a PhD in Physics in 2009, from the same University. During her PhD she was student member of CESAM (Center for Environmental and Marine Studies). Her research activities focused on modelling processes in estuarine and coastal regions with emphasis on hydrodynamics and biogeochemical processes. She is currently working on environmental problems covering local and regional scale of Portugal coastal ecosystems: (i) ecological modelling in estuarine and costal systems; (ii) water quality processes in estuarine systems. She has 3 peer-reviewed scientific papers and 6 publications in conference proceedings, on subjects related to computational environmental fluid mechanics

Philip Chatwin has been Emeritus Professor of Applied Mathematics at the University of Sheffield, United Kingdom, since 2005. He was a student in DAMTP at the University of Cambridge (BA and Part III from 1960 to 1964, PhD from 1964 to 67, supervised by George Batchelor), a Royal Society PG Fellow at the University of Grenoble from 1967 to 1968, Lecturer and Senior Lecturer in Applied Mathematics at the University of Liverpool from 1968 to 1985, Professor of Mathematics at Brunel University from 1985 to 1990, and Professor of Applied Mathematics at the University of Sheffield from 1991 to 2005. His early research (until about 1975) was on longitudinal (Taylor) dispersion. His most important research has been on turbulent dispersion (fundamentals and atmospheric applications), much of it in collaboration with Paul Sullivan and Nils Mole.

Ana Cirisan is a postgraduate student of Meteorology and Environmental Modelling at the Center for Interdisciplinary and Multidisciplinary Studies and Developmental Research (ACIMSI) at the University of Novi Sad, Serbia. She is a young researcher - scholarship holder on the project *Modelling and numerical simulations of complex physical sistems* ON141035 funded by Ministry of Science and Environmental Protection. Ana Cirisan is also included on the FP6 project (RRP-CMEP, ON043670) at the Center for Meteorology and Environmental Predictions, Faculty of Sciences, University of Novi Sad. Her main research interest is the numerical weather forecast and modeling of atmospheric processes.

Athanassios A. Dimas is currently Associate Professor in Coastal Hydraulics at the Department of Civil Engineering, University of Patras, Greece. He received his Diploma in Naval Architecture and Marine Engineering at the National

Technical University of Athens. He received his M.Sc. in Ocean Engineering and M.Sc. in Mechanical Engineering both at MIT, where he also received his Ph.D. in Ocean Engineering. He was a post-doctoral research associate at the Levich Institute, City College of New York, and an assistant professor at the Department of the Mechanical Engineering, University of Maryland. He is the author or co-author of 63 refereed scientific articles, including 16 publications in scientific journals, 3 book chapters, 29 publications in international conference proceedings, and 15 publications in national conference proceedings in subjects related to coastal hydraulics, turbulence simulation, flow instability and computational fluid dynamics. He contributed as reviewer in several scientific journals (e.g. *Journal of Fluid Mechanics, Physics of Fluids, Journal of Fluids and Structures, International Journal for Numerical Methods in Fluids, Journal of Biomechanics*). He is active as expert reviewer for research funding in the European Commission. He is member of the *American Society of Mechanical Engineers* (ASME). He was awarded the Offshore Mechanics Prize by the Offshore Mechanics and Polar Engineering Council. He is the co-holder of a U.S. Patent.

Sergio De Felice was graduated with honours in 2004 in Civil Engineering, major Hydraulics, at the University of Napoli *Federico II*, Italy, defending a research thesis about *Interactions between turbulent boundary layer and vegetated surfaces through LDA technique*. He received his PhD in 2009 from the same University defending the thesis *Experimental study on the hydrodynamic characteristics of a vegetated channel*. His publication record holds 11 products, including: 4 textbooks, 1 paper in a national journal, 4 papers in international conferences, and 2 papers in national conferences. His research interests are: tuning and implementation of LDA devices and Labview systems, turbulent flows, boundary layer flows, vegetated beds flows.

Vicente S. Fuertes Miquel is an Industrial Engineer (specialized in Energy) from the Universidad Politécnica de Valencia (UPV), Spain (1992). He received a PhD in Hydraulic Engineering from the Department of Hydraulic and Environmental Engineering, UPV, Spain (2001). He has been a Civil Servant of the Spanish Administration in different categories (since 1994). Currently, he is an Assistant Professor of Fluids Mechanics. His activity in *Centro Multidisciplinar de Modelación de Fluidos* is threefold and falls completely within the fluids, water and environmental fields. He is the author and editor of several books; he has written a number of research papers published in high impact journals and presented many contributions to international events; he is co-author of several commercial computer simulation tools in the water field; he periodically co-organizes international events in his own University; he has participated in numerous R&D projects, and served as project coordinator in some of them that were financed by several official institutions. In terms of consulting, he has carried out frequent design, project, management, audits, and studies in water resources for private and public Companies and Administrations.

Vladimír Fuka is currently a Ph.D. student of meteorology at Department of Meteorology and Environment Protection, Faculty of Mathematics and Physics, Charles University in Prague, Czech Republic. His dissertation topic deals with the airflow modelling in the complex geometry areas with the main stress given to turbulence. He presented results of his work at several international meetings.

Timothy Ginn graduated from Purdue University, USA, with a Ph.D. in Civil Engineering in 1988 and is now Professor in the Faculty of Civil and Environmental Engineering at University of California, Davis, USA. Before joining the UC Davis Faculty in 1996, Dr. Ginn was employed as a senior research scientist at the Pacific Northwest National Laboratory of the U.S. Department of Energy. His research integrates modeling approaches into collaborative efforts targeting multicomponent and multiphase fate and transport, in the contexts of contaminant hydrology, population dynamics, water disinfection, colloid filtration, groundwater age, and inverse problems.

Peter Green is from the University of California, Davis, USA, where has been a researcher in the Department of Civil and Environmental Engineering since 2000. His degrees are in Chemistry from Stanford University and the Massachusetts Institute of Technology, and he works on a variety of water quality and air quality projects, including collaborations with ecologists and clinical researchers. Current research covers organic pollutants (volatile and non-volatile) as well as metals other than mercury. Ongoing projects involve mercury and children's health. A 2006 publication on the microbial methylation of mercury has been widely cited.

Branko Grisogono got his B.S. and M.S. in physics, geophysics with meteorology at the University of Zagreb, Croatia, in 1983 and 1987, respectively. He obtained his Ph.D. from Desert Research Institute, Univ. of Nevada, Reno, USA, in physics, physics of the atmosphere, in 1992. After a short postdoc there, he moved to the Department of Meteorology at Uppsala Univ., Sweden, combining his experience in boundary-layer (BL) physics and wave dynamics through mesoscale modeling and asymptotic methods. He became Assoc. Prof. in 1997 and moved to the Dept. of Meteorology at Stockholm Univ., Sweden, and remained until 2003 working on waves and turbulence in stable boundary layers and downslope windstorms. There he taught a few courses, supervised students, and became a Member of Tellus Advisory Board. He continued working on coastal and mountain meteorology at the Dept. of Geophysics, Univ. of Zagreb, Croatia, at the end of 2003, teaching a few courses and tutoring several Ph.D. students at the Dept. and the Croatian Weather Service. After co-organizing a few conferences and workshops and reviewing for peer-journals and funding agencies, he was a guest coeditor for special issues of *Meteorol. Zeitschrift* and *Boundary-Layer Meteorology* in 2006 and 2007, respectively. He is in Editorial Boards for a couple of other journals and in ICAM Steering Committee. His main contributions relate to airflow over double bell-shaped ridges, inclusion of the WKB method in geophysical BL studies, Coriolis effects on sloped BLs and behind orographic-wave breaking, mixing length-scale over sloped terrain, etc.

Zoran Grsic is a research fellow and the head of the Meteorological Group of Radiation and Environmental Protection Department in the Institute of Nuclear Sciences *Vinca*, Belgrade, Serbia. He received a B.Sc and M.Sc in meteorology at the Department of Meteorology, Faculty of Physics, Belgrade University of Belgrade. His Professional interests are in air pollution modeling, air quality monitoring networks, automated systems for real time air pollution distribution assessment and determining zones of influence of chemical and radiation sources. As an expert in meteorological aspects of air pollution, he participated the international project: *Council of Economic Advisers (CEA) Meteorological Aspects of Air Pollution* (theme U.4. *Research on atmospheric radioactivity*). At national level he was involved in several projects related to the modelling and monitoring of chemical and radioactive pollution: *Radiation Protection of the Nuclear Reactor RA, Scientific Basis of Environmental Protection, Developing monitoring system in cement industry of Serbia (HTR257)* and *Investigation and development procedures for radioactive and hazard waste disposal and environmental impact assessment (HOI 1985).* Currently he is leading meteorologist in three national projects: *The decommission of the nuclear* *reactor in Vinca, Automated air quality system in the industrial zone of Pancevo* and *Design of continual observing system for the assessment of the influence of thermo-power plant Nikola Tesla*. He has four papers in the international journals and 27 papers presented at the various international conferences. He is a member of Air Protection Society of Serbia, Meteorological Society of Serbia and Balkan Environmental Association.

Paola Gualtieri, graduated in Hydraulic Engineering in 1989 and specialized in Hydraulic and Environmental Engineering in 1991, received her PhD in 1995 at the University of Napoli *Federico II*, Italy. She teaches courses in hydraulics, fluid mechanics, hydraulic measurements and models. Since 2006 she is Associate Professor in Hydraulics at the University of Napoli *Federico II*. She was member of the organizing committee of *Riverflow 2004* conference and is now member of the Board of an engineering PhD School at the University of Napoli *Federico II*. She was reviewer for international conferences, INTAS and Georgia National Science Foundation. Twelve times she was member of national research groups. She produced 53 scientific papers, including 2 textbooks, 3 chapters in international scientific books; 6 papers in international journals, 19 papers in international conferences, 13 papers in national conferences. Her research interests are: hydraulic measurements and models, turbulent flows, boundary layers flow, air entrainment, cavitation, reaeration and volatilization, vegetated open channels flows.

András Zénó Gyöngyösi is currently PhD student of Earth Sciences, Meteorology at the Eötvös Loránd University, Budapest, Hungary. He received MSc in Meteorology in 2000. His Master Thesis was the Development of the NRIPR Mesoscale Meteorological Model at the Eötvös University. He wrote his thesis at the Meteorological Institute at the Uppsala University, Sweden on a scholarship from the European Union in the framework of the ERASMUS program. He is working mainly in projects involved in micrometeorological issues. His research topic besides is environmental modeling.

Milena Horvat is senior researcher and the Head of the Department of Environmental Studies at the Jožef Stefan Institute, Slovenia. She coordinates a number of projects related to mercury research including development of analytical methods, biogeochemical cycling and modeling in the environment, and collaborates in health related studies. She is a coordinator of several projects in China, Portugal, Japan, France, and Croatia. She has published over 120 papers in the peer-reviewed literature and proceedings of international conferences and books, mainly related to research on mercury. She has supervised a number of MS and Ph.D. theses. She served as the chairperson for the organization of the 7th International Conference on Mercury as Global Pollutant, Ljubljana, 2004, and was a member of the steering committee for the 8th ICMGP in Madison, WI, in 2006.

László Horváth is currently principal councilor of the Hungarian Meteorological Service and titular professor of Szent István University, Hungary. He received M.Sc. (chemistry) and dr. univ. degree (meteorology) in Eötvös Loránd University, Hungary, Candidate of Sciences degree and Doctor of Hungarian Academy of Sciences title (earth sciences) at the Hungarian Academy of Sciences and dr. habil. degree in Szent István University. He is author or co-author of 270 publications in the field of atmospheric chemistry. He is a member of Biological and Earth Sciences' PhD School at Szent István University and Eötvös Loránd University, respectively. He has been national leader of 4 terminated and 1 current EU framework projects and several national research projects. His main interest is the exchange of trace materials within the biosphere and atmosphere especially the fluxes of nitrogen and sulfur compounds and greenhouse gases. He is a member of various international and national organizations, societies, actions such as IUPAC (International Union of Pure and Applied Chemistry, national representative), 3 COST (European Cooperation in Science and Technology) actions as MC member (actions 729, ES 0804 and 735), ACCENT (Atmospheric Composition Change, The European Network of Excellence, associated), IAPSO International Association for the Physical Sciences of the Oceans, INI (International Nitrogen Initiative) European Centre, advisory group etc. .

Aleš Jirk is a Ph.D. student of meteorology at the Department Meteorology and Environment Protection, Faculty of Mathematics and Physics, Charles University in Prague, Czech Republic. In his diploma thesis he focused on the problem of stratified flow and he continues in work on this problem in the framework of his dissertation.

Darko Kapor is the Professor of the Theoretical Physics at the Department of Physics, Faculty of Sciences, University of Novi Sad, Serbia. He teaches various

Theoretical and Mathematical Physics courses to all Physics students, a course on Atmospheric radiation to the students of Physics and Meteorology and History of Physics for Physics teachers. He received a B.S. in Physics at the University of Novi Sad, his M.Sc. in Theoretical Physics at the University of Belgrade, Serbia and defended his Ph.D.Thesis in Theoretical Physics at the University of Novi Sad. Along with his teaching activities in Physics, he also teaches at the multidisciplinary studies of the Center for Meteorology and Environmental Modelling (CMEM) which is the part of the Association of Centers for Multidisciplinary and Interdisciplinary Studies (ACIMSI) of the University of Novi Sad. His main research interest is the Theoretical Condensed Matter Physics, where he is the head of the project financed by the Ministry for Science of the Republic of Serbia. During the last 15 years, he has developed an interest in the problems of theoretical meteorology and worked with the Meteorology group at the Faculty of Agriculture and Faculty of Sciences. He is engaged in transferring the concepts and techniques of the theoretical physics to meteorology, for example to the problem of agregated albedo for heterogeneous surfaces and more recently the problem of chaos.

Gerasimos A. Kolokythas is currently a PhD candidate at the Department of Civil Engineering, University of Patras, Greece. He received his Diploma in Civil Engineering at the University of Patras, where he also received his M.Sc. in Civil Engineering, Water Resources and Environmental Engineering. Mr. Kolokythas is the author or co-author of 6 refereed scientific articles in scientific journals, international and national conferences.

José F. Lopes is Professor at the Department of Physics of the University of Aveiro, Portugal, where he currently teaches several undergraduate and graduate courses, namely General Physics, Mechanics, Continuum Media Physics and advanced theoretical and numerical modeling subjects in the area of Oceanography. He holds a B.Sc. on Physics and a M.Sc. on Physics/Energetics (Fluid Mechanics and Heat Transfer) from the University of Aix-Marseille, France, where he also defended his Ph.D thesis in Physics/Energetics. He is member of CESAM (Centre for Environmental and Marine Studies) an interdisciplinary R&D Unit of the University of Aveiro, a government funded Laboratory, with the mission of developing research in the coastal sciences and providing specialized services in the coastal environment. His research activities have focused on modeling processes in estuarine and coastal processes with emphasis on hydrodynamics and transport processes. Presently he is working on environmental problems covering local and regional scale of coastal ecosystems: (i) ecological modeling in estuarine and costal systems integrating the atmosphere, hydrosphere and biosphere; (ii) water quality processes in estuarine systems. He has 84 peer-reviewed scientific papers, including 18 publications in scientific journals, 55 publications in conference proceedings, and he is co-author of 8 textbooks on subjects related to aerodynamics and computational environmental fluid mechanics.

Petra Amparo López Jiménez is currently Associate Professor in Hydraulic Engineering at the Department of Hydraulic and Environmental Engineering of the Universidad Politécnica de Valencia (UPV), Spain. She was graduated with a Bachelor of Engineering (Honours), from Coventry University (UK) in 1992. She received her Industrial Engineer degree from UPV in 1994, and she finally received a PhD in Hydraulic Engineering from UPV in 2001. Dr. López has more than a decade of experience in teaching in engineering fields, related to hydraulic and environmental topics. She is author or editor of more than twenty books about hydraulic engineering, water quality and atmospheric pollution. She has participated and led some national and international R&D projects financed by public and private funds. She has also co-organized national and international seminars. She is an experienced University teacher, an active researcher and a former practicing engineer, also very active in Continuous Professional Development activities. She has 71 reviewed scientific contributions, including 10 publications in scientific journals and peer-reviewed scientific books. She has been Academic Director of the School of Industrial Engineering (UPV) from 2006 to 2008. Currently she is the Secretary of the Department of Hydraulic and Environmental Engineering in her University.

F. Javier Martínez Solano is PhD in Engineering by the Universidad Politécnica de Valencia (UPV), Spain, since 2002. He is also lecturer of Fluid Mechanics in the Department of Hydraulic and Environmental Engineering of this University since 1994. He is teaching Hydraulic Machinery, Industrial Ventilation and Water Supply Network Modeling at the School of Industrial Engineering, UPV. He has authored 2 text books about Hydraulic Machinery and other 3 books regarding drinking water and storm water network modeling. He develops his research in Centro Multidisciplinar de Modelación de Fluidos. He is author of more than 20 research papers in reviewed scientific journals and 30 communications to international scientific events. Besides, he has participated both as an author and as editor in more than twenty books related to hydraulic engineering. Most of these books have been published by international editorials, such as John Wiley & Sons, McGraw-Hill, ASCE, Elsevier, Kluwer, or Balkema. His research concentrates in the modeling of both drinking and waste water networks, as well as the integration of simulation models in GIS environments.

Arash Massoudieh is currently Assistant Professor in Environmental Engineering at the Civil Engineering Department of the Catholic University of America, USA. He received a B.Sc. in Civil Engineering at the Sharif University, Tehran, Iran, and his Masters degree in Civil Engineering at the University of Tehran. He received a Ph.D. in Environmental Engineering at the University of California, Davis. Prof. Massoudieh's main interest is contaminant fate and transport in aquatic systems including surface water, groundwater and vadose zone with a focus on modeling the role of the moving solid phase such as suspended sediments or colloidal particles on the transport of sorbing contaminants. Dr. Massoudieh's other interests include modeling actively mobile organisms including fish and micro-organism in aquatic systems, inverse-modeling using heuristic methods, as well as stochastic methods in reactive transport of contaminants in porous media.

Lubos Matejicek is Assistant Professor in environmental modeling and GIS at the Institute for Environmental Studies of the Charles University in Prague, Czech Republic. He holds a Dipl. Ing. in automatic control from Czech Technical University in Prague and Ph.D. in applied and landscape ecology from Charles University in Prague. He served as a research fellow founded by the Royal Society in London in the framework of the project: application GIS in environmental systems (1996-1997). His research interests include spatio-temporal modeling, GIS applications in ecology, and spatial databases. He is author of 50 publications in conference proceedings, scientific journals, and books. His teaching is focused on environmental informatics, environmental modeling, and GIS. He is member of the *International Society for Ecological Modeling (ISEM)*, of the *Geographic Society* and of the *International Environmental Modelling & Software Society (iEMSs)*.

Nils Mole has been a Lecturer and Senior Lecturer in Applied Mathematics at the University of Sheffield, United Kingdom, since 1991. He was a student in the Department of Applied Mathematics and Theoretical Physics (DAMTP) at the University of Cambridge (BA and Part III from 1978 to 1982) and in the Department of Meteorology at the University of Reading (PhD from 1982 to 1986). He was then a Research Assistant in the Department of Mathematics and Statistics at Brunel University (from 1986 to 1990, supervised by Philip Chatwin) and a Lecturer in Mathematics at the University of Essex, from 1990 to 1991. Most of his research has been on turbulent dispersion, often in collaboration with Philip Chatwin and Paul Sullivan.

José de Jesús Mora Rodríguez is a Civil Engineer who graduated in 2002 from the Michoacán University of San Nicolás of Hidalgo, (UMSNH), Mexico. In 2005 he received a M.Eng. degree in Hydraulics from the National Autonomous University of Mexico (UNAM). He has been a Ph.D. student in Spain, from 2005) of Hydraulic Engineering and Environment of the Universidad Politécnica de Valencia (UPV). His main research area is related to model leakage supply networks and their impacts on water quality. He was an academic technician of the UMSNH on 2001. From 2003 to 2005, he investigated groundwater hydrology at the Mexican Institute of Water Technology (IMTA). He has been a member of Centro Multidisciplinar de Modelación de Fluidos (CMMF, since 2005). Since 2007, he has been a recipient of a research grant (FPI) of Genelitat Valenciana, Spain.

Zoltán Nagy is meteorologist at the Hungarian Meteorological Service. He is the head of the Surface Observations Division where his responsibility is managing and supervising the automatic surface weather network, solar radiation and stratospheric ozone measurements, radioactivity network and the Calibration Laboratory of Hungarian Meteorological Service. His responsibility also is to operate the abovementioned networks and activity in accordance of requirements of ISO 9001/2000 quality assurance system. His activity covers the reconstruction of the long time series of global and UV-B radiation, and developing statistical methods that give possibility for real-time checking of data coming from the automatic surface weather network. He is leader of the WMO Regional Solar Radiation Centre for RA VI in Budapest. Recently he is the manager of the project at Hungarian Meteorological Service that aims to establish a special climate network to follow the effects of climate change in Hungary with the accuracy on the possible highest level. Connecting to this project he, recently, studies the effect of thermometer screens on accuracy of temperature measurements and effects of wind on the accuracy of precipitation measurements.

Guelfo Pulci Doria, graduated in Electronic Engineering on 1966, is Fluid Mechanics Full Professor since 1986 at the University of Napoli *Federico II*, Italy. He was Head of Department during 6 years, and President of Naples Water Supply Authority during 2 years. He was six times Head of the Board and four times external reviewer of PhD courses. He contributed in organizing 16 scientific conferences. He was reviewer for international conferences and journals. Ten times he was coordinator of national research projects and one time local coordinator of a EU project. His publication record holds 171 products, including 2 scientific books, 2 textbooks, 5 chapters in international scientific books, 5 papers in international journals, 5 papers in italian journals, 3 invited lectures in conferences, 37 papers in international conferences, 41 papers in national conferences, and 10 multimedia DVD for a television broadcasted *Hydraulics* course. His research interests are: hydraulic measurements and models, turbulent flows, boundary layers, vortex shafts, air entrainment, groundwater, cavitation, biological hydrodynamics, dispersion and particle deposition, reaeration and volatilization, vegetated open channels flows, history of science.

Borivoj Rajkovic is employed at the Institute of Meteorology, Faculty of Physics, Belgrade University, Serbia, in the position of assistant professor. He teaches courses in Dynamic Meteorology, Micrometeorology, Numerical and Physical Modeling of the Atmosphere. He graduated from Faculty of Mathematics, Mechanics and Astronomy, Department of Meteorology. He received a M.Sc and Ph.D. in meteorology at Princeton University, USA. His professional interests are in numerical modeling of the atmosphere and ocean, micrometeorology and parameterization of physical processes in the atmosphere. As an expert in numerical modeling he participated in several international and national projects: *Adriatic integrated coastal areas and rivers basin management system project (ADRICOSM-EXT)*, under IOC-UNESCO, *Simulations of climate change in the Mediterranean Area (SINTA)*, *Adricosm integrated river basin and coastal zone management system: Montenegro coastal area and river Bojana catchment (ADRICOSM-STAR)*, *Implementation of the coupled Ocean-Atmosphere model* and *Extreme weather condition over Serbia*. During 1993-95 he was a visiting scientist at the World Laboratory in LAND-3 project *Protection of the Coastal Marine Environment in the Southern Mediterranean Sea*. Selected publications include papers on several international conferences and workshops. He has published 16 papers in the international journals from the fields of micrometeorology, physical processes and oceanography. He is the author of the text-book *Micrometeorology*.

Robert J. Schindler's research centers on the micro-scale dynamics of aquatic flows. He is currently a postdoctoral scientist in the Physical Ecology Laboratory, Department of Integrative Biology at the University of Guelph, Canada. He received his PhD in the School of Geography, University of Leeds, United Kingdom, where in conjunction with the UK Centre for Ecology & Hydrology, he examined how macrophytes (aquatic plants) change the hydrodynamics of river channels. In his MSc research, he examined fundamental linkages between flow structure, form and sediment transport to develop an improved model for the ripple-dune transition. His current research at the University of Guelph examines fundamental aspects of boundary roughness, turbulent flow and scalar dispersion. Other interests include the development of acoustic and particle-tracking techniques to examine the complex linkages between hydrodynamics, channel morphology and aquatic species, and the denoising of turbulent signals.

Paul Sullivan has been Emeritus Professor of Applied Mathematics at the University of Western Ontario, Canada, since 2004. He was a student of Engineering at the University of Waterloo (BSc 1964, MSc 1965) and in DAMTP at the University of Cambridge (PhD 1968). He was a Research Fellow in Applied Science at the California Institute of Technology from 1969 to 1970, and Professor of Applied Mathematics at the University of Western Ontario from 1969 to 2004. His research has mostly been on turbulent dispersion, often in collaboration with Philip Chatwin and Nils Mole.

Carlos Cabrera Toledo obtained a Bachelor's degree in Civil Engineering from the University *Diego Portales* in Santiago de Chile, Chile. He is currently finishing his Master's degree in Civil and Environmental Engineering at the University of California, Davis, performing research on modeling the mercury fate, transport and speciation in rivers.

Mirjam Vujadinovic is currently a second year Ph.D student at the Institute of Meteorology, Faculty of Physics, Belgrade University, Serbia where she is also a teaching assistant for courses Modeling of the Atmosphere 1 and 2. She received a B.Sc and M.Sc in meteorology at the University of Belgrade. Her main research interests are modeling of the atmosphere, modeling of regional climatology, air-sea interaction and modeling of the hydrology. She holds a fellowship of Serbian Ministry of Science and Technological Development. She was involved in two international and two national projects: *Simulations of climate change in the Mediterranean Area (SINTA), Adricosm integrated river basin and coastal zone management system: Montenegro coastal area and river Bojana catchment (ADRICOSM-STAR), Modeling and numerical simulations of complex physical systems* and *Weather and climate forecast in Serbia*. She presented 5 papers in international conferences.

Tess Weathers is currently a graduate student at the Department of Civil and Environmental Engineering at the University of California, Davis, USA. She obtained a BS in Civil Engineering from the California Polytechnic State University, San Luis Obispo in 2007. Her work focuses on multicomponent reactive transport applied to ureolytic-induced calcite precipitation and strontium co-precipitation, and streamtube-ensemble techniques for well-to-well reactive transport modeling.

Tamás Weidinger is currently Associate Professor at the Department of Meteorology, Eötvös Loránd University, Budapest, Hungary. He received his M.Sc. and dr. univ. degree in Meteorology at the Faculty of Sciences of Eötvös Loránd University, and Candidate of Sciences degree (same as Ph.D.) in Geography (Meteorology) at the Hungarian Academy of Sciences. He teaches dynamic meteorology, micrometeorology and basic meteorology courses to Earth Sciences, Environmental Sciences and Meteorology students. Prof. Weidinger has more than 140 peer-reviewed scientific papers, including 21 publications in SCI scientific journals, 54 publications in conference proceedings, and 30 other refereed journal publications in subjects related to micrometeorology, measurement and modeling surface-biosphere-atmosphere turbulent exchanges, wind energy modeling and analysis of local effects of global climate changes. He is also a co-author of 2 textbooks on Dynamic meteorology and Micrometeorology (in Hungarian). He is a member of the Editorial Board of *Theoretical and Applied Climatology (Springer)* and *Időjárás (Quaterly Journal of the Hungarian Meteorological Service)*. He contributed as reviewer in several scientific journals (e.g. *Environmental Research Letters, Environmental Modelling and Software, Journal of Applied Meteorology, Meteorology and Atmospheric Physics*). He is a member of the Management Committee of the *COST ES0804* program and of the *American Meteorological Society (AMS)* and has been participating in different EU research projects in the field of micrometeorology and trace gases flux calculations in the framework of *GRAMINAE, GREENGASS* and *NitroEurope*.

Dušan Žagar is Assistant Professor in the Faculty of Civil and Geodetic Engineering of the University of Ljubljana, Slovenia. Dr. Žagar obtained his Ph.D. in 1999 with research specialization on multidimensional hydrodynamic and pollutant transport and fate modeling in non-uniform flow fields. His original research provided the development of quantitative tools for high-resolution, three-dimensional simulation of mercury transport and transformation processes in the marine coastal environment. In particular, Dr. Žagar has contributed to the development of coupled models of mercury fate and transport in the Idrijca/Soča river system, the Gulf of Trieste, and in the Mediterranean overall.

xxxi

PART ONE

THEORETICAL AND MODELING ASPECTS

Chapter 1

TURBULENT DISPERSION: HOW RESULTS FOR THE ZERO MOLECULAR DIFFUSIVITY CASE CAN BE USED IN THE REAL WORLD

NILS MOLE[*], PHILIP CHRISTOPHER CHATWIN[†] and
PAUL J. SULLIVAN[‡]

We consider the dispersion of contaminants in turbulent flows at high Péclet number. Except at the smallest scales, molecular diffusion acts slowly by comparison with turbulent advection. Molecular diffusion is still important because it is the only means by which the concentration in a fluid particle can be changed. Exact solution of the full problem, including both turbulent advection and molecular diffusion, is not possible because of the well-known turbulence closure problem. But exact results for moments and the probability density function (pdf) of concentration can be derived for the hypothetical case of zero diffusivity. For high Péclet number these results can be expected to hold in certain ranges of space and time with only slight modification. We outline the results in the absence of molecular diffusion, and consider how the results for the moments can be modified to account for the presence of slowly acting diffusion. The model expressions for the moments involve the mean concentration, and a set of parameters which vary slowly with distance from the source, or with time since release. The corresponding form of the pdf is considered, and results for large concentrations are presented. It is shown that many of the results obtained with this approach agree well with experimental observations. Areas needing further work are also suggested.

Keywords: Concentration Fluctuations; Concentration Moments; Intermittency; Maximum Concentration; Molecular Diffusion; Probability Density Function; Turbulent Dispersion; Generalised Pareto Distribution

[*]Department of Applied Mathematics, University of Sheffield, Hicks Building, Sheffield S3 7RH, U.K.,   Email: n.mole@sheffield.ac.uk

[†]Department of Applied Mathematics, University of Sheffield, Hicks Building, Sheffield S3 7RH, U.K.,   Email: p.chatwin@sheffield.ac.uk

[‡]Department of Applied Mathematics, The University of Western Ontario, London N6A 5B7, Ontario, Canada,   Email: pjsul@uwo.ca

1. Introduction

We present a review of a novel approach to the modelling of turbulent dispersion which has been developed over the last twenty years or so. Rather than attempting to solve the Navier-Stokes and advection-diffusion equations for the full problem, or applying closures directly to those equations, this approach involves deriving exact results for the hypothetical case of no molecular diffusion, and then making physical arguments for how these results should be modified in the presence of diffusion.

Two fundamental physical processes govern the dispersion of a passive conserved scalar in a turbulent flow: advection by the turbulent velocity field, and molecular diffusion. The relative importance of these processes can be roughly quantified using the Péclet number $Pe = ul/\kappa$, where u and l are appropriate velocity and length scales for the turbulent fluctuations, and κ is the molecular diffusivity. In most environmental and engineering applications the Péclet number is very large[a]. In such cases turbulent advection acts on a much shorter timescale than molecular diffusion, the ratio of the timescales being given by Pe^{-1} if the length scales for the velocity and concentration fields are comparable. Advection acts to stretch the scalar cloud or plume out into thin sheets and strands [4], as observed experimentally [2, 5–7]. Molecular diffusion is a much slower process, but it is nevertheless an important one, since it is the only means by which the scalar concentration in a fluid particle can be altered, and it limits the smallest scales which can be present in the scalar field. For Schmidt number ν/κ of order 1 or greater, where ν is the kinematic viscosity, this smallest scale is of the order of the conduction cut-off length $\lambda_c = (\nu\kappa^2/\epsilon)^{1/4}$ [8], where ϵ is the turbulent energy dissipation rate per unit mass. Molecular diffusion also has the effect of dissipating the variance and higher moments of the concentration [8–10].

Nevertheless, it seems likely that at high Pe, and for times t and positions $\mathbf{x}$ not too far from the source values, the effect of molecular diffusion can be described by appropriate modification of the results when there is

---

[a]For example, for heat, water vapour, carbon monoxide, methane or propane dispersing in air, κ is of order $10^{-5}\,\mathrm{m^2 s^{-1}}$ [1–3], so even values as small as $u = 0.1\,\mathrm{ms^{-1}}$ and $l = 0.1\,\mathrm{m}$ would give Pe of order 10^3. For salt, methane, ethanol, toluene or urea dispersing in water, κ is of order $10^{-9}\,\mathrm{m^2 s^{-1}}$, and for heat dispersing in water it is of order $10^{-7}\,\mathrm{m^2 s^{-1}}$. So even for scales as small as $u = 10^{-3}\,\mathrm{ms^{-1}}$ and $l = 0.1\,\mathrm{m}$ we would have Pe of order 10^5 and 10^3, respectively. In many applications the typical velocity and length scales would be larger than the values given here, leading to even larger values of Pe.

no molecular diffusion. This approach was initiated in [11], and taken further in many papers, especially [12–14]. This article is an appraisal of the present position.

In Sec 2 we derive results for the moments and probability density function (pdf) of concentration for the case of no molecular diffusion. In Sec 3 we show how the results for the moments can be modified to include the effects of diffusion, and in Sec 4 we discuss the implications for the pdf, including its behaviour at high concentrations. Finally, in Sec 5, we discuss the successful aspects of this approach, and suggest areas which require further work.

2. The case of no molecular diffusion

2.1. *Uniform source*

In [11] the following results for the case of no molecular diffusion and a uniform concentration source were derived. The only possible concentrations are then zero and the source concentration θ_1. The pdf, $p(\theta; \mathbf{x}, t)$, of concentration $\Gamma(\mathbf{x}, t)$ is defined by

$$p(\theta; \mathbf{x}, t) = \frac{\mathrm{d}}{\mathrm{d}\theta} \mathrm{Prob}(\Gamma(\mathbf{x}, t) \leq \theta),$$

and can then be written as

$$p(\theta) = (1 - \pi)\delta(\theta) + \pi\delta(\theta - \theta_1), \tag{1}$$

where $\pi(\mathbf{x}, t)$ is the probability that $(\mathbf{x}, t)$ is in fluid which originated in the source, and δ is the Dirac delta-function. Note that $\pi(\mathbf{x}, t)$ is completely determined by the statistical properties of the turbulent velocity field.

The mean concentration $C(\mathbf{x}, t) = \mathrm{E}\{\Gamma(\mathbf{x}, t)\}$, where $\mathrm{E}\{\cdot\}$ denotes the expected value, or ensemble mean, is then given by

$$C = (1 - \pi)\int \theta\delta(\theta)\,\mathrm{d}\theta + \pi\int \theta\delta(\theta - \theta_1)\,\mathrm{d}\theta = \pi\theta_1, \tag{2}$$

and the nth central moment

$$\mu_n(\mathbf{x}, t) = \mathrm{E}\{[\Gamma(\mathbf{x}, t) - C(\mathbf{x}, t)]^n\}$$

is given by

$$\begin{aligned}
\mu_n &= (1 - \pi)\int (\theta - C)^n\delta(\theta)\,\mathrm{d}\theta + \pi\int (\theta - C)^n\delta(\theta - \theta_1)\,\mathrm{d}\theta \\
&= (1 - \pi)(-C)^n + \pi(\theta_1 - C)^n.
\end{aligned}$$

This can be rewritten, using Eq. (2), as

$$\frac{\mu_n}{\theta_1^n} = \frac{C}{\theta_1}\left(1 - \frac{C}{\theta_1}\right)^n + (-1)^n \left(1 - \frac{C}{\theta_1}\right)\left(\frac{C}{\theta_1}\right)^n. \tag{3}$$

2.2. *Non-uniform source*

We now consider the more general case of random, non-uniform, source concentration. [15] included random, but stationary and homogeneous, source concentration, while [12] derived results for a non-random, spatially varying instantaneous source. Here we generalise to the case of random source concentration whose distribution may vary in space and time. A smokestack is a practical example of a case where the concentration of effluent at the source will be a random function of spatial position (within the stack exit cross-section) and of time. The development given below follows that of [12], but is applied to these more general source conditions.

Let the source concentration be $\Gamma_S(\mathbf{x}, t)$, with Γ_S random, and let the probability that fluid at $(\mathbf{x}, t)$ came from a volume-time element $dV(\mathbf{y})\, d\tau$ about $(\mathbf{y}, \tau)$ be $P(\mathbf{y}, \tau; \mathbf{x}, t)\, dV(\mathbf{y})\, d\tau$. Then the pdf of $\Gamma(\mathbf{x}, t)$ can be written as

$$p(\theta; \mathbf{x}, t) = \int_{a.s.t.} p_S(\theta; \mathbf{y}, \tau)\, P(\mathbf{y}, \tau; \mathbf{x}, t)\; dV(\mathbf{y})\, d\tau,$$

where $a.s.t.$ denotes that the integration takes place over all space and for all times satisfying $\tau \leqslant t$, and $p_S(\theta)$ is the pdf of Γ_S. This is essentially equation (10.5) of [16], applied to the pdf of concentration rather than to the concentration itself.

We assume that the source occupies a finite volume V_0. Then

$$p(\theta; \mathbf{x}, t) = \int_{\mathbf{y} \notin V_0, \tau \leqslant t} \delta(\theta)\, P(\mathbf{y}, \tau; \mathbf{x}, t)\; dV(\mathbf{y})\, d\tau$$

$$+ \int_{\mathbf{y} \in V_0, \tau \leqslant t} p_S(\theta; \mathbf{y}, \tau)\, P(\mathbf{y}, \tau; \mathbf{x}, t)\; dV(\mathbf{y})\, d\tau$$

$$= [1 - \pi(\mathbf{x}, t)]\,\delta(\theta) + \int_{\mathbf{y} \in V_0, \tau \leqslant t} p_S(\theta; \mathbf{y}, \tau)\, P(\mathbf{y}, \tau; \mathbf{x}, t)\; dV(\mathbf{y})\, d\tau,$$

$$\tag{4}$$

where

$$\pi(\mathbf{x}, t) = \int_{\mathbf{y} \in V_0, \tau \leqslant t} P(\mathbf{y}, \tau; \mathbf{x}, t)\; dV(\mathbf{y})\, d\tau \tag{5}$$

is the probability that fluid at $(\mathbf{x}, t)$ came from the source.

Far from the source, the initial location of a fluid particle within the source is "forgotten". If we take the origin to be within V_0, we then have

$$P(\mathbf{y}, \tau; \mathbf{x}, t) \approx P(\mathbf{0}, \tau; \mathbf{x}, t)$$

for $\mathbf{y} \in V_0$, and

$$\pi(\mathbf{x}, t) \approx V_0 \int_{\tau \leqslant t} P(\mathbf{0}, \tau; \mathbf{x}, t)\, \mathrm{d}\tau. \tag{6}$$

This gives

$$p(\theta; \mathbf{x}, t) \approx [1 - \pi(\mathbf{x}, t)]\, \delta(\theta) + \int_{\tau \leqslant t} \mathrm{d}\tau\, P(\mathbf{0}, \tau; \mathbf{x}, t) \int_{\mathbf{y} \in V_0} \mathrm{d}V(\mathbf{y})\, p_S(\theta; \mathbf{y}, \tau). \tag{7}$$

If $p_S(\theta; \mathbf{y}, \tau)$ does not depend on τ then the time and space integrals decouple, and we are able to derive a result analogous to Eq. (1). This is the case if either (a) the source concentration distribution is stationary, or (b) the release is instantaneous. In both of these cases we find, using Eq. (6), that

$$p(\theta; \mathbf{x}, t) \approx [1 - \pi(\mathbf{x}, t)]\, \delta(\theta) + \frac{\pi(\mathbf{x}, t)}{V_0} \int_{\mathbf{y} \in V_0} \mathrm{d}V(\mathbf{y})\, p_S(\theta; \mathbf{y}). \tag{8}$$

In this formulation, all $(\mathbf{x}, t)$ dependence is accounted for by the variation of $\pi(\mathbf{x}, t)$. Thus, for a wide class of source conditions, very simple results can be derived in the zero molecular diffusivity case.

For a uniform source, $p_S(\theta; \mathbf{y}) = \delta(\theta - \theta_1)$ and Eq. (1) is recovered from Eq. (8). It can also be shown that Eq. (8) is consistent with results in [12, 15] – for details see Appendix A.

Here we consider the case of a random source concentration $\Gamma_S(\mathbf{x}, t)$ which varies in space and time, but has a pdf p_S which does not depend on time, so that Eq. (8) applies. The absolute moments m_n of Γ therefore satisfy

$$m_n(\mathbf{x}, t) = \int \theta^n p(\theta; \mathbf{x}, t)\, \mathrm{d}\theta = \frac{\pi(\mathbf{x}, t)}{V_0} \int_{\mathbf{y} \in V_0} m_{Sn}(\mathbf{y})\, \mathrm{d}V(\mathbf{y}), \tag{9}$$

where

$$m_{Sn}(\mathbf{x}) = \int \theta^n p_S(\theta; \mathbf{x})\, \mathrm{d}\theta$$

is the nth absolute moment of $\Gamma_S(\mathbf{x}, t)$. If we let the mean source concentration be $\theta_1(\mathbf{x}) = m_{S1}(\mathbf{x})$, then the mean concentration $C(\mathbf{x}, t)$ is given by

$$C(\mathbf{x}, t) = m_1(\mathbf{x}, t) = \frac{\pi(\mathbf{x}, t)}{V_0} \int_{\mathbf{y} \in V_0} \theta_1(\mathbf{y}) \, dV(\mathbf{y}). \tag{10}$$

Define the constant concentration scale θ_0 by

$$\theta_0 = \frac{\displaystyle\int_{\mathbf{y} \in V_0} m_{S2}(\mathbf{y}) \, dV(\mathbf{y})}{\displaystyle\int_{\mathbf{y} \in V_0} \theta_1(\mathbf{y}) \, dV(\mathbf{y})}. \tag{11}$$

Then the central moments $\mu_n = \mu_n(\mathbf{x}, t)$ for $n \geqslant 2$ can easily be shown to be given by

$$\begin{cases} \mu_2(\mathbf{x}, t) = C(\mathbf{x}, t) \left\{ \theta_0 - C(\mathbf{x}, t) \right\} \\ \mu_3(\mathbf{x}, t) = C(\mathbf{x}, t) \left\{ \lambda_3^2 \theta_0^2 - 3\theta_0 C(\mathbf{x}, t) + 2C(\mathbf{x}, t)^2 \right\} \\ \mu_4(\mathbf{x}, t) = C(\mathbf{x}, t) \left\{ \lambda_4^3 \theta_0^3 - 4\lambda_3^2 \theta_0^2 C(\mathbf{x}, t) + 6\theta_0 C(\mathbf{x}, t)^2 - 3C(\mathbf{x}, t)^3 \right\} \\ \qquad \vdots \end{cases}$$
$$\tag{12}$$

where the non-dimensional constants λ_n are defined for $n \geqslant 3$ by

$$(\lambda_n \theta_0)^{n-1} = \frac{\displaystyle\int_{\mathbf{y} \in V_0} m_{Sn}(\mathbf{y}) \, dV(\mathbf{y})}{\displaystyle\int_{\mathbf{y} \in V_0} \theta_1(\mathbf{y}) \, dV(\mathbf{y})}. \tag{13}$$

(Note that Eq. (11) and Eq. (13) for $n = 2$ are consistent with taking $\lambda_2 = 1$.) Apart from trivial changes of notation, Eq. (12) is the same as equation (10) of [12], and Eq. (13) is the generalisation of equation (11) of [12].

3.　The effects of molecular diffusion

For the real case with molecular diffusion, we consider the analogue of Eq. (1) and Eq. (4). Suppose we let $f_s(\theta; \mathbf{x}, t)$ be the pdf of $\Gamma(\mathbf{x}, t)$ conditional on being in fluid that came from the source, and $f_a(\theta; \mathbf{x}, t)$ be the pdf of $\Gamma(\mathbf{x}, t)$ conditional on being in ambient fluid that did not come from the source. As outlined by [15], it follows from the law of total probability that for any source the pdf of concentration can then be written as

$$p(\theta; \mathbf{x}, t) = \pi(\mathbf{x}, t) f_s(\theta; \mathbf{x}, t) + \{1 - \pi(\mathbf{x}, t)\} f_a(\theta; \mathbf{x}, t), \tag{14}$$

where $\pi(\mathbf{x}, t)$, as before, is the probability that fluid at $(\mathbf{x}, t)$ came from the source. This probability is not affected by molecular diffusion, but depends only on the velocity field, so Eq. (5) still applies.

3.1. *Intermittency*

In turbulent dispersion applications the concept of intermittency is used by many authors[b]. The aim is to characterise the contrast between concentrations close to zero (found mainly in ambient, non-source, fluid) and relatively large concentrations (which can often be many standard deviations larger than the mean) found mainly in fluid originating from the source. Such authors usually define intermittency to be the probability of non-zero concentration (see, for example, [17]). There are two problems associated with this definition. Firstly, [15] pointed out that the solution of the advection-diffusion equation for Γ will have non-zero values for all $\mathbf{x}$ for all times after release, so this definition of intermittency would strictly give a value of 1 everywhere. Secondly, small measured concentrations will be contaminated with noise. The most usual methods for dealing with this have been the use of a threshold, or the fitting of a Gaussian distribution to small concentrations. The measured value of intermittency depends strongly on the arbitrary choice of threshold, or on the method chosen to fit the Gaussian.

[15] proposed instead that intermittency be defined as $\pi(\mathbf{x}, t)$, the probability that fluid at $(\mathbf{x}, t)$ comes from the source. Intermittency is then well-defined, and conveys valuable information about the effect of the velocity field. To measure π experimentally one can make the reasonable assumption that the mean concentration is little affected by molecular diffusion. Eq. (10) then gives

$$\pi(\mathbf{x}, t) \approx \frac{V_0 C(\mathbf{x}, t)}{\displaystyle\int_{\mathbf{y} \in V_0} \theta_1(\mathbf{y}) \, dV(\mathbf{y})}. \tag{15}$$

---

[b]In the theory of turbulence itself, "intermittency" is used in a different sense, to denote the spatial patchiness of dissipation.

This is consistent with equation (15) of [12], which applies to a homogeneous source for which $\theta_1(\mathbf{x})$ is a constant. A practical advantage of this (albeit approximate) method of estimating the intermittency $\pi(\mathbf{x}, t)$ is that it will be little affected by measurement errors associated with noise and/or lack of resolution.

3.2. *Concentration moments*

[11] argued that, since molecular diffusion is a slow process, the structure of the moments will be close to that in the absence of molecular diffusion. They proposed that for self-similar dispersion in stationary turbulent flows, downwind of a uniform source Eq. (3) could be modified in two ways to take account of the effect of molecular diffusion.

Firstly, they replaced the constant source concentration θ_1 by a representative local value αC_0, where α is a constant and C_0 is the mean concentration on the plume centreline for a steady release (or at the cloud centre for an instantaneous release [12]). Secondly, constants of proportionality were introduced to take account of dissipation of concentration moments and of the increased background concentration resulting from diffusion out of the sheets and strands of source fluid with high concentrations:

$$\frac{\mu_n}{(\alpha C_0)^n} = \beta^n \left\{ \hat{C}(1 - \hat{C})^n + (-1)^n (1 - \hat{C})\hat{C}^n \right\}, \tag{16}$$

where

$$\hat{C} = \frac{C}{\alpha C_0}.$$

(Note that [11] used $\beta^{1/2}$ where we use β in Eq. (16), *i.e.* we follow the more convenient usage of [12].) Because of dissipation, β would be expected to be less than or equal to 1 and, to avoid negative variance, we need $\alpha \geqslant 1$. [11] found that observations from a number of experiments could be fitted well by Eq. (16) with a constant value of α, and β approximately constant in the crosswind direction.

Normalised moments, in particular the skewness K_3, kurtosis K_4 and higher order equivalents K_n, defined by

$$K_n = \frac{\mu_n}{\mu_2^{n/2}} \qquad \text{for} \quad n = 3, 4, \ldots, \tag{17}$$

are useful in many contexts. [13] showed that Eq. (16) implies that

$$\begin{cases} K_4 = K_3^2 + 1 \\ K_5 = K_3^3 + 2K_3, \end{cases} \tag{18}$$

and gave a general expression for K_n as a function of K_3. Analysis of experimental data from a steady release close to the ground in the field suggested that Eq. (18) ought to be replaced by

$$\begin{cases} K_4 = a_4 K_3^2 + b_4 \\ K_5 = a_5 K_3^3 + b_5 K_3, \end{cases} \tag{19}$$

where a_4, b_4, a_5 and b_5 are constants. Eq. (19) has been shown to be a good approximation in a variety of experiments, including plumes in atmospheric boundary layers under various stability classes [13, 18], clouds in the wind tunnel with varying density and with different forms of fence [19], and plumes in wind tunnels [20, 21]. Figure 1 shows an example for the steady line source, grid turbulence, releases of [22], which were analysed further by [12].

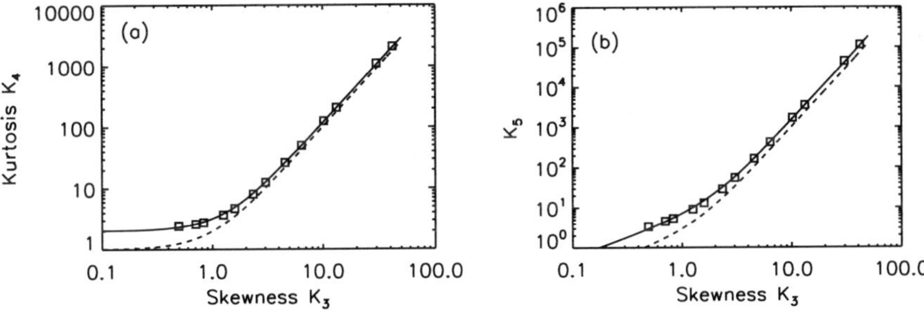

Fig. 1. An example of the fit of Eq. (19) to data from the steady line source, grid turbulence, releases of [22] at downwind distance 50 mm from the source. The fitted values are $a_4 = 1.161$, $b_4 = 2.025$, $a_5 = 1.485$, and $b_5 = 5.647$. The squares are the measured values, and the solid curves are the fits of Eq. (19). (a) Kurtosis K_4 against skewness K_3, (b) K_5 against K_3. The dashed curves show the relationships given by Eq. (18).

[12] argued that Eq. (16) would describe the lateral moment structure, with α and β varying with time for an instantaneous release, or with downwind distance for a steady release. For the case of non-uniform source concentration they applied the same argument that led to Eq. (12), to

obtain

$$
\begin{cases}
\dfrac{\mu_2}{(\alpha\beta C_0)^2} = \hat{C}(1 - \hat{C}) \\[2ex]
\dfrac{\mu_3}{(\alpha\beta C_0)^3} = \hat{C}\left(\lambda_3^2 - 3\hat{C} + 2\hat{C}^2\right) \\[2ex]
\dfrac{\mu_4}{(\alpha\beta C_0)^4} = \hat{C}\left(\lambda_4^3 - 4\lambda_3^2\hat{C} + 6\hat{C}^2 - 3\hat{C}^3\right) \\[2ex]
\qquad\qquad\vdots
\end{cases}
\tag{20}
$$

for some parameters $\lambda_3, \lambda_4, \ldots$ If $\lambda_n = 1$ for all n then Eq. (20) reduces to the uniform source case Eq. (16). Since we expect a non-uniform source to increase the spread of concentration values by comparison with a uniform source, and since $\hat{C} \leqslant 1$, we expect $\lambda_n \geqslant 1$ in general. This is supported by the fitted values given in Table 1 of [21]. The λ_n would be expected to vary in time for an instantaneous release, or with downwind distance for a steady release, in the same way as for α and β. [12] fitted these parameters to data from a steady line source experiment in wind tunnel grid turbulence [22], finding that they varied very slowly with downwind distance (see Table 1 of [21]). We would also expect the parameters a_4, b_4, a_5, b_5 and higher order equivalents to be approximately constant in the crosswind direction, but to vary slowly with time or downwind distance. This has been confirmed for the experiments of [22] by [20, 21]. [20] also showed that the expression for K_4 in Eq. (19) followed approximately from Eq. (20). The relationships between the moments proposed by [11, 12] have been found to agree reasonably well with measurements from a range of experiments, including jets, wakes, plumes, uniformly sheared flow and buoyant jets [11, 12, 23, 24]. Figure 2 shows an example from the steady line source releases of [22].

[14] argued that β is mainly a measure of the dissipation accomplished by molecular diffusion, so that β tends to zero in the limit of large diffusion time. Therefore β would not be expected to be constant across the whole plume cross-section. Any source material arriving far from the centreline will have taken a long time to get there, because of the large distance involved. So it will have been acted on by diffusion for a long time, implying that β tends to zero far from the centreline. But the distance from the centreline at which β becomes small compared with its centreline value will be much greater than the downwind distance from the source and, hence, the plume width, which is the distance at which C becomes small. Where C is very small, it is usually difficult to obtain reliable measurements, so we

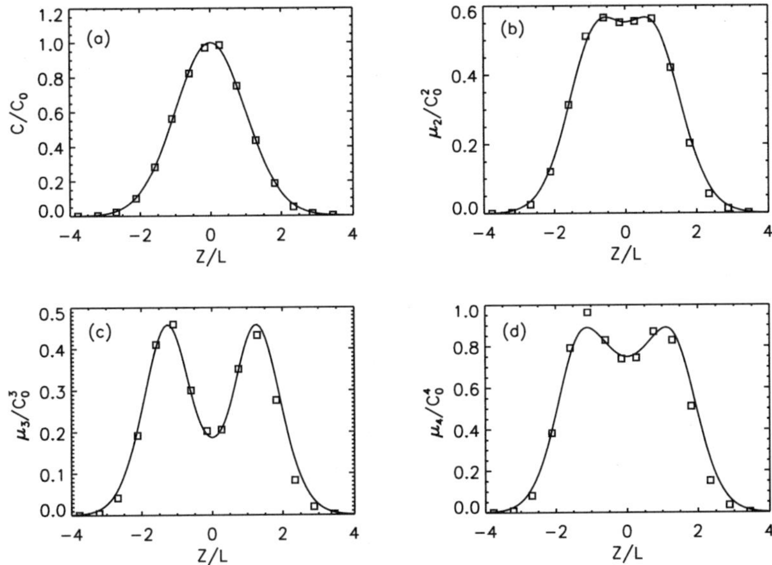

Fig. 2. Fits of Eq. (20) to data from [22] at 50 mm downwind of the source. The fitted values are $\alpha = 1.729$, $\beta = 0.871$, $\lambda_3^2 = 1.161$ and $\lambda_4^3 = 1.511$. The horizontal axis is the distance Z from the mean plume centreline divided by the mean plume width L. The squares are the measurements, and the curves are the fits. (a) Gaussian fit to C/C_0, (b) Second moment μ_2/C_0^2, (c) Third moment μ_3/C_0^3, (d) Fourth moment μ_4/C_0^4.

expect *measured* values of β to be approximately constant in the crosswind direction over the range of distances for which measurements are usually made, consistent with the results described above.

The successful agreement of Eq. (19) and Eq. (20) with data from a wide variety of experiments encourages us to develop the argument further, especially to the pdf itself.

4. The probability density function of concentration

4.1. *The overall pdf*

The pdf is the principal function describing the distribution of concentration in turbulent diffusion. This fact is now, but belatedly, well recognised (and there is also recognition that, in itself, the mean concentration is of little value, theoretically and practically). The pdf is also practically important; for example, if the lower and upper flammable limits of a flammable gas are θ_L and θ_U, the probability that the mixture at $(\mathbf{x}, t)$ is flammable is

$$\int_{\theta_L}^{\theta_U} p(\theta; \mathbf{x}, t) \, \mathrm{d}\theta.$$

Many other examples can be given.

For a uniform source the exact pdf of concentration in the absence of molecular diffusion is given by Eq. (1), and consists of delta-functions located at zero concentration (corresponding to non-source fluid), and at the source concentration θ_1. The corresponding model moments when diffusion is included, given by Eq. (16), satisfy the first of Eq. (18). But the theory of probability distributions shows that this implies that the pdf must consist of two delta-functions (see, for example, the appendix of [13]). [24, 25] showed that this pdf, with moments given by Eq. (16), is

$$p(\theta) = (1 - \hat{C})\delta(\theta - \theta_{\min}) + \hat{C}\delta(\theta - \theta_{\max}), \tag{21}$$

where

$$\theta_{\min} = (1 - \beta)C \geqslant 0 \quad \text{and} \quad \theta_{\max} = (1 - \beta)C + \alpha\beta C_0 \leqslant \alpha C_0, \tag{22}$$

with equality in both cases when $\beta = 1$, which corresponds to being at the source, where we also have $\alpha = 1$ and $C_0 = \theta_1$. In the absence of diffusion, $\alpha C_0 = \theta_1$, $\hat{C} = C/\theta_1 = \pi$ and $\beta = 1$, so Eq. (1) is recovered. (It takes some time for the cumulative effects of molecular diffusion to become important, so at small times after release, or at small downwind distances from the source, the $\kappa = 0$ results will still give a good approximation.)

This model captures the effect of molecular diffusion in reducing the largest concentrations and increasing the smallest concentrations. Since it only has two possible concentrations, $\theta_{\min}$ and $\theta_{\max}$, it fails to represent the continuous distribution of concentrations in real turbulent dispersion. However, Eq. (21) can give a good indication of the structure of the real pdf. [24, 25] analysed experimental data for line and point sources in grid turbulence, and for a buoyant jet in a boundary-layer cross flow, and showed that in cases when the pdf had 2 peaks, the location and height of the peaks corresponded quite well to the location and size of the delta-functions in Eq. (21).

If we non-dimensionalise the concentration by $\hat{\Gamma} = \Gamma/(\alpha\beta C_0)$ then we can rewrite Eq. (21) for a uniform source as

$$\hat{p}(\hat{\theta}) = (1 - \hat{C})\delta(\hat{\theta} - D) + \hat{C}\delta(\hat{\theta} - D - 1), \tag{23}$$

where $\hat{p}$ is the pdf of $\hat{\Gamma}$ and

$$D = \left(\frac{1}{\beta} - 1\right)\hat{C}.$$

For a non-uniform source we can write the pdf corresponding to the moments Eq. (20), and analogous to Eq. (23), in the form

$$\hat{p}(\hat{\theta}) = \hat{p}_D(\hat{\theta} - D),$$

where $\hat{p}_D(\hat{\theta})$ is a pdf with mean $\hat{C}$, second absolute moment $\hat{C}$, and nth absolute moment $\lambda_n^{n-1}\hat{C}$ for $n = 3, 4, \ldots$ (these results are a consequence of the choice of θ_0 in Eq. (11)). Details of the derivation of these results are given in Appendix B. At first sight it might appear that $\hat{p}_D$ should be given by the appropriately normalised version of p in Eq. (8). In that case the normalisation $\hat{\theta} = \theta/\theta_0$ and $\hat{C} = C/\theta_0$ gives first and second absolute moments equal to $\hat{C}$, and higher absolute moments equal to $\lambda_n^{n-1}\hat{C}$. But this is inconsistent with our present normalisation, which is $\hat{\theta} = \theta/(\alpha\beta C_0)$ and $\hat{C} = C/(\alpha C_0)$. This means that, in general, $\hat{p}_D$ cannot be obtained from Eq. (8).

If $\lambda_n = 1$ for all n, then $\hat{p}_D(\hat{\theta} - D) = (1 - \hat{C})\delta(\hat{\theta} - D) + \hat{C}\delta(\hat{\theta} - D - 1)$ and Eq. (23) is recovered. Given the dependence of the moments of $\hat{p}_D$ on $\hat{C}$, it would be natural to look for a general solution in the form $\hat{p}_D(\hat{\theta} - D) = (1 - \hat{C})\delta(\hat{\theta} - D) + \hat{C}\hat{f}(\hat{\theta} - D)$ for some pdf $\hat{f}$, so that $\hat{C}$ was scaled out of the moments. Thus $\hat{f}$ would have first and second absolute moments equal to 1, and higher moments equal to λ_n^{n-1}. However, this gives zero variance for $\hat{f}$, so $\hat{f}$ would have to be a single delta-function, and this form is only possible if $\lambda_n = 1$ for all n, i.e. the uniform source case. The fitted values of λ_n in [12, 21] are close to 1, so we would expect $\hat{p}_D$ to be bimodal, with molecular diffusion having the effect of smoothing out the delta-functions of the no-diffusion case.

4.2. *Large concentrations*

For a general, non-uniform, source, while we are unable to produce a simple exact closed form expression for $\hat{p}_D$, and hence for $\hat{p}$ or p, there are some questions of practical interest for which we do not need this. One such is applications to toxic and malodorous gases where we want to know the distribution of large concentration values, but we do not necessarily need to know the form of the distribution for smaller concentrations.

Statistical extreme value theory provides a framework for analysing the distribution of large concentration values. Of particular use is the work of [26] on values above a high threshold. Let $P_c(\theta; \theta_T)$ be the distribution function of the conentration Γ, conditional on Γ being greater than a threshold θ_T, i.e.

$$P_c(\theta; \theta_T) = \text{Prob}(\Gamma < \theta | \Gamma > \theta_T) = \frac{\int_{\theta_T}^{\theta} p(\phi)\, d\phi}{1 - \int_0^{\theta_T} p(\phi)\, d\phi} \qquad \text{for } \theta \geqslant \theta_T.$$

The pdf $p_c(\theta; \theta_T)$ of Γ, conditional on $\Gamma \geqslant \theta_T$, is then

$$p_c(\theta; \theta_T) = \frac{d}{d\theta} P_c(\theta; \theta_T) = \frac{p(\theta)}{1 - \int_0^{\theta_T} p(\phi)\, d\phi} \qquad \text{for } \theta \geqslant \theta_T.$$

For large θ_T, [26] showed that, subject to some conditions which are satisfied in most cases,

$$p_c(\theta; \theta_T) \approx g(\theta - \theta_T; k, a), \tag{24}$$

where $g(\theta; k, a)$ is the pdf of the generalised Pareto distribution (GPD) and is given by

$$g(\theta; k, a) = \frac{1}{a}\left(1 - \frac{k\theta}{a}\right)^{1/k - 1}. \tag{25}$$

Here k is the shape parameter and $a\,(> 0)$ is the scale parameter. In the present application the maximum possible concentration $\theta_{\max}$ is finite (less than or equal to the largest possible source concentration), in which case we expect $k > 0$ so that $g(\theta; k, a)$ has a finite upper endpoint a/k. We then have

$$\theta_{\max} = \theta_T + \frac{a}{k}. \tag{26}$$

It is straightforward to show from Eq. (25) that increasing the threshold θ_T does not change the form of the asymptotic distribution nor the value of the shape parameter k, but does change the value of the scale parameter a. Eq. (26) shows that this change can be calculated by $\Delta a = -k\Delta\theta_T$, where Δa and $\Delta\theta_T$ are the changes in a and θ_T. Thus a higher threshold gives a smaller value of the scale parameter.

The standard statistical approach is to fit Eq. (25) to excesses over a high threshold, using maximum likelihood (for examples of applications to turbulent dispersion see [27–30]). This does not, however, lend itself to modelling based on expressions for concentration moments, like Eq. (20). [14] presented an alternative method, which allows k, a and, hence, $\theta_{\max}$ to be derived from the moments. In [14] the overall pdf was expressed as

$$p(\theta) = (1 - \eta)f(\theta) + \eta g(\theta; k, a),$$

for some function f and parameter $\eta\,(>0)$, with f assumed to make a negligible contribution for large θ. The upper endpoint of the distribution given by $p(\theta)$ is then a/k, so

$$\theta_{\max} = \frac{a}{k}.$$

If $p(\theta) \approx \eta g(\theta; k, a)$ for all θ above some threshold θ_T, then it is straightforward to show that

$$p_c(\theta; \theta_T) \approx \frac{\left(1 - \frac{k\theta}{a}\right)^{1/k-1}}{a\left(1 - \frac{k\theta_T}{a}\right)^{1/k}} = g(\theta - \theta_T; k, a - k\theta_T) \qquad \text{for } \theta \geqslant \theta_T,$$

consistent with Eq. (24). [14] found for a line source experiment that the upper 50% or more of the concentration range is well-approximated by the GPD.

For large n, the absolute moment m_n could be expected to be dominated by the contribution from large concentrations, so that

$$m_n \approx \eta \int_0^{\theta_{\max}} \theta^n g(\theta; k, a)\, \mathrm{d}\theta. \tag{27}$$

[14] showed that this implied that, for sufficiently large n,

$$\frac{m_{n-1}}{m_n} \approx \frac{1}{a}\left(\frac{1}{n}\right) + \frac{k}{a}. \tag{28}$$

Thus, the parameters k and a can be identified from the linear relationship between the ratio of successive moments, m_{n-1}/m_n, and $1/n$. This method shares with the standard statistical approach the advantage that the fits are hardly affected by baseline and/or noise problems at low measured concentrations. It also has the advantage that it can be used when only moments are known, as in the model described in Sec. 3.2. Figure 3 shows examples of some fits of the GPD to concentration tails, carried out in this way, for the data of [22].

[14] also considered the behaviour in Eq. (20) far from the plume centreline, for a steady line source release (the same arguments would also apply for a steady point source). They deduced that

$$\lambda_n^{n-1} = a_n \lambda_3^{2(n-2)}, \tag{29}$$

where the a_n are defined by Eq. (19). By combining this result with Eq. (28) far from the centreline, they derived the relationship

$$\frac{a_{n-1}}{a_n} = \frac{1}{r}\left\{1 - \frac{20}{n}\left(\frac{a_4^2 - a_5}{5a_4^2 - 4a_5}\right)\right\}, \tag{30}$$

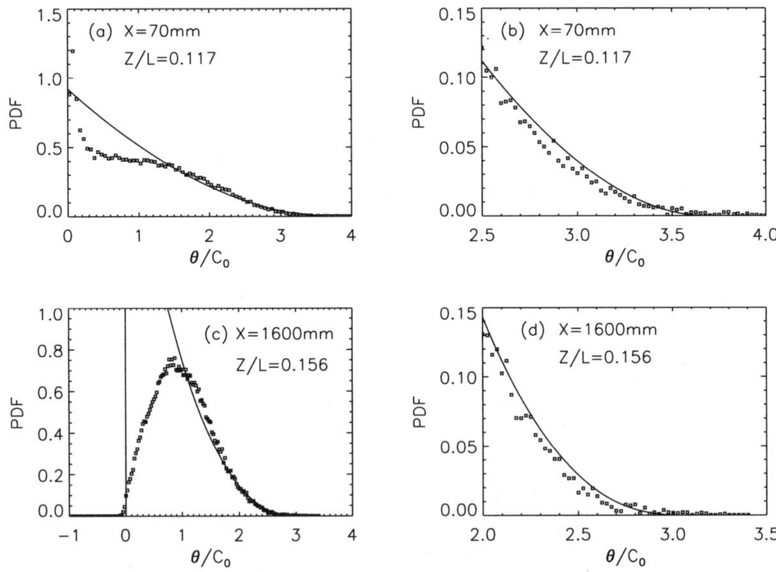

Fig. 3.   Examples of the measured pdf of θ/C_0 (points), and GPD (curves) for fitted k and a values, for the data of [22]. The right-hand panels show blow-ups of the tails. X is the downwind distance from the source.

where

$$r = \frac{a_4 a_5}{5a_4^2 - 4a_5},$$

and also showed that

$$\frac{\theta_{\max}}{C_0} \approx \alpha \left\{ \beta \lambda_3^2 r + (1 - \beta)\hat{C} \right\} = \alpha\beta(\lambda_3^2 r + D). \tag{31}$$

Note that for a uniform source we have $\lambda_n = 1$ and $a_n = 1$ so that, as desired, Eq. (31) reduces to Eq. (22). Equation (31) has strictly only been shown to be valid far from the centreline, *i.e.* as $\hat{C} \to 0$, but since α, β, λ_3, a_4 and a_5 are expected to be roughly constant over the distances over which $\hat{C}$ becomes small, it seems likely that Eq. (31) can also be applied near the centreline. The results of [14] support this.

Equation (31) suggests that as we go very far from the centreline, ultimately $\theta_{\max}/C_0 \to 0$, since we expect that $\beta \to 0$ in that limit. Since, as noted earlier, $\beta \to 0$ more slowly than $\hat{C}$, being controlled by diffusion rather than advection, the first term in Eq. (31) will give the leading or-

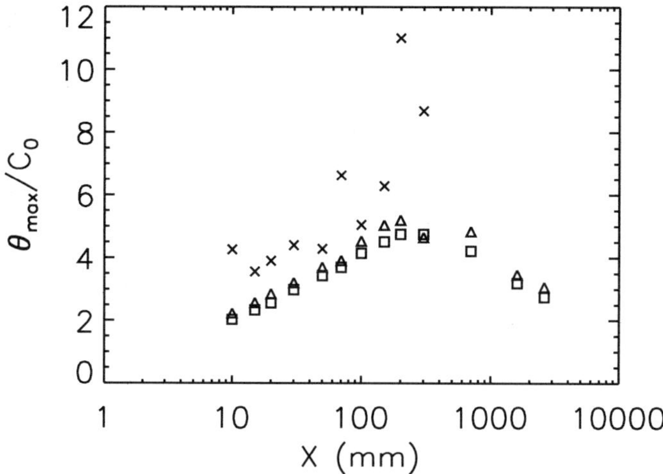

Fig. 4.　Downwind variation of $\theta_{\max}/C_0$ for the data of [22]. X is the downwind distance from the source. The crosses are the model values calculated from Eq. (31); the squares are calculated directly from the measurements by least-squares fits to the moment ratios for $n = 4$ to $n = 8$; and the triangles are calculated directly from the measurements by a linear fit to the moment ratios for $n = 7$ and $n = 8$.

der behaviour. Far downwind we expect $\beta \to 0$, since diffusion has had a long time to act, so Eq. (31) suggests that $\theta_{\max}/C_0 \to \alpha\hat{C} = C/C_0$, *i.e.* $\theta_{\max} \to C$. [14] argued that both of these results would be expected on physical grounds.

Figure 4 compares the downwind variation of $\theta_{\max}/C_0$ estimated from Eq. (31) using fitted values of the parameters, with estimates obtained directly from the concentration data, for the data of [22]. Results of two different estimation methods are shown — one where Eq. (28) was fitted by least-squares to the points for $n = 4$ to $n = 8$, and one which used a straight line fit through the points for $n = 7$ and $n = 8$. The model values are larger than the direct estimates, and are not shown at the furthest downwind distances (where they become unreliable, because the denominator in r becomes close to zero). The direct estimate using only $n = 7$ and $n = 8$ is a little larger than that using $n = 4$ to $n = 8$, suggesting that n is not yet large enough to have attained the asymptotic result Eq. (27), and that $\theta_{\max}/C_0$ may well be underestimated. So the model may be in better agreement with the true values of $\theta_{\max}/C_0$ than indicated by this figure. Given the difficulty of making accurate estimates of the upper endpoint of a distribution, the degree of agreement here is very promising. Further investigation into the optimal fitting of Eq. (28) is required, and also of the standard errors and biasses of the estimates.

5. Discussion

The models described here were derived by considering the hypothetical case of no molecular diffusion, for which many exact results can be obtained, and then taking account of the physical effects of molecular diffusion. These models have proved highly successful at reproducing a number of aspects of turbulent dispersion, thus allowing results for the zero molecular diffusivity case to be used in the real world.

The models, given by Eq. (16) and Eq. (20), for the crosswind variation of concentration moments have been shown to fit experimental data very well. For example, excellent fits were found by [11] for point, jet and wall sources in boundary layers and grid turbulence, and by [12] for a line source in grid turbulence. One feature of the latter experiment which is captured particularly well is the downwind evolution of the cross-plume structure of the variance. Near the source the variance is bimodal (corresponding to $\alpha < 2$), with peaks located symmetrically about the mean plume centre-line. Further downwind the variance becomes unimodal (corresponding to $\alpha > 2$), with a single peak on the centreline, and far downwind it becomes bimodal again. The relationship between skewness and kurtosis (and also higher order equivalents), given by Eq. (19), approximates a variety of experiments well, including steady sources and clouds, and cases with and without fences (further details were given in Sec. 3.2). The most recent work, in [14], shows very promising results when these models are extended to tackle the highly challenging problem of large concentrations, in particular that of estimating the largest possible concentration.

There are a number of pieces of work which would improve these models and make them even more useful for real world problems. The first is to develop better models for the evolution (with time since release or with distance from the source) of the parameters involved in the models. This will require further investigation of suitable closure schemes, similar to those discussed by [31–35]. The second is to carry out a thorough analysis of the moment-based method (i.e. that using Eq. (28)) for fitting the generalised Pareto distribution (GPD) for large concentrations. This will aim to answer questions about the optimum choice of moments to use, and about biasses and standard errors of the resulting estimates. Finally, further work on identifying the form of the pdf $\hat{p}$ corresponding to the moments Eq. (20) would give valuable insight into the underlying structure of the model.

Acknowledgements

We are grateful for the financial support we have received for this work over the years, from Canada NSERC, UK Research Councils (in particular SERC, now EPSRC), the Royal Society, the UK Ministry of Defence, the UK Health and Safety Executive, the European Union and NATO.

Appendix A.

We show here that Eq. (8) is consistent with the results obtained previously by [12, 15].

In [15] it was assumed that the source was stationary and homogeneous, i.e. that the probability distribution of the source concentration $\Gamma_S(\mathbf{x}, t)$ was independent of $\mathbf{x}$ and t. This case corresponds to applying Eq. (8) with $p_S(\theta; \mathbf{y}) = p_S(\theta)$. Thus

$$\int_{\mathbf{y} \in V_0} d\mathbf{y}\, p_S(\theta; \mathbf{y}) = V_0\, p_S(\theta),$$

so Eq. (8) becomes

$$p(\theta; \mathbf{x}, t) \approx [1 - \pi(\mathbf{x}, t)]\, \delta(\theta) + \pi(\mathbf{x}, t)\, p_S(\theta),$$

which is equation (14) of [15].

In [12] it was assumed that the release occurred instantaneously at $t = 0$, with spatially-varying but non-random concentration $\Gamma_S(\mathbf{x})$. This corresponds to taking $p_S(\theta; \mathbf{y}) = \delta(\theta - \Gamma_S(\mathbf{y}))$, so that Eq. (8) immediately gives equation (8) of [12].

Appendix B.

Let us define a pdf $q(\hat{\theta})$ for $0 \leqslant \hat{\theta} \leqslant \infty$ by $q(\hat{\theta}) = \hat{p}_D(\hat{\theta} - D)$, where $D = \left(\dfrac{1}{\beta} - 1\right)\hat{C}$ and $\hat{p}_D(\hat{\theta})$ has mean $\hat{C}$, second absolute moment $\hat{C}$, and nth absolute moment $\lambda_n^{n-1}\hat{C}$ for $n = 2, 3, \ldots$ Let $q(\hat{\theta})$ have mean μ_q, and nth central moment μ_{qn} for $n = 2, 3, \ldots$ Then

$$\mu_q = \hat{C} + D = \frac{1}{\beta}\hat{C} = \frac{C}{\alpha\beta C_0}$$

and

$$\mu_{qn} = \int_0^\infty \left(\hat{\theta} - \mu_q\right)^n q(\hat{\theta})\, d\hat{\theta} = \int_{-D}^\infty \left(s - \hat{C}\right)^n \hat{p}_D(s)\, ds$$

$$= \sum_{i=0}^n \binom{n}{i} \left(-\hat{C}\right)^{n-i} \hat{m}_{Di},$$

where $\hat{m}_{Di}$ is the ith absolute moment of $\hat{p}_D(\hat{\theta})$.

Thus we have

$$\mu_{q2} = \hat{C}(1 - \hat{C}),$$

and for $n \geqslant 3$

$$\mu_{qn} = \hat{C}\left\{ (n-1)(-\hat{C})^{n-1} + \binom{n}{2}(-\hat{C})^{n-2} + \sum_{i=3}^n \binom{n}{i}\lambda_i^{i-1}(-\hat{C})^{n-i} \right\}.$$

Thus the central moments of $q(\hat{\theta})$ are just the right-hand sides of Eq. (20). The left-hand sides of Eq. (20) are the central moments of $\hat{p}(\hat{\theta})$ (since the normalisation is $\hat{\Gamma} = \Gamma/(\alpha\beta C_0)$). Since $\hat{p}(\hat{\theta})$ also has mean $C/(\alpha\beta C_0) = \mu_q$, we can identify $\hat{p}$ with q, and hence obtain

$$\hat{p}(\hat{\theta}) = \hat{p}_D(\hat{\theta} - D).$$

References

[1] G. K. Batchelor, *An Introduction to Fluid Dynamics*. (Cambridge University Press, Cambridge, 1967).

[2] K. A. Buch and W. J. A. Dahm, Experimental study of the fine-scale structure of conserved scalar mixing in turbulent shear flows. Part 2. $Sc \approx 1$, *J. Fluid Mech.* **364**, 1–29, (1998).

[3] D. R. Lide, Ed., *CRC Handbook of Chemistry and Physics*. (Taylor & Francis, Boca Raton, Florida, 2005), 86th edition.

[4] G. K. Batchelor, The effect of homogeneous turbulence on material lines and surfaces, *Proc. R. Soc. Lond. A.* **213**, 349–366, (1952).

[5] W. J. A. Dahm, K. B. Southerland, and K. A. Buch, Direct, high resolution, four-dimensional measurements of the fine scale structure of $Sc \gg 1$ molecular mixing in turbulent flows, *Phys. Fluids A.* **3**, 1115–1127, (1991).

[6] A. F. Corriveau and W. D. Baines, Diffusive mixing in turbulent jets as revealed by a pH indicator, *Exps. Fluids.* **16**, 129–136, (1993).

[7] K. A. Buch and W. J. A. Dahm, Experimental study of the fine scale structure of conserved scalar mixing in turbulent shear flows. Part 1. $Sc \gg 1$, *J. Fluid Mech.* **317**, 21–71, (1996).

[8] G. K. Batchelor, Small-scale variation of convected quantities like temperature in turbulent fluid. Part 1. General discussion and the case of small conductivity, *J. Fluid Mech.* **5**, 113–133, (1959).

[9] P. C. Chatwin and P. J. Sullivan, The relative diffusion of a cloud of passive contaminant in incompressible turbulent flow, *J. Fluid Mech.* **91**, 337–355, (1979).

[10] P. C. Chatwin and P. J. Sullivan, Cloud-average concentration statistics, *Math. Computers Sim.* **32**, 49–57, (1990).

[11] P. C. Chatwin and P. J. Sullivan, A simple and unifying physical interpretation of scalar fluctuation measurements from many turbulent shear flows, *J. Fluid Mech.* **212**, 533–556, (1990).

[12] B. L. Sawford and P. J. Sullivan, A simple representation of a developing contaminant concentration field, *J. Fluid Mech.* **289**, 141–157, (1995).

[13] N. Mole and E. D. Clarke, Relationships between higher moments of concentration and of dose in turbulent dispersion, *Boundary-Layer Met.* **73**, 35–52, (1995).

[14] N. Mole, T. P. Schopflocher, and P. J. Sullivan, High concentrations of a passive scalar in turbulent dispersion, *J. Fluid Mech.* **604**, 447–474, (2008).

[15] P. C. Chatwin and P. J. Sullivan, The intermittency factor of scalars in turbulence, *Phys. Fluids A.* **1**, 761–763, (1989).

[16] A. S. Monin and A. M. Yaglom, *Statistical Fluid Mechanics, Volume 1.* (The MIT Press, Cambridge, 1971).

[17] D. J. Wilson, A. G. Robins, and J. E. Fackrell, Intermittency and conditionally-averaged concentration fluctuation statistics in plumes, *Atmos. Environ.* **19**, 1053–1064, (1985).

[18] D. M. Lewis, P. C. Chatwin, and N. Mole, Investigation of the collapse of the skewness and kurtosis exhibited in atmospheric dispersion data, *Il Nuovo Cimento.* **20 C**, 385–398, (1997).

[19] P. C. Chatwin and C. Robinson, The moments of the pdf of concentration for gas clouds in the presence of fences, *Il Nuovo Cimento.* **20 C**, 361–383, (1997).

[20] T. P. Schopflocher and P. J. Sullivan, The relationship between skewness and kurtosis of a diffusing scalar, *Boundary-Layer Met.* **115**, 341–358, (2005).

[21] T. P. Schopflocher, C. J. Smith, and P. J. Sullivan, Scalar concentration reduction in a contaminant cloud, *Boundary-Layer Met.* **122**, 683–700, (2007).

[22] B. L. Sawford and C. M. Tivendale. Measurements of concentration statistics downstream of a line source in grid turbulence. In *Proc. 11th Australasian Fluid Mechanics Conf.*, Hobart, Australia (Dec., 1992).

[23] D. J. Moseley. A closure hypothesis for contaminant fluctuations in turbulent flow. Master's thesis, University of Western Ontario, Canada, (1991).

[24] H. Ye. *A New Statistic For The Contaminant Dilution Process In Turbulent Flows.* PhD thesis, University of Western Ontario, London, Ontario, Canada, (1995).

[25] P. J. Sullivan and H. Ye, Moment inversion for contaminant concentration in turbulent flows, *Canadian Applied Mathematics Quarterly.* **4**, 301–310, (1996).

[26] J. Pickands, Statistical inference using extreme order statistics, *Ann. Statist.* **3**, 119–131, (1975).

[27] N. Mole, C. W. Anderson, S. Nadarajah, and C. Wright, A generalized Pareto distribution model for high concentrations in short-range atmospheric dispersion, *Environmetrics.* **6**, 595–606, (1995).

[28] C. W. Anderson, N. Mole, and S. Nadarajah, A switching poisson process model for high concentrations in short-range atmospheric dispersion, *Atmos. Environ.* **31**, 813–824, (1997).

[29] T. P. Schopflocher, An examination of the right-tail of the PDF of a diffusing scalar in a turbulent flow, *Environmetrics.* **12**, 131–145, (2001).

[30] R. J. Munro, P. C. Chatwin, and N. Mole, The high concentration tails of the probability density function of a dispersing scalar in the atmosphere, *Boundary-Layer Met.* **98**, 315–339, (2001).

[31] L. Clarke and N. Mole, Modelling the evolution of moments of contaminant concentration in turbulent flows, *Environmetrics.* **6**, 607–617, (1995).

[32] F. Labropulu and P. J. Sullivan, Mean-square values of concentration in a contaminant cloud, *Environmetrics.* **6**, 619–625, (1995).

[33] N. Mole, The α-β model for concentration moments in turbulent flows, *Environmetrics.* **6**, 559–569, (1995).

[34] N. Mole, The large time behaviour in a model for concentration fluctuations in turbulent dispersion, *Atmos. Environ.* **35**, 833–844, (2001).

[35] P. J. Sullivan, The influence of molecular diffusion on the distributed moments of a scalar PDF, *Environmetrics.* **15**, 173–191, (2004).

Chapter 2

HIERARCHY AND INTERACTIONS IN ENVIRONMENTAL INTERFACES REGARDED AS BIOPHYSICAL COMPLEX SYSTEMS

DRAGUTIN T. MIHAILOVIC[1], IGOR BALAZ[2]

[1]*Faculty of Agriculture, University of Novi Sad, Novi Sad, Serbia*
[2]*Scientific Computing Laboratory, Institute of Physics, Belgrade, Serbia*

The field of environmental sciences is abundant with various interfaces and is the right place for the application of new fundamental approaches leading towards a better understanding of environmental phenomena. For example, following the definition of environmental interface by Mihailovic and Balaž [23], such interface can be placed between: human or animal bodies and surrounding air, aquatic species and water and air around them, and natural or artificially built surfaces (vegetation, ice, snow, barren soil, water, urban communities) and the atmosphere. Complex environmental interface systems are open and hierarchically organised, interactions between their constituent parts are nonlinear, and the interaction with the surrounding environment is noisy. These systems are therefore very sensitive to initial conditions, deterministic external perturbations and random fluctuations always present in nature. The study of noisy non-equilibrium processes is fundamental for modelling the dynamics of environmental interface systems and for understanding the mechanisms of spatio-temporal pattern formation in contemporary environmental sciences, particularly in environmental fluid mechanics. In modelling complex biophysical systems one of the main tasks is to successfully create an operative interface with the external environment. It should provide a robust and prompt translation of the vast diversity of external physical and/or chemical changes into a set of signals, which are "understandable" for an organism. Although the establishment of organisation in any system is of crucial importance for its functioning, it should not be forgotten that in biophysical systems we deal with real-life problems where a number of other conditions should be reached in order to put the system to work. One of them is the proper supply of the system by the energy. Therefore, we will investigate an aspect of dynamics of energy flow based on the energy balance equation. The energy as well as the exchange of biological, chemical and other physical quantities between interacting environmental interfaces can be represented by coupled maps. In this chapter we will address only two illustrative issues important for the modelling of interacting environmental interfaces regarded as complex systems. These are (i) use of algebra for modelling the autonomous establishment of local hierarchies in biophysical systems and (ii) numerical investigation of coupled maps representing exchange of energy, chemical and other relevant biophysical quantities between biophysical entities in their surrounding environment.

Keywords: environmental interface; nonlinearity; chaos; logistic equation; energy balance equation; coupled maps; hierarchy; complex system

1. Introduction

The field of environmental sciences is abundant with various interfaces and is the right place for the application of new fundamental approaches leading towards a better understanding of environmental phenomena. We defined the environmental interface as an interface between two either abiotic or biotic environments which are in relative motion exchanging energy through biophysical and chemical processes and fluctuating temporally and spatially regardless of its space and time scale [23]. This definition broadly covers the unavoidable multidisciplinary approach in environmental sciences and also includes the traditional approaches in sciences that deal with environmental space. The environmental interface as a complex system is a suitable area for the occurrence of irregularities in temporal variations of some physical or biological quantities describing their interactions [28] [30] [33]. For example, such interface can be placed between: human or animal bodies and surrounding air, aquatic species and water and air around them, and natural or artificially built surfaces (vegetation, ice, snow, barren soil, water, urban communities) and the atmosphere. The environmental interface of different media was considered in different contexts [11] [18] [26] [27]. Complex environmental interface systems are open and hierarchically organised and interactions between their parts are nonlinear, while their interaction with the surrounding environment is noisy. These systems are therefore very sensitive to initial conditions, deterministic external perturbations and random fluctuations always present in nature. The study of noisy non-equilibrium processes is fundamental for (i) modelling the dynamics of environmental interface systems and (ii) understanding the mechanisms of spatio-temporal pattern formation in contemporary environmental sciences, particularly in environmental fluid mechanics [8]. Recently, considerable effort has been invested to develop an understanding of how different fluctuations arise from the interplay of noise, forcing, and nonlinear dynamics. The understanding of complexity in the framework of environmental interface systems may be enhanced by starting from the so-called simple systems in order to catch the phenomena of interest and then adding details that introduce complexity at many levels. In general, the effects of small perturbations and noise, which is ubiquitous in real systems, can be quite difficult to predict and can often yield counterintuitive behaviour. Even low-dimensional systems exhibit a huge variety of noise-driven phenomena, ranging from a less ordered to a more ordered system dynamics. Before proceeding further, several terms require detailed clarification. The term *complex system* we use in Rosen's sense [28] as explicated in the comment by Colier [6]: "In Rosen's sense a complex system cannot be decomposed non-trivially into a set of parts for which it is the logical sum. Rosen's modelling relation requires this. Other notions of modelling would allow complete models of Rosen style complex systems, but the models would have to be what Rosen calls *analytic*, that is, they would have to be a logical product. Autonomous systems must be complex. Other types of systems may be complex, and some may go in and out of complex phases". Also, the term *complexity* can entail a lot of ambiguities, since there is a great variety of its uses. Sometimes [e.g., 28] it just refers to systems that cannot be modelled precisely in all respects. However, following [2], the term "complexity" has three levels of meaning: (a) there is self-organization and emergence in complex systems [9], (b) complex systems are not organized centrally but in a distributed manner—there are many connections between the system's parts [9] [17] and (c) it is difficult to model complex systems and to predict their behaviour, even if one knows to a large extent the parts of such systems and the connections between the parts [9] [12].

In the past years the study of deterministic mathematical models of environmental systems has clearly revealed a large variety of phenomena, ranging from deterministic chaos to the presence of spatial organization. The chaos in higher dimensional system is one of the focal subjects of physics today. Along with the approach starting from modelling physical systems with many degrees of freedom, there emerged a new approach, developed by Kaneko [16], to couple many one-dimensional maps to study the behaviour of the system as a whole. However, this model can only be applied to study the dynamics of a single medium such as the pattern formation in a fluid. What happens if two media border on each other like environmental interface? One may naturally lead to the model of coupled logistic maps with different logistic parameters. Even two logistic maps coupled to each other may serve as the dynamical model of driven coupled oscillators [21]. It has been found that two coupled identical maps possess several characteristic features which are typical of higher dimensional chaos. This model of coupling can be applied, for example, on the modelling of energy exchange between two interacting environmental interfaces [22]. In modelling the processes on environmental interfaces we should keep in mind that in such interacting biophysical systems hierarchical relations are always established. Practically it means that we cannot directly compare interactions from different hierarchical levels. Their mutual relations are always mediated through particular segments of underlying processes, which serve as inputs/outputs of functional regulations. In order to formally represent this, we cannot use standard tools from mathematical analysis. Instead we need to use a more general algebraic approach under which we can construct subsystems with different local rules [4].

In this chapter we will address only two illustrative issues important for the modelling of interacting environmental interfaces regarded as complex systems. These are (i) use of algebra for modelling the autonomous establishment of local hierarchies in biophysical systems and (ii) numerical investigation of coupled maps representing (ii-a) an approach in analysis of the energy balance equation for environmental interface and energy exchange between interacting environmental interfaces and (ii-b) substance exchange between biophysical entities in their surrounding environment.

2. Algebraic Representation of Local Hierarchies in Biophysical Systems

As it has already been implied, one of the pillars of establishing and maintaining functionality in biophysical systems is the successful formation of an operative interface with the external environment. It should provide a robust and prompt translation of the vast diversity of external physical and/or chemical changes into a set of signals that are "understandable" for the system [1] [19]. Through physiological processes, such signals are then processed, resulting in the activation of adaptive reactions to observed changes. However, the process of creating understandable signals from raw environmental material is far from simple and unambiguous. There have been several attempts to analyse it, either from a wider perspective, by developing the general theory of signs by C.S. Peirce [for an overview see: 3], or from more particular perspectives, taking human language [29] and animal communication [13] as focal points. Although they provide us with some very useful concepts and make a solid path for further investigations, these attempts are primarily developed as philosophical considerations and as such have to be further operationalised for our present purpose. Therefore, in this chapter we will take the most general concepts developed in the aforementioned literature and we will postulate them as starting points for our investigation, where: (i) each individual is considered as an observer who resides within some environment and (ii) observation is an act of assimilating a segment of external changes with the appropriate set of internal operations in order to produce a reaction. Therefore, we can state that we see the outside world, i.e. the world of phenomena (*ambience*), from the observer's perspective (the inner world). In the ambience, there are *systems* of different levels of complexity and their environments. The *system* in the ambience is a collection of perceptions, while whatever lies outside, like the component of a set, constitutes the environment [28] [31]. The fate of science lies in the fact that it is focused on the system [28]. Furthermore, the system is described by states (determined by observations), while the environment is characterized through its effects on the system. Moreover, all of these concepts (system, state, effect), when taken in biological context, presuppose the establishment of hierarchies between interacting parts. Therefore, our goal is to construct a formal system where hierarchical levels can be formed by the system itself without explicit action of some external agent (e.g. modeller, controller...). Beyond the existence of explicit regulative mechanisms, the preconditions for performing such a task are: (i) decomposition of the system's process(es) into n subprocesses, where the performance of each subprocess can be influenced by other systemic elements (either regulators or outputs of other processes/subprocesses) and (ii) existence of mutual distribution of priority and interdependence of actions.

Both of these characteristics can be obtained by using the strategy of constructing a formal system as a decision-making system, developed by Mesarovic et al. [20]. If we have input/output relation $S \subseteq P \times E$, where the population of systemic elements is denoted by $P = \{p_1, p_2, p_3, ..., p_n\}$, and the set of external influences by $E = \{e_1, e_2, e_3, ..., e_n\}$, it can be regarded as decision-making if there is given a family of decision problems $D_p | p \in P$, and the solution set Γ, where mapping $T : \Gamma \to E$ is such that $\forall (p \in P \wedge e \in E), (p, e) \in S$ only if $\exists \gamma \in \Gamma$ such that γ is solution to D_p and $T(\gamma) = e$. Having such background, the decision making hierarchy can be easily represented through the use of partially ordered sets. In other words, if Ψ is a set of systemic elements, and binary relation $>$ is a partial ordering over that set, then we can say that $(\Psi, >)$ is a partially ordered set (poset) with hierarchical ordering. Further, if we introduce Ψ as a finite family of systems $S_i | i \in I$ where $I = \{i, j, k, ...\}$ is a finite index set and if $>$ is a strict partial ordering of I, then $(\Psi, >)$ is a hierarchy of systems. If $(\Psi, >)$ is a hierarchy of decision making system such that $i > j$ iff $\psi_j \in \Psi$ have priority of action over $\psi_j \in \Psi$, then we can say that (Ψ, >) is a decision making hierarchy. From such hierarchy, following Mesarovic et al. [20], we can construct a simple multi-level hierarchical system by separating set I into an arbitrarily chosen number of subsets such that $\Psi^1 = \{S_i \mid i \in I_1\}$ represent the first level where $I_1 = \{i \mid i$ is a minimal element of $I\}$. Accordingly, the i-th level is

$$\Psi^i = \{S_k \mid k \in I_i\} \tag{1}$$

where

$$I_i = \{k \mid k \text{ is a minimal element of } I - [I_1 \cup I_2 \cup ... \cup I_{i-1}]\}. \tag{2}$$

Here, for a given set Ψ, the index set is associated with its elements in accordance with their properties and following (1) and (2) multi-level hierarchy can be established. However, if we want to represent hierarchy in biological systems, we will face two situations where such approach would fail, i.e. external, unentailed action would be necessary in order to re-establish the proper hierarchy. Firstly, in different contexts, properties of a single element can be evaluated in different ways. More precisely, if $S_i < S_j$ and $S_j < S_k$, it is not necessary that $S_i < S_k$. At first sight, it may look like a violation of the rule of transitivity, but as we will see later it is not exactly the case. Secondly, a new element can be introduced, and its place in the hierarchy (eventually) should be autonomously determined. This is not a problem when new elements are part of a formal system implicitly introduced through the already given elements of Ψ (e.g. succession of real numbers, or succession of metabolic transformation of substrate in the form $...A_3 \rightarrow A_4 \rightarrow A_5 \rightarrow A_6...$). However, if a new element is not part of the previously introduced formal system (e.g. introduction of complex number, or introduction of metabolic transformation $A_4 \rightarrow B_1$), the organization of multi-level hierarchy will be restricted to the character of imposed binary relation, or, in the best case, to a family of relations for which the system is designed.

It seems that Mesarovic's scheme is the most optimal solution for externally designed systems where goal and purpose precede organization itself. In that case, distribution of levels and corresponding compartmentalization into subprocesses are created having some particular purpose in mind. If the goal is changed, the system can only be re-organized by purposeful external action. On the other hand, biophysical systems are organized in a manner that enables the incorporation of additional elements and processes without necessarily disturbing the already established functional structures. It is sure that such organization demands random searching through potentially vast state-space, which is an undesired strategy in contemporary process design where one of the crucial requirements is the need for instant optimality (i.e. the performance of a process should be optimal from the moment of putting that process into work). However, regardless of its non-optimality, it is also the only strategy which enables autonomous evolvability of the system.

We propose that the main strategy toward the proposed goal is the ability of a constructed formal system to establish local hierarchies, instead of one, general hierarchy for the whole system. In that way, we will bypass validity of transitivity for all systemic elements without violating the transitivity principle *per se.*

In order to introduce locally created hierarchies, we should change the manner in which index set I is created, since it should not be applicable to the whole set Ψ. Keeping parallelism with empirical situation, where e.g. each enzyme in accordance with its structure determines the set of external compounds with which it can react, each $\psi \in \Psi \equiv S \subseteq P \times E$ should posses a certain set of attributes through which it will define its subjective environment. In order to do that, we should clarify the functioning of the input/output relation $S \subseteq P \times E$. At each moment in time (for the sake of simplicity we will apply the notion of objective time external to the system presented as a succession of evenly distributed successions of temporal points), the state of the population P is determined by the collection of mappings $\omega : e \rightarrow p | e \in E, p \in P$, while the succession of states can be obtained by introducing temporal sequences of events, such that $E = \{e : T \rightarrow I\}$ and $P = \{p : T \rightarrow R\}$, where T is a set of time points t, I is a set of external stimuli on a particular agent such that at each time point the system receives stimulus $i(t)$ and R there is a set of appropriate responses, $r(t)$. Therefore, we should explicitly formalize preconditions for creating set I. As it has already been stated, what can be defined as functional external stimuli is determined by the structure of functional elements and basically is of binary nature: only those changes in vicinity which activate functional operations will be perceived, so any part of the environment can be either "invisible" or can act as a negative/positive activator of some functional process. Formally, if we define set $O = \{o_1, o_2, o_3 ... o_n\}$ as a set of all environmental objects, and set $M = \{m_1, m_2, m_3 ... m_n\}$ as a set of attributes (dependent on structure of functional elements) such that $\Lambda \subseteq O : \{ \forall \lambda \in \Lambda \,|\, (\lambda, m) \in \Theta \}$, where Θ is a binary relation between M and Λ, then set Λ is dependent and restricted to the available scope of set M. In other words, set Λ is a set of all objects which can be perceived by systemic functional elements. Further, we can iterate the same process again and apply it on the elements of Λ:

$$\Lambda_{\lambda_i} \subseteq \Lambda : \{ \forall m \in M, \lambda_i \in \Lambda \,|\, (\lambda_i, m) \in \Theta \} \tag{3}$$

in order to obtain a set of observable elements Λ_{λ_i} for each $\lambda_i \in \Lambda$. If we incorporate this finding into the definition of multi-level hierarchical systems described above, we can se that

$$\Lambda_{\lambda_i} = \{ \psi_i \in \Psi \} \cup \{ o_i \in O \}. \tag{4}$$

In terms of real life systems, it means that the observable vicinity of each functional element of a biological system is composed of other systemic functional elements (e.g. enzymes), as well as of extra-systemic compounds (e.g. substrates). However, if we now create a poset $(\Psi, >)$ where partial ordering relation over the set Λ_{λ_i} is derived from the structure of each observer $\lambda_i \in \Lambda$, because of (3), hierarchy will be only local, applicable merely to a subset of elements. Also, index set I, is derived from M using appropriate mapping. Finally, if we now apply (1) and (2), we obtain:

$$\Lambda_{\lambda_i} = \{ S_k \,|\, k \in I_i \} \cup \{ o_i \in O \}. \tag{5}$$

By this, each functional element is able to create its own subjective environment with locally established hierarchies dependent on a portion of the totality of available attributes. Therefore, for each Λ_{λ_i} hierarchy is defined by the applied set of attributes and can be represented by a simple linear ordering of priorities.

3. Energy Balance Equation for Environmental Interface

Although the establishment of organisation in any system is of a crucial importance for its functioning, it should not be forgotten that we are dealing with real-life problems in biophysical systems where a number of other conditions should be reached in order to put the system to work. Undoubtedly, one of the key conditions is the proper supply of the system by the energy. In biological systems as part of biophysical ones, for example, this can be achieved by various mechanisms like assimilation, transpiration, chemical transformations, etc. In all of these cases, the survival of individual entities depends on the balance between energy reached and energy spent. Therefore, in this section we will investigate the dynamics of energy flow based on the energy balance equation. In its basic form it includes temperature differences between the underlying surface and surrounding environment. However, it can be used in a more general form for analysis of energy balance of any environmental interface. We keep it in the basic form, since it is suitable not only for investigation of biophysical systems but also for differently created environmental interfaces. Since all the energy transfer processes occur in the finite time interval, we shall immediately write this equation in terms of finite differences, i.e. in the form of difference equation

$$DT_i = F_n \tag{6}$$

where D is the finite difference operator defined as $DT_i = (T_{i,n+1} - T_{i,n})/Dt$, T_i is the environmental interface temperature, n is the time level, Dt is the time step, $F_n = (R_n - H_n - E_n - S_n)/c_i$ is defined at the nth time level, R is net radiation, H and E are sensible and latent heat, respectively, transferred by convection, and S is the heat transferred by conduction into deeper layers of underlying matter while c_i is the environmental interface soil heat capacity per unit area. Eq. (6) is also written in the finite difference form for an additional reason. It can be explained if we follow the comprehensive consideration provided by van der Vaart [32] about replacing given differential equations by appropriate difference equations in modelling of phenomena in physical and biological world. According to him many mathematical models for physical and biological problems have been and will be built in the form of differential equations or systems of such equations. With the advent of computers, one has been able to find (approximate) solutions for equations that used to be intractable. Many of the mathematical techniques used in this area amount to replacing the given differential equations by appropriate difference equations, so extensive research has been done into how to choose appropriate difference equations whose solutions are "good" approximations to the solutions of the given differential equations.

In Eq. (6) the sensible heat is calculated as $C_H(T_i - T_a)$, where C_H is the sensible heat transfer coefficient and $T_a(t)$ is the gas temperature given as the upper boundary condition. The heat transferred into underlying soil matter is calculated as $C_D(T_i - T_d)$, where C_D is the heat conduction coefficient, while

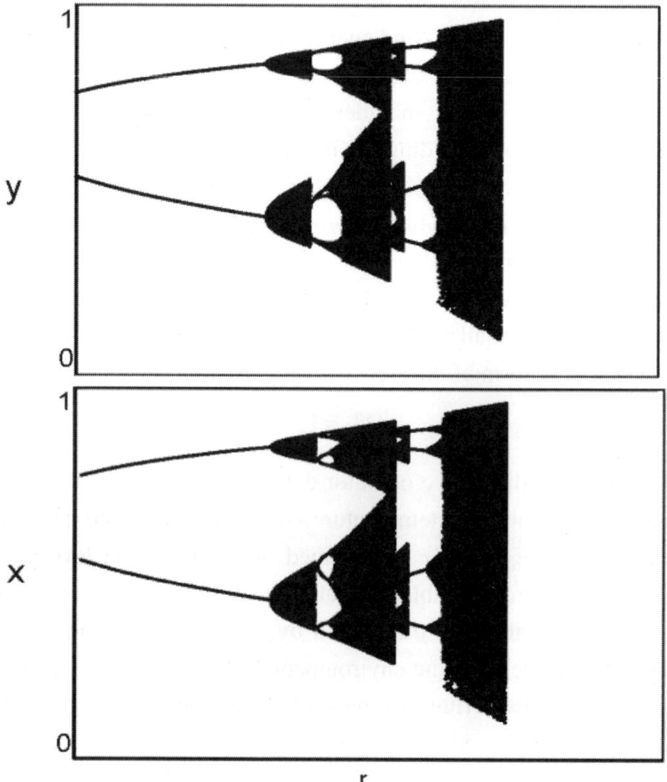

Figure 1. Bifurcation diagram for coupled maps with r ranging from 3 to 4 and $\varepsilon = 0.06$. For each values of r, we used the final point of the previous r value and 1500 iterates are plotted. This shows the period-doubling sequence as well as quasi-periodic and chaotic regions.

$T_d(t)$ is the temperature of the deeper layer of underlying matter that is given as the lower boundary condition. Following Bhumralkar [5] and Holtslag and van Ulden [15], the net radiation term can be represented as $C_R(T_i - T_a)$, where C_R is the radiation coefficient. For small differences of T_a and T_d, according to Mihailovic et al. [24], the expression for the latent heat can be written in the form $C_L f(T_a)[b(T_i - T_a) + b^2(T_i - T_a)^2]/2$. Here, C_L is the latent heat transfer coefficient, $f(T_a)$ is the gas vapour pressure at saturation and b is a constant characteristic for a particular gas. Calculation of time dependent coefficients C_R, C_H, C_L and C_D can be found in Monteith and Unsworth [25]. After collecting all terms in Eq. (6) we get

$$DT_i = A_1(T_{i,n} - T_{a,n}) - A_2(T_{i,n} - T_{a,n})^2 - A_3(T_{d,n} - T_{a,n}) \tag{7}$$

where $A_1 = [C_R - C_H - C_L bf(T_a)]/c_i$, $A_2 = b^2 f(T_a)/(2c_i)$ and $A_3 = C_D/c_i$ are coefficients also depending on Dt. With $Dt_p = 1/(A_1 - C_D/c_i)$ we indicate the scaling time range of energy exchange at the environmental interface including coefficients that express all kinds of energy reaching and departing from the environmental interface. For any chosen time interval, for solving Eq. (7), there always exists $Dt_{p,l} = Min[Dt_p(c_i, C_R, C_H, C_L)]$ when energy at the environmental interface is exchanged in the fastest way by radiation, convection and conduction. If we define dimensionless time $\tau = Dt/Dt_{p,l}$ and use the lower boundary condition $T_{d,n} = T_{a,n} - (c_i/C_D)DT_a$, then Eq. (7), after some transformations, takes the form of the logistic equation, i.e.

$$\Gamma_{n+1} = \beta \Gamma_n (1 - \Gamma_n) \tag{8}$$

where the symbols introduced have the following meaning: Γ is the dimensionless quantity [24]. while $\beta = 1 + \tau$ is taking values in the interval $1.0 < \beta < 4.0$.

4. Interaction Between Environmental Interfaces

Under the aforementioned conditions, Eq. (8) represents energy exchange at a uniform environmental interface. In nature, however, we usually encounter a mixture of two or more environmental interfaces, for example, a surface covered by spots consisting of different plant communities and barren soil. In this case there exist a number of interacting environmental interfaces. Therefore, the energy exchange between them is more complex because it has to be described with more equations having the form of Eq. (8). Besides the energy, these environmental interfaces can also exchange many other biological, chemical or physical quantities. Like many other interesting physical problems [14], interaction between environmental interfaces can be described by the dynamics of coupled oscillators. In order to study their behaviour as a function of coupling strength and nonlinearity, we consider the dynamics of two coupled maps belonging to the same universality class as oscillators. Here we first consider techniques that can be used to characterize orbit of discrete maps as periodic, quasiperiodic or chaotic. These methods are then applied to investigate the effect of coupling two period-doubling maps. Consider the general vector mapping

$$\vec{x}_{n+1} = \vec{F}(\vec{x}_n), \qquad n = 0,1,\ldots \tag{9}$$

and its *Nth* iterate $\vec{F}^{(N)}(\vec{x}) \equiv \vec{F}^{(N-1)}(\vec{F}(\vec{x}))$ with $\vec{F}^{(1)}(\vec{x}) \equiv \vec{F}(\vec{x})$. The asymptotic behaviour of a series of iterates of the map can be characterized by the largest

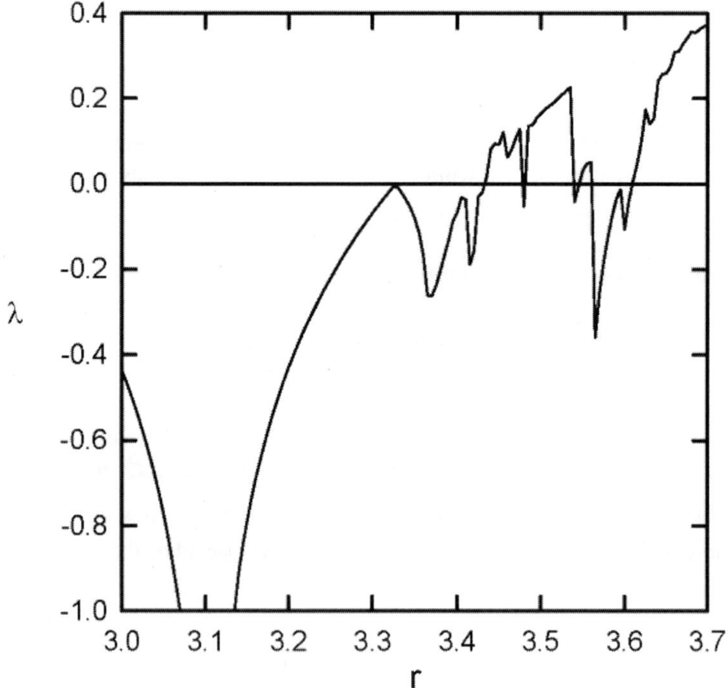

Figure 2. Lyapunov exponent for the coupled maps as a function of r ranging from 3.0 to 3.7 and with $\varepsilon = 0.06$. Each point was obtained by iterating many times from the initial condition to eliminate transient behaviour and then averaging over another 50 000 iterations. Initial conditions were $x = 0.2, y = 0.4$ with $1000r$ values.

Lyapunov exponent, which, for an initial point $\vec{x}_0$ is an attracting region, is defined to be

$$\lambda = \lim_{N \to \infty} \{ \ln[\| D^{(N)}(\vec{x}_0) \|] / N \} \tag{10}$$

where $\| D^{(N)} \|$ is the norm of the derivative matrix

$$D_{ij}^{(N)}(\vec{x}) = \frac{\partial \vec{F}_i^{(N)}(\vec{x})}{\partial x_j}. \tag{11}$$

This exponent measures how rapidly two nearby orbits in attracting region converge or diverge. It can be evaluated by noting that $D^{(N)}(\vec{x}_0) = D^{(N-1)}(\overrightarrow{F(\vec{x}_0)})D(\vec{x}_0)$; so if $\vec{x}_0, \vec{x}_1, \vec{x}_2, \ldots$ are successive iterates of the map, then

$$D^{(N)}(\vec{x}_0) = D(\vec{x}_{N-1}) \cdots D(\vec{x}_1) D(\vec{x}_0). \tag{12}$$

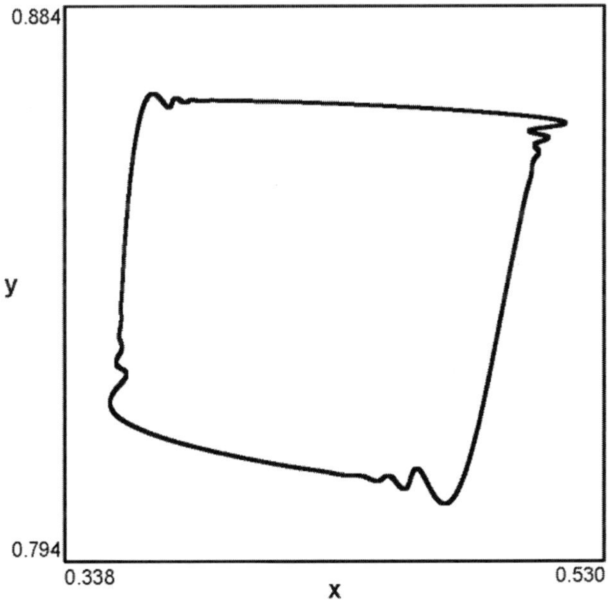

Figure 3. Plot of the iterates of the map (x_n, y_n) for quasiperiodic motion when $r = 3.39$, $\varepsilon = 0.06$, and initial point $x = 0.2, y = 0.4$.

In practice, λ is computed by initially iterating the map many times to eliminate transient behaviour and then using a large number N of successive points to compute the derivative matrix as indicated in Eq. (11). Finally, the quantity $\ln[\|D^{(N)}(\vec{x}_0)\|]/N$ is used as an approximate value of the Lyapunov exponent for the attracting region [10]. This exponent provides a way to distinguish among periodic, quasiperiodic, and chaotic motion. Specifically, if $\vec{x}_0$ is part of a stable periodic orbit of length K, then the norm of the derivative matrix $\|D^{(K)}(\vec{x})\|$ will be less than one for every $\vec{x}$ in the K cycle. Thus the exponent will be negative and will characterize the rate at which small perturbations from the fixed cycle decay. A zero value for the exponent indicates quasiperiodic behaviour in which nearby paths maintain their distance on average. Finally, when λ becomes positive, nearby points in the attracting region diverge from each other giving chaotic motion. In general, the exponent will depend on the

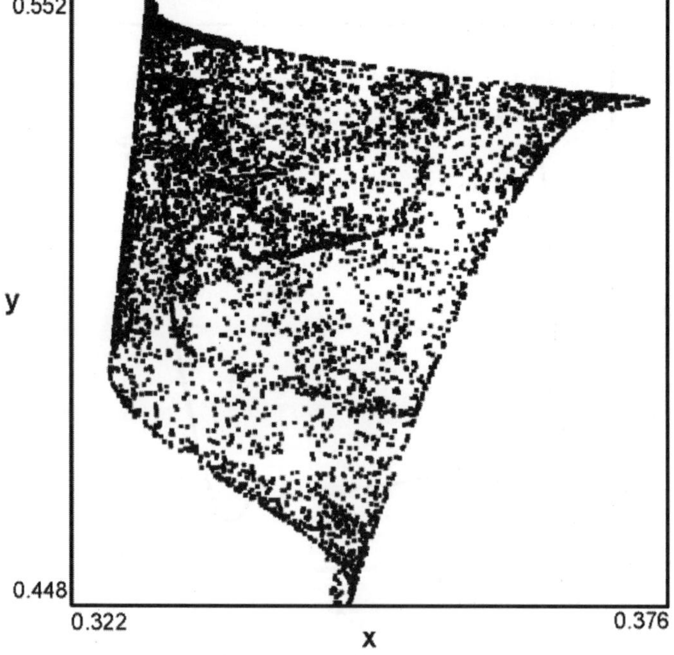

Figure 4. Plot of the iterates of the map (x_n, y_n) for chaotic motion when $r = 3.615$, $\varepsilon = 0.06$, and initial point $x = 0.2$, $y = 0.4$.

initial point used in the iteration because there may be several stable attractors, each with a separate basin of attraction [14].

We now use these techniques to examine the effect of coupling two nonlinear maps that display period doubling. In particular, we consider two-dimensional mapping

$$x_{n+1} = rx_n(1 - x_n) + \varepsilon(y_n - x_n) \tag{13a}$$

$$y_{n+1} = ry_n(1 - y_n) + \varepsilon(x_n - y_n) \tag{13b}$$

where the nonlinearity parameter (logistic parameter) r is such that $0 \le r \le 4$ and the coupling parameter ε satisfies $0 \le \varepsilon \le 1$. This map displays a wide range of behaviour, as the parameters r and ε are varied including periodic, quasiperiodic, and chaotic motion. Note that the derivative matrix D of Eq. (12) at the point (x, y) is given by

$$D(x, y) = \begin{bmatrix} r - \varepsilon - 2rx & \varepsilon \\ \varepsilon & r - \varepsilon - 2ry \end{bmatrix}. \tag{14}$$

In particular, the norm of the derivative matrix shows that there is a stable fixed point for the region in which $r < 3 - 2\varepsilon$. The in-phase 2 cycle is stable for $1 + 2(1 + \varepsilon + \varepsilon^2)^{1/2} \le r \le 1 + 6^{1/2} = 3.449$ with $\varepsilon \le (3^{1/2} - 1)/2 = 0.37$. Similarly, the out-of-phase case is stable for $3 - 2\varepsilon \le r \le 1 + (6 - 10\varepsilon + 4\varepsilon^2)^{1/2}$. Note that these regions overlap, indicating that the coupled maps have multiple basins of attraction for some parameter values. For nonzero values of the coupling constant, instead of simple period doubling, successive periods can be separated by regions of quasiperiodic motion in which the frequencies of the two oscillators incommensurate. These transitions regions also contain various locked orbits of high period, including an asymmetric period in which x and y cycle over a distinct set of values.

A more detailed picture of this map can be obtained by investigating the behaviour across section of the phase diagram. Specifically, we chose the value $\varepsilon = 0.06$, which gives typical results from this map. A bifurcation diagram for a particular choice of initial conditions $(x = 0.2, y = 0.4)$ is given in Figure 1. This figure was obtained by starting with $r = 3$ and incrementing in small steps up to $r = 4$. At each new r value the final (x, y) iterate of the previous values was used as the initial point. This corresponds to continuously changing the system's parameters without restoring the original state of the system. The behaviour of the Lyapunov exponent for the above initial condition is shown in Figure 2 for r in the range 3.0 - 3.7 and $\varepsilon = 0.06$. The value of the second Lyapunov exponent for these parameters is so close to that of the largest exponent that they are not distinguishable on the scale of the plot.

Note that the second Lyapunov exponent is computed by using the minimum value of $|A\vec{v}|/|\vec{v}|$, instead of the norm $\|A\| \equiv \max_{\{\vec{v}\}}\{|A\vec{v}|/|\vec{v}|\}$, where $\vec{v}$ ranges over all nonzero N vectors. In particular, the plot in Figure 2 shows that the transition from out-of-phase period 2 to 4 is separated by a quasiperiodic region in which the exponent is zero, as well as a chaotic region with positive exponent. The detailed transition out of the quasiperiodic region involves a complicated series of periodic orbits [14].

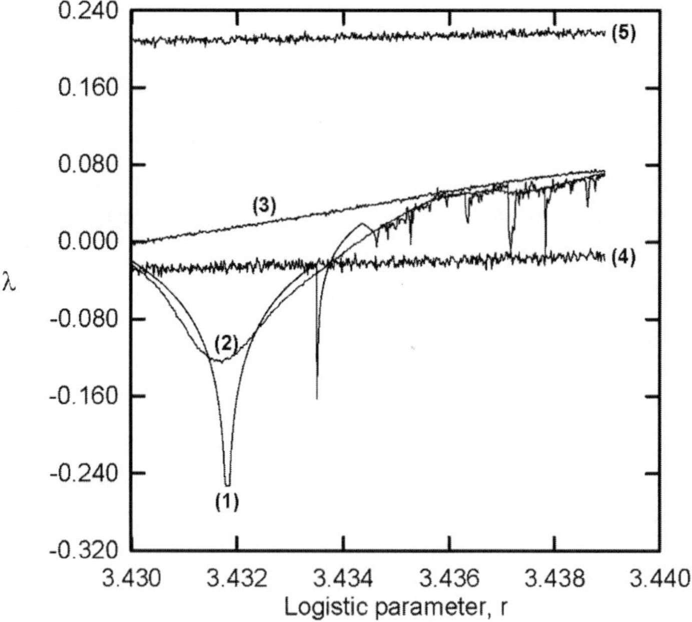

Figure 5. Effect of additive noise on the scale structure of the Lyapunov coefficient for r ranging from 3.430 to 3.440, $\varepsilon = 0.06$ and initial point $x = 0.2, y = 0.4$. Here (1) refers absence of noise while the noise amplitude takes values $\alpha = 0.0001$ (2), 0.001 (3), 0.01 (4) and 0.1 (5). Plot consists of $501r$ values each of which was obtained by iterating 100000 times from the initial condition and then averaging over another 50 000 iterations.

Many successive frequencies locking with high periods appear before the period-4 orbits. A plot of these iterates of the map for a locked periodic orbit nearby quasiperiodic motion is given in Figure 3. Note that the iterates of the map in this region actually form two groups of points located symmetrically around the diagonal $x = y$. For clarity, only one of these groups is actually shown in the figure. Figure 4 shows the iterates of a particular chaotic orbit.

We already emphasized that the study of noisy non-equilibrium processes is fundamental for modelling the dynamics of environmental interface systems. Here we consider the behaviour of the coupled maps in the presence of thermal fluctuations or other noise. As can be shown in the case of uncoupled non-linear

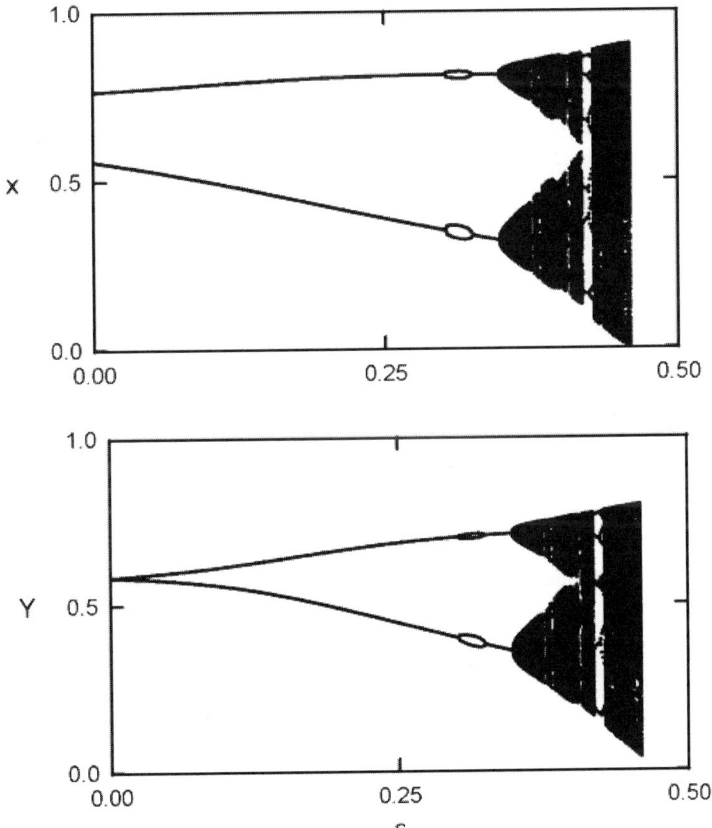

Figure 6. Bifurcation diagram for the map of Eqs. (16) with $r^{(1)} = 3.1$ and $r^{(2)} = 2.4$, and $0 \le \varepsilon \le 0.5$. For each value of ε, the map was iterated 500 times from the initial point $x = 0.2, y = 0.4$ to eliminate transients, and the next 500 iterate was plotted.

                    D.T. Mihailovic and I. Balaz

oscillators, the addition of external or parametric fluctuations has a pronounced effect on the dynamics of such systems [7]. The effect of noise was modelled by adding uniformly distributed random numbers to the map (13). Specifically, we considered the map

$$x_{n+1} = rx_n(1-x_n) + \varepsilon(y_n - x_n) + \alpha\delta_n^{(1)} \tag{15a}$$

$$y_{n+1} = ry_n(1-y_n) + \varepsilon(x_n - y_n) + \alpha\delta_n^{(2)} \tag{15b}$$

where $\delta_n^{(1)}$ and $\delta_n^{(2)}$ are random numbers uniformly distributed in the interval $[-1,1]$ and α is the amplitude of the noise. Figure 5 shows the effect of different amounts of noise on the fine scale structure of the Lyapunov exponent. An additional case occurs when the parameters of the oscillators have small, random variations due, for example, to external noise. These so-called parametric fluctuations can be simulated by modulating the values of the logistic parameters by uniform random numbers in the s small interval.

The preceding discussion considered the effect of coupling two identical maps. However, it is also of interest to examine the case in which the logistic parameters are unequal, corresponding to the coupling of two different oscillators. Specifically, we considered the map

$$x_{n+1} = r^{(1)}x_n(1-x_n) + \varepsilon(y_n - x_n) \tag{16a}$$

$$y_{n+1} = r^{(2)}y_n(1-y_n) + \varepsilon(x_n - y_n) \tag{16b}$$

with $r^{(1)} \neq r^{(2)}$. The general kind of behaviour seen is similar to the previous case, that is, the coupling produces a new quasiperiodic motion and can lead to chaos. This can be seen in the bifurcation diagram of Figure 6, which displays the behaviour of Eqs. (16) for $r^{(1)} = 3.1, r^{(2)} = 2.4$, and with $0 \le \varepsilon \le 0.5$.

In modelling complex environmental interface systems, it is also interesting to consider the behaviour of the following system of two linearly coupled maps

$$x_{n+1} = (1-\varepsilon)f(r^{(1)}x_n) + \varepsilon f(r^{(1)}y_n) \tag{17a}$$

$$y_{n+1} = (1-\varepsilon)f(r^{(2)}y_n) + \varepsilon f(r^{(2)}x_n) \tag{17b}$$

where the map $f(r,x) = rx(1-x)$ is taken to be the logistic map with logistic parameters $r^{(1)}$ and $r^{(2)}$. In the case of $r^{(1)} = r^{(2)}$, two maps soon become synchronized no matter what the initial conditions may be, i.e., coupled maps are identical with a single logistic map. Interesting is the case of $r^{(1)} \neq r^{(2)}$. In

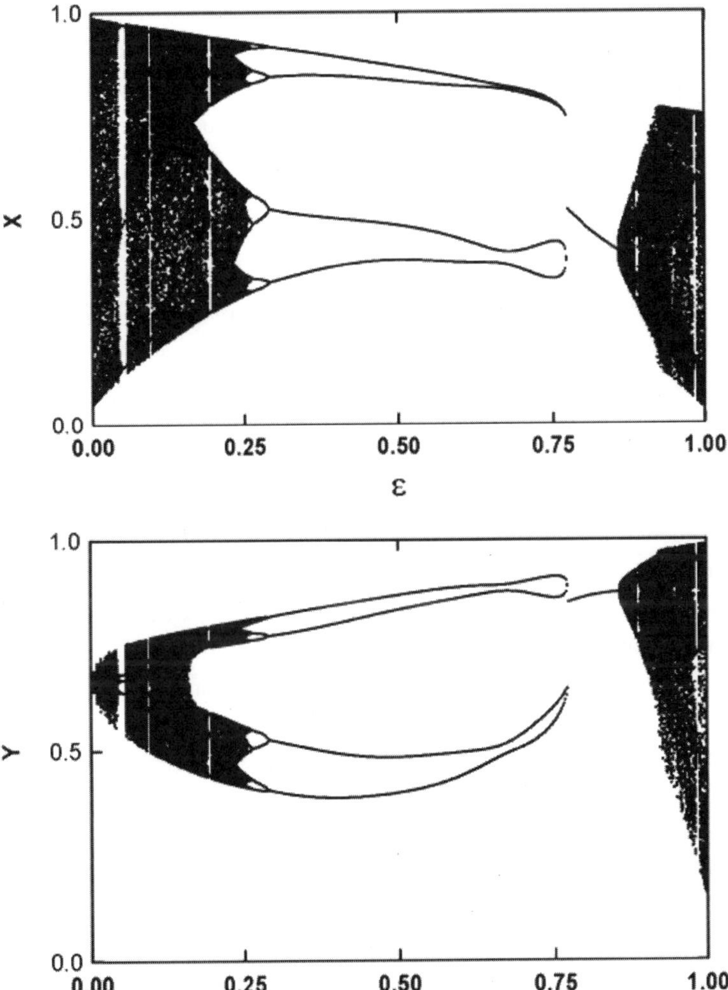

Figure 7. Bifurcation diagram for the map of Eqs. (17) with $r^{(1)} = 4.0$ and $r^{(2)} = 3.0$, and $0 \leq \varepsilon \leq 1$. For each value of ε, the map was iterated 500 times from the initial point $x = 0.2, y = 0.4$ to eliminate transients, and the next 500 iterate was plotted.

the following we fix the logistic parameters above and below the critical value $r^{(1)} = 3.56994$ for $r^{(1)}$ and $r^{(2)}$ respectively. We choose the logistic parameters $r^{(1)}$ and $r^{(2)}$ and regard the coupling parameter ε as the controlling parameter. In Figures 7 and 8, the attractors of the coupled-map are displayed as functions of coupling ε. Figure 7 shows the result of $r^{(1)} = 4$ and $r^{(2)} = 3$ while Figure 8 shows that of $r^{(1)} = 4$ and $r^{(2)} = 2$. In both cases, for each value of ε we used the final value of the previous ε and 1500 iterations were plotted. They are two typical examples of the various values of $r^{(1)}$ and $r^{(2)}$. One immediately notices several interesting features. The fact that there are two chaotic regions in both $\varepsilon = 0$ and $\varepsilon = 1$ ends seems odd at first sight, but after some reflection, one realizes that very weak ε means very strong $(1-\varepsilon)$, which brings chaos first to the variable x and then to y, however weak the coupling term may be. The most salient feature is the appearance of a stable period four cycle right after the period one around $\varepsilon = 0.77$ in Figure 8. Another case, found both in Figure 7 and Figure 8, is the sudden filling of the x and y space around $\varepsilon = 0.85$ and above. The broad window-like region with period four around $\varepsilon = 0.9$ in the case of Figure 8 is also noteworthy.

5. Conclusion

We considered a combined approach to the modelling of environmental interfaces regarded as biophysical complex systems. They are higher dimensional complex systems where both of their parts, organization and temporal dynamics, demand different kinds of formalism. Therefore, we constructed an outline of establishing local hierarchies and then we reported the results of numerical investigation on the systems of two coupled maps, representing the exchange of energy, chemical and biophysical quantities of two interacting environmental interfaces. It has been done by calculating the phase and bifurcation diagrams of the coupled maps for different values of the logistic and coupling parameters as well as by the calculation of the Lyapunov exponent. It seems that further analysis of this system will be useful for understanding the processes of the exchange of different quantities between two interacting environmental interfaces.

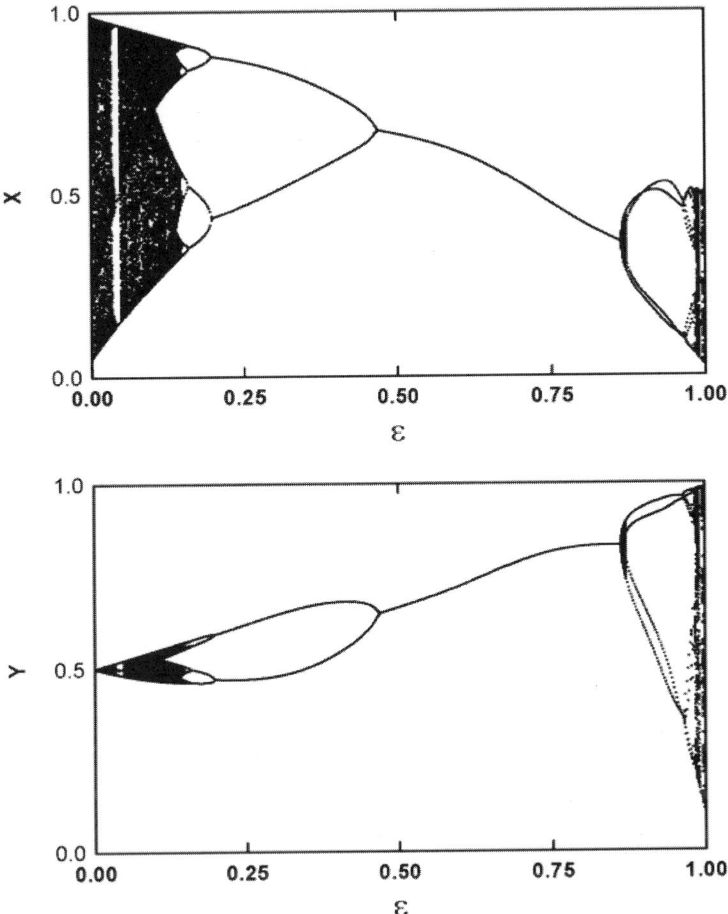

Figure 8. Bifurcation diagram for the map of Eqs. (16) with $r^{(1)} = 4.0$ and $r^{(2)} = 2.0$, and $0 \leq \varepsilon \leq 1$. For each value of ε, the map was iterated 500 times from the initial point $x = 0.2, y = 0.4$ to eliminate transients, and the next 500 iterate was plotted.

Acknowledgement

The research work described in this paper has been funded by the Serbian Ministry of Science and Technology under the project "Modelling and numerical simulations of complex physical systems", No. ON141035 for 2006-2010.

            D.T. Mihailovic and I. Balaz

Appendix - list of symbols

List of Symbols		
Symbol	**Definition**	**Dimensions or Units**
C_D	heat conduction coefficient	[Wm^{-2}K^{-1}]
C_H	sensible heat transfer coefficient	[Wm^{-2}K^{-1}]
C_L	latent heat transfer coefficient	[Jm^{-2}K^{-1}Pa^{-1}]
C_R	radiation coefficient	[Wm^{-2}K^{-1}]
D	derivative matrix	
D	finite difference operator	
$D^{(N)}$	norm of derivative matrix	
E	latent heat	[Wm^{-2}]
$\vec{F}(\vec{x}_n)$	general vector mapping	
F$_n$	nth time level	
H	sensible heat	[Wm^{-2}]
R	net radiation	[Wm^{-2}]
S	heat transferred by conduction into deeper layers of underlying matter	[Wm^{-2}]
T_a	gas temperature given as the upper boundary condition	[K]
T_d	temperature of deeper layer of underlying matter given as the lower boundary condition	[K]
T_i	environmental interface temperature	[K]
b	constant characteristic for a particular gas	[K^{-1}]
c_i	environmental interface soil heat capacity per unit area	[Jm^{-2}s^{-1}]
$f(T_a)$	gas vapour pressure at saturation	[Pa]
r	logistic parameter	
α	amplitude of the noise	
$\delta_n^{(1)}, \delta_n^{(2)}$	random numbers	
ε	coupling parameter	
λ	Lyapunov exponent	
τ	dimensionless time	

References

1. R. Amils, C. Ellis-Evans, and H. Hinghofer-Szalkay, *Life in Extreme Environments*, (Springer, 2007).
2. V. Arshinov, C. Fuchs, *Causality, Emergence, Self-Organisation;* Arshinov V., Fuchs C. Eds.; NIA-Priroda: Moscow, pp. 1-18 (2003).
3. A. Atkin, Zalta EN Ed., http://plato.stanford.edu/entries/peirce-semiotics/ (2006).
4. Balaz, and D.T. Mihailovic, Evolvable biological interfaces: Outline of the new computing system. *Fourth Biennial Meeting: International Congress on Environmental Modelling and Software,* Barcelona, Spain, 104-113 (2008).
5. C.M. Bhumralkar, *J. Appl. Meteorol.* **14**, 1246-1258 (1975).
6. J.D. Collier, *Causality, Emergence, Self-Organisation*; Arshinov V.; Fuchs C., Eds.; NIA-Priroda: Moscow, pp. 150-166 (2003).
7. J.P. Crutchfield, D. Farmer and B.A. Huberman, *Phys. Rep.* **92**, 45 (1982).
8. B. Cushman-Roisin, C. Gualtieri and D.T. Mihailovic, Environmental Fluid Mechanics: Current Issues and Future Outlook. Gualtieri C., Mihailovic D.T. Eds., Taylor & Francis, 1-16 (2008).
9. B. Edmonds, *The Evolution of Complexity*; Heylighen F., Bollen, J., Riegler, A. Eds. Kluwer: Dordrecht, pp. 1-17 (1999).
10. D. Feit, *Commun. Math. Phys.* **61**, 249 (1978).
11. J. Glazier, J. and F. Graner, *Phys. Rev. E*, **47**, 2128 – 2154 (1993).
12. F. Heylighen, *The Evolution of Complexity*; Heylighen F., Bollen, J., Riegler, A. Eds. Kluwer: Dordrecht, pp. 17-47 (1999).
13. J. Hoffmeyer, *Eur. J. Semiot. Stud.*, **9**, 355-376 (1997).
14. T. Hogg and B.A. Huberman, *Phys. Rev. A* **29**, 275-281 (1984).
15. A.A. Holtslag, and A.P. van Ulden, *J. Appl. Meteor.*, **22**, 517-529 (1975).
16. K. Kaneko, *Prog. Theor. Phys.* **69**, 1427-1442 (1983).
17. S. Kauffman, *The Origins of Order: Self-Organization and Selection in Evolution* (Oxford University Press, 1993).
18. M.L. Martins, G. Ceotto, G. Alves, C.C.B. Bufon, J.M. Silva, and F.F. Laranjeira, *Phys. Rev. E*, **62**, 7024 – 7030 (2000).
19. R.M. May, and A.R. McLean, *Theoretical Ecology: Principles and Applications* (Oxford University Press Inc. 2007).
20. M.D. Mesarovic, D. Macko, Y. Takahara, *Theory of Hierarchical, Multilevel, Systems* (Academic Press, 1970).
21. S. Midorikawa, K. Takayuki, and C. Taksu, *Prog. Theor. Phys.*, **94**, 571-575 (1995).

22. D.T. Mihailovic, Two interacting environmental Interfaces: Folded bifurcation in coupled asymmetric logistic maps. *Fourth Biennial Meeting: International Congress on Environmental Modelling and Software (iEMSs 2008)*, Barcelona, Spain, 134-140 (2008).

23. D.T. Mihailovic, and I. Balaž, *Idojaras*, **111**, 209-220 (2007).

24. D.T. Mihailovic, D.V. Kapor, B. Lalic, and I. Arsenic, The chaotic time fluctuations of ground surface temperature resulting from energy balance equation for the soil-surface system, *26th General Assembly of European Geophysical Society*, Nice, France, (2001).

25. J.L. Monteith, M. Unsworth, *Principles of Environmental Physics* (Elsevier, 1990).

26. N. Nikolov, W. Massman, A. Schoettle, *Ecol. Modelling*, **80**, 205-235 (1995).

27. D.S. Niyogi, and S. Raman, *Numerical modelling of gas deposition and bidirectional surface–atmosphere exchanges in mesoscale air pollution systems*. Boybeyi Z Ed. WIT Publications, Advances in Air Pollution, Vol. **9**, 1-51, (2001).

28. R. Rosen, *Life itself: A comprehensive inquiry into the nature, origin, and fabrication of life* (Columbia University Press, 1991).

29. F. Saussure, de, *Course in General Linguistics* (McGraw-Hill, 1965).

30. A.M. Selvam, and S. Fadnavis, *Meteor. Atmos. Phys.*, **66**, 87-112 (1998).

31. T.H. Sivertsen, *Phys. Chem. Earth.* **30**, 35-43 (2005).

32. H.R. van der Vaart, *Bull. of Math. Biol.*, **35**, 195-211 (1973).

33. M. Zupanski, and I.M, Navon, *Bull. Amer. Meteor. Soc.*, **88**, 1431–1433 (2007).

Chapter 3

SOME RECENT ADVANCES IN MODELING STABLE ATMOSPHERIC BOUNDARY LAYERS

BRANKO GRISOGONO

*Department of Geophysics, Faculty of Science & Mathematics,
University of Zagreb, Zagreb, Croatia*

The atmospheric boundary layer (ABL) is the lowest part of the atmosphere that is continuously under the influence of the underlying surfaces through mechanical (roughness and shear) and thermal effects (cooling and warming), and the overlying, more free layers. Such boundary layers and the related geophysical turbulence exist also in oceans, seas, lakes and rivers. Here we focus on those in the atmosphere; however, similar reasoning as presented here also applies to the other geophysical flows mentioned. Since most of human activities and overall life take place in the ABL, it is easy to grasp the need for an ever better understanding of the ABL: its nature, state and future evolution. In order to provide a reasonable and reliable short- or medium-range weather forecast, a decent climate scenario, or an applied micrometeorological study (for e.g. agriculture, road construction, forestry, traffic), etc., the state of the ABL and its turbulence should be properly characterized and marched forward in time in concert with the other prognostic fields. This is one of many tasks of numerical weather prediction and climate models. Many of these models have problems in handling rapid surface cooling under weak or without synoptic forcing (e.g. calm nighttime mountainous or even hilly conditions).

Overall research during the last ~ 10 years or so, strongly suggests that the evolution of the stable ABL is still poorly understood today. There we make a contribution by assessing some recent advances in the understanding of nature, theory and modeling of the stable ABL (SABL). In particular, we address inclined very (or strongly) stratified SABL in more details. We show that a relatively thin and very SABL, as recently modeled using an improved "z-less" mixing length scale, can be successfully treated nowadays; the result is quietly extended to other types of the SABL. Finally, a new generalized "z-less" mixing length-scale is proposed. At the same time, no major improvements in modeling weak-wind strongly-stable ABL is reported yet.

Keywords: convective ABL; mixing length-scale; Monin-Obukhov length; numerical modeling; parameterization; Prandtl number; Richardson number(s); stratified turbulence; "z-less" turbulence; very stable boundary layer.

1. Introduction

Most of human life takes place in the atmospheric boundary layer (ABL). The ABL behaves as an active intra- and inter-layer among various underlying surfaces on one side (e.g. sea, inclining terrain, urban areas), and the rest of the capping atmosphere. There the ABL exchanges, modulates or even alters a multitude of information ranging from radiative and moisture fluxes, to momentum, heat and species fluxes, etc. After a very brief review of a few recent advancements related to the convective ABL (CABL), we focus on the (very) stable ABL (SABL). This is justified because in the very SABL (VSABL) progressively smaller eddies still play significant roles in the nature and ultimate fate of the layer, which is usually not the case for the CABL where the largest eddies determine most of turbulent flow properties. Almost needless to say, small eddies are difficult to measure at statistically meaningful levels, as well as to calculate the related turbulent fluxes due to these relatively small eddies [21] [20] [22], which may not only exist but also significantly contribute to the fluxes. Hence, it is generally less known about the SABL than about the CABL today [3] [27]. Turbulent structures associated with the CABL and SABL are under various influences due to e.g. surface fluxes, near-surface temperature inversions, low-level jets (LLJ), wind meandering, unsteadiness, internal boundary layers, etc. [10] [4] [6] [5] [12]. These features strongly affect and often determine the ABL turbulence; therefore, these structures should be included in new ABL turbulence parameterizations for numerical weather prediction (NWP), i.e. meteorological, air-chemistry and even climate models.

A contemporary research overview of the CABL structures, with the emphasis on shear affecting the CABL evolution, is in [9]. The attention there is paid to the surface layer, mixed layer and entrainment layer – all in the view of both barotropic and baroclinic effects, recalling observations, numerical modeling and analytic works. The CABL evolution is explained in terms of Monin-Obukhov length, L, friction and convective velocity scales, u_* and w_*, respectively, and the inversion height; various results are often inferred there from the authors successful large-eddy simulations, LES. The latter technique is generally better suited for the CABL than for the SABL [3]. Although Fedorovich and Conzemius, [9], display several results, we stress here only a few. As the contribution of wind shear to the turbulence production increases, the coherent structure in the CABL alters from quasi-hexagonal cells to horizontal convective rolls oriented parallel to the mean flow vector. The CABL may behave as a single layer only in the shear-free case. Finally, most of modeled CABL structures can be described by Richardson numbers. A few other intriguing features of the CABL, e.g. dispersion of air pollutants, nonlinear interplays between advection and diffusion, etc. are in e.g. [1]. Certain ABL effects on NWP model initialization and data assimilation are in e.g. [23]. These aspects deserve a few independent review articles or even a new book.

Several specific questions will be discussed related to the SABL's stratified turbulence. We focus on e.g. a proper treatment of L over sloped surfaces, and the corresponding "z-less" length-scale [24] [25] which is active above the surface layer overlying the sloping surface (if the latter sub-layer even exists in a VSABL). These features should help preventing NWP and air-chemistry models' problems like runaway cooling, frictional decoupling [28] [11] [15], or a systematic thermal bias [28] [12]. One of the VSABL types is that due to weak-wind stable conditions [19] [5], another is katabatically driven SABL [11], etc. [22] [31] [10] [4] [6]; furthermore, one often talks about the SABL in plural (different types of SABLs). Among several reasons for the importance of these usually thin and very stratified boundary-layer flows, one of them being a more proper weather and climate simulations over e.g. Antarctica and Greenland, we also explain the vertical diffusion of the slope parallel wind component (the one induced by katabatic flows and Coriolis force). Through this vertical diffusion, in principle, the long-lived VSABL might interact with the polar vortex. A few recommendations for modeling purposes will be provided as well. This study continues on a few other recent works of the author and the collaborators [3] [11] [14] [22]. We will end up with a couple of plausible derivations, brief enough but hopefully inviting for a scientific contemplation and scrutiny while reflecting on how much of misinterpretation and parameterization mishaps we have had during the last few decades. Apparently, nature continues yet to be nice to us, to our clumsy modeling approaches and sophisticated visions that often end up in the middle of the nowhere.

In this Introduction a necessary background and an incomprehensive overview have been just given. The next (largest) section is the core of this contribution going through certain details of the SABL mentioned above. The main results are outlined in the concluding section.

2. Recent improvements in modeling the SABL

2.1. *Mixing length-scale*

Figure 1 illustrates the starting point of this study: an over-diffusive SABL in a typical mesoscale numerical model (solid curves); the profiles are taken from [11], based on their Figure 1, simulated by using MIUU model [2] [7] [8] [1] [23] [13]. The overall perspective to this problem has been also elaborated elsewhere recently [3] [28] [22]. The simulation details will be given in the next subsection; here we first wish to boldly introduce the problem of the SABL over-diffusion and its remedy in Figure 1. The dashed-dotted curves, shown on both panels in Figure 1 for the downslope velocity U and potential temperature θ, respectively, represent the corresponding simulation with the problem alleviated. The latter simulation (dash-dotted) is a more trustful one because it also corresponds to another model, i.e. a calibrated analytic Prandtl model result [11]. Of course, both models, MIUU and Prandtl, had been previously checked independently against various observations. These totally independent models both qualify as valuable tools for studying role types of SABL flows, their different level of complexity, and the sets of underlying approximations; that give viability and credibility to our approach.

The "z-less" mixing length scale has been most often defined as a local quantity (e.g. a MIUU model default):

$$l_{STAB} = a\frac{(TKE)^{1/2}}{N}.$$

Its modification is [11]:

$$l_{STAB} = \min[a\frac{(TKE)^{1/2}}{N}, b\frac{(TKE)^{1/2}}{\$}], \tag{1}$$

where *TKE* is turbulent kinetic energy, *N* and *\$* is buoyancy and shearing frequency (based on the absolute shear: $\$ = |S|$), respectively, $a \approx 0.5$ and $b = a/2$, valid for the gradient Richardson number $0 < Ri \leq 1$, $Ri = (N/S)^2$; otherwise, for $Ri > 1$, only the 1st term in (1) is kept. If (1) is applied for all $Ri > 0$, then the old formulation above (1), will be valid only for $Ri \geq 4$, provided again $b = a/2$ (namely, this validity goes in (1) as the square of a/b due to $Ri = (N/\$)^2$). Note that in the SABL, *TKE* usually scales as a local u_*^2. (Often-present factor of two around TKE due to related higher-order closures is not tracked down here for simplicity; this should be done properly while coding.) Plausibly define: the weakly stratified SABL exhibiting everywhere $0 < Ri \ll \infty$ (typically $Ri \leq 1$), and the VSABL characterised by containing (sub)regions with $Ri \gg 1$. Dash-dotted curves in Figure 1 are obtained using (1). While the over-diffusive SABL modeled, Figure 1 (solid), is much too deep, its properly modeled behavior, i.e. the VSABL (dash-dotted), is in agreement with another model (that of calibrated Prandtl, see below), and it is also numerically and physically stable (e.g. it does not show a sign of frictional decoupling as that in e.g. [15]). It is expected that (1) ought to improve simulations for other types of SABL flows too [11] [10] [4] [6] because the overall turbulence scheme deployed, a higher-order one, so called level 2.5 [2] [7], is slope insensitive. Hence, this scheme as such does not care whether a particular flow is katabatic or not. Since wind shear is generally more changeable than buoyancy frequency in the stable atmosphere, it makes sense to try to use (1).

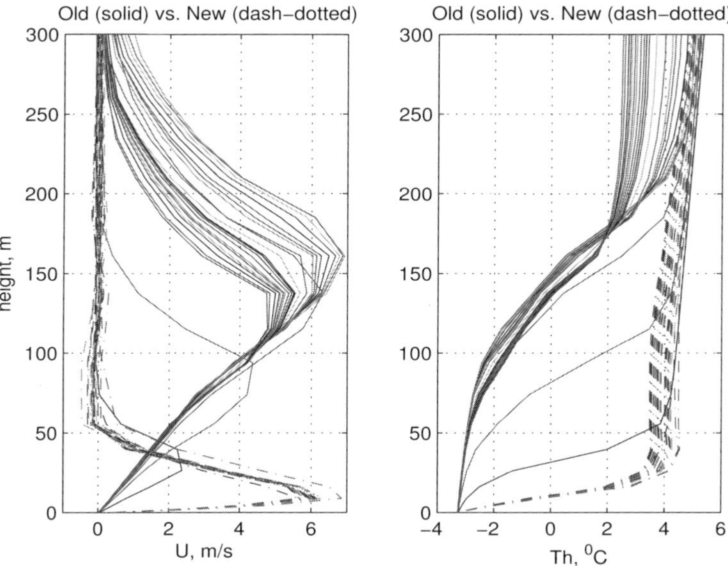

Figure 1. Two numerical simulations of same katabatic flow using two different parameterizations for the "z-less" mixing length-scale in MIUU model [11]. The profiles of the downslope wind component U (left) and potential temperature θ (right), are shown for 24 h of simulations. Over-diffusive SABL (solid) consists of an elevated LLJ and a capping inversion spreading over the lowest ~ 200 m. Using a newly proposed mixing length-scale eqn. (1), the SABL becomes much thinner (dash-dotted) and in agreement with calibrated analytic Prandtl model (see below).

The main advantage of (1) seen in Figure 1 is the prevention (dash-dotted) of an excessive vertical diffusion of the SABL in time; such over-diffusive behavior is clearly shown in both panels, U and θ, respectively (solid). We shall return to a re-derivation of (1) and its eventual new modification later. Testings for other types of (more realistic, etc.) flows are being performed elsewhere while this paper is being written.

2.2. *Coriolis effect in the VSABL*

Next, we display a few detailed, additional katabatic flow fields from MIUU model, some of these were shown to correspond very well to the calibrated Prandtl model [11]. Of course, all the fields modeled are coupled among themselves in the dynamically consistent way through the governing equations [2] [7] [8]. Figure 2 displays U and θ, Figures 2a and 2b, also from Figure 1, dash-dotted, for a constantly sloped terrain of -2.2°, under a calmly stratified background atmosphere of $\Delta\theta/\Delta z$ = 5K/km with the surface potential temperature deficit of 6.5 K (i.e. same as in [11]). An intriguing feature is in Figure 2c displaying the slope-parallel wind component, V, i.e. the component continuously diffusing upward [17] [27] [29]. Apparently, this diffusive-like behaviour of a SABL flow component has been theoretically known for a few decades in the school of Lev N. Gutman [27], but it was poorly accessible in the peer-review English literature. It is only this flow component that exhibits the overall vertical diffusion, Figure 2c, but the essential fields corresponding directly to the katabatic forcing, Figures 2a and 2b, remain confined in the lowest few tens of meters. While the simulation results shown up to Figure 2c are very recent [11], the latter figure is one of the new results of this study (shown only in conferences and the proceedings). Incidentally, the poorly modeled fields in Figure 1 (solid), are qualitatively somewhere between Figures 2a and 2b on one side, and Figure 2c, on the other side. We shall return to the V-component, Figure 2c, later.

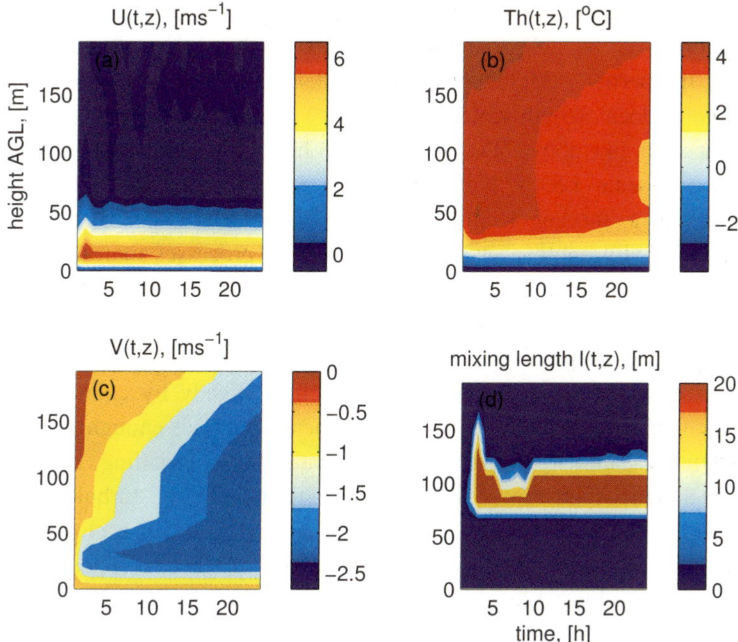

Figure 2. Displaying details from Figure 1, dash-dotted, concurring to pure katabatic wind: (a) the downslope wind-component U, (b) the potential temperature, (c) the slope parallel (Coriolis-induced) wind component V, (d) the turbulent mixing length-scale. Note that (a) and (b) correspond to dash-dotted curves on the left and right panels in Figure 1, respectively.

Figure 2d shows the relevant mixing length-scale based on (1), allowing for the whole flow field in Figure 2 (concurring to the calibrated Prandtl model [11]), i.e. its switch in Figure 1 from solid to dash-dotted solutions. Typical values of this master length-scale in Figure 2d are less than a couple of meters, often only a few decimetres, except in the upper part of the SABL, $z \sim (100 \pm 20)$ m, that is weakly stratified (see Figure 2b) in the presence of sheared flow (Figures 2a and 2c); there the mixing length is ≤ 20 m, Figure 2d. It is the LLJ and its shear determining the turbulence properties, not e.g. a distance from the surface; this is also in a qualitative agreement with [4] [6]. Those modelers

                               B. Grisogono

trying to describe the VSABL with e.g. Blackadar type of the mixing length-scale, will never be able to represent katabatic flows that often govern the VSABL properties. Moreover, the modelers deploying even a more sophisticated local length-scale, e.g. "z-less" length-scale based or related to Ozmidov scale, like above (1), will also often fail because of excluding the most relevant time-scale, i.e. the wind shear, explicitly in the length-scale. In other words, even a mixing length-scale based on Ozmidov scale should be accompanied with another scale sensing shearing processes more explicitly in the SABL, e.g. as in (1).

Some authors assess the turbulence energy parameterization, by e.g. involving the concept of total turbulent energy [22] [31]; some others find sufficient improvements in the SABL modeling already by changing only a mixing length-scale formulation [11] [14]. The author finds enough amusements and scientific challenge just by reformulating and rescaling the relevant turbulent length-scale, Figure 1, almost as adjustable as a "turbulence rubber gum", within the accepted concepts of atmospheric turbulence closure single-point modeling. After returning to Figure 2c, we will discuss a possibility that an NWP model's lowest level is (still) higher than the LLJ. Finally, we will re-derive and propose a slightly different but more generalized expression than (1) for the "z-less" mixing length-scale.

Simple katabatic flows, e.g. as those already shown (i.e. hydrostatic and Boussinesq, quasi-1D, without large-scale pressure gradient, all for constant: slope, surface potential temperature deficit and roughness), if persistent enough, like those over long glaciers during the polar night, may produce a permanent effect on the whole troposphere [30] [17], also see Figure 2c. Under such persistent katabatic forcing, the cross-slope wind component V is induced due to the Coriolis effect [17] [27] [29]; V diffuses upwards without a well-defined spatio-temporal scale, Figure 2c. Hence, this might affect, in principle, the whole troposphere, all the way up to the polar vortex (after ~ 180 days of polar night), which is not an intuitive result. To make this statement more convincing, we compare the V component from MIUU model with the analytical solution [17], Figure 3. The latter asymptotic solution is based on the *WKB* method (the letters coming from the last names of the method's promoting scientists), which is an elegant singular perturbation technique, checked for the Prandtl model and against appropriate observations [11] [17] [26]. This asymptotic solution, shown on the left panel in Figure 3, consists of a suitable combination of the error function and exponentially decaying cosine, both having a dimensionless similarity variable given with an integral of height, time and a gradually varying eddy diffusivity/conductivity [17] [29]. The latter feature is checked recently against a new set of measurements and LES data with promising results [16]. Of course, the models cannot agree in certain details, simply because of intrinsic differences in their respective nature, ranging from turbulence parameterizations, spatial dimensions involved, grid-point distribution to the boundary conditions, etc. [11]. Nonetheless, it appears that both models converge in their message about the spatio-temporal evolution of the V-component shown in Figure 3; likewise, the models agree in the other fields (not shown, also see [11] [17]).

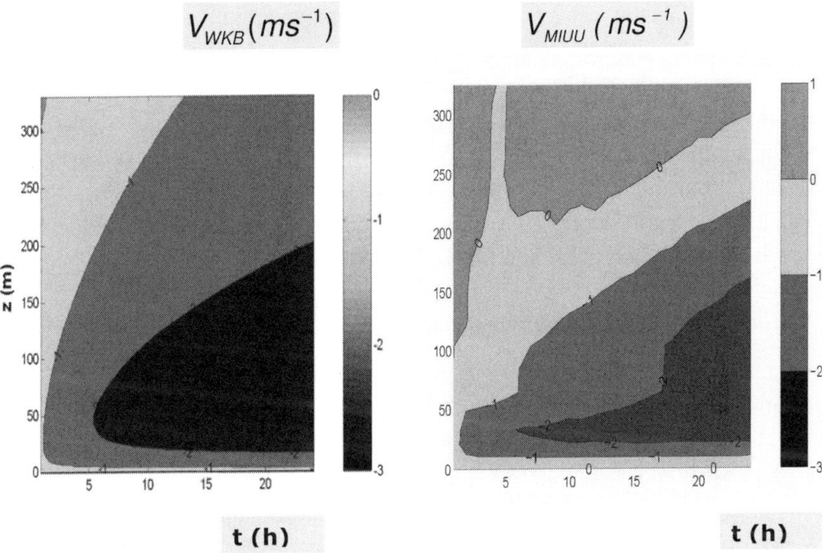

Figure 3. The slope-parallel wind component V as obtained analytically (left) using *WKB* method [17] and (right, as in Figure 2c) by MIUU mesocale model [11, this study]. Both models show a vertical diffusion of the V component. Certain quantitative differences between the analytical and numerical model arrive from two classes of a multitude of reasons. One is in the underlying model basic assumptions, another is in their technical formulations, e.g. spatio-temporal resolution, etc. [5]. A few details about the simulations: the Coriolis parameter, slope angle, surface potential temperature deficit and background temperature gradient are $(f, \alpha, C, \Delta\theta/\Delta z) = (10^{-4} \text{ s}^{-1}, -2.2°, -6.5 °C, 5 \cdot 10^{-3} \text{ K(km)}^{-1})$. Moreover, in the analytic model only (left), the additional parameters are the Prandtl number, height and the maximum eddy conductivity: $(Pr, h, Kmax) = (1.1, 200 \text{ m}, 2 \text{ m}^2\text{s}^{-1})$; these model details are in [17].

                    B. Grisogono

In modeling cases inevitably deploying much too poor spatial resolution, to resolve the LLJ with at least three to four vertical gridpoints, another point is that a katabatic LLJ is also likely to appear below the lowest NWP model level. At the same time, most of NWP models use some version of Monin-Obukhov length, L, to parameterize the near-surface fluxes; this length does not sense any terrain slope (which is increasingly resolved with ever finer resolution in NWP models), i.e. it assumes horizontal homogeneity for the near-surface flow variables. Hence, L, as such, cannot accommodate any direct influence of katabatic (or anabatic) flows. For such situations, a modified Monin-Obukhov length, L_{MOD}, has been recently proposed [14]. It includes a possible bulk effect of katabatic LLJ; the latter height may be estimated from the background flow variables and underlying terrain parameters. This work, as well as a few other contemporary findings [20] [19] [14] [22] [31] [10], give strong evidence that there is no critical Ri pertaining to shifts back and forth between turbulent and laminar geophysical flow regimes. Various turbulent forms exist at various Ri values. To put it simply, historically we did not relate Ri and Pr values for the SABL flows appropriately; this seems now to be quite a settled issue which also discards the existence of critical Ri [3] [22] [31] [18].

2.3. *Simplified TKE equation and a new generalized "z-less" length-scale*

An extension of (1) follows together with future work remarks. A few plausible derivations for new "z-less" mixing length-scales stem from a few recent works [3] [11] [31], thus allowing for another new result of this study. Let us start with the prognostic equation for *TKE* under typical simplifying conditions such as horizontal homogeneity, Boussinesq and hydrostatic approximation and the absence of a mean vertical motion:

$$\frac{\partial(TKE)}{\partial t} = -\overline{u'w'}\frac{\partial \bar{u}}{\partial z} + \frac{g}{\theta_0}\overline{w'\theta'} - \frac{\partial}{\partial z}\left[\overline{w'(\frac{p'}{\rho_0} + TKE)}\right] - \varepsilon \cdot \tag{2}$$

The terms have their usual meaning in this well known equation: the local rate of change of *TKE* on the LHS is balanced by the consecutive terms on the RHS: the shear production, buoyant destruction (in the SABL, while in the CABL this is a source term), transport and redistribution due to pressure- and turbulence-correlations ("fluxes of turbulent fluxes") and viscous dissipation, respectively. Next, we assume a steady-state and neglect transport and redistribution terms.

The steadiness assumed also implies here that the mixing length-scale will not remember its own history, i.e. it will be a diagnostic quantity that immediately adjusts to the conditions imposed. Transport and redistribution terms, the square brackets on the RHS of (2), are notoriously difficult both to measure and to model; sometimes these are treated as diffusive-like processes, sometimes are simply leftovers from a bulk budget of the other terms in (2). Furthermore, we parameterize the momentum and heat fluxes in (2) as $K_m\$$ and $K_h\$$, where K_m and K_h are eddy diffusivity and conductivity and $\$$ is (again) the absolute shear. The last term in (2) is parameterized as $\varepsilon = b(TKE)^{3/2}/\Lambda$, where b is an empirical constant and Λ is a new mixing length-scale (Λ replaces l_{STAB} from (1)). Under these simplifications (2) becomes:

$$0 = K_m\$^2 - K_h N^2 - \frac{b}{\Lambda}(TKE)^{3/2}, \tag{3}$$

signifying that the buoyant destruction and viscous dissipation (last two terms) compete in partitioning *TKE* after the mechanical/shear production of *TKE*.

There are a few ways to proceed from (3) in order to estimate Λ, the goal of this subsection. A simple, 1st order closure would assume, based on the absolute shear $\$$: $K_m = a_1\Lambda^2\$$ and likewise $K_h = a_1\Lambda^2\$/Pr$, where a_1 is a model constant and Pr is again turbulent Prandtl number; typically $Pr \geq 1$ in the SABL [14] [17] [22] [31]. A more advanced and, arguably, better parameterization would be a higher-order closure, with a simplest form as $K_m = a_2\Lambda(TKE)^{1/2}$ and likewise $K_h = a_2\Lambda(TKE)^{1/2}/Pr$. When either of these parameterizations is plugged in (3), the following expression for Λ ensues (the first index will be for the 1st order closure, the second index will be for the higher-order closure in $\Lambda_{1,2}$):

$$\Lambda_{1,2} = c_{1,2} \frac{(TKE)^{1/2}}{\$(1 - Ri/Pr)^{1/(3,2)}}, \tag{4}$$

where $c_{1,2}$ are appropriate coefficients obtained from b, a_1 or a_2, respectively; moreover, the root exponent in the denominator in (4) is either *1/3* or *1/2* for the 1st or the higher-order closure, respectively. While in higher-order closures *TKE* is most often forecasted, in 1st order schemes it may be only diagnosed. After including an important recent finding about the SABL that

$$Pr \approx 0.8 + 5\,Ri\,, \tag{5}$$

from [31] into (4), it appears that the denominator in (4) may be justifiably expanded into binomial series since for the SABL (5) gives *max(Ri/Pr) ≤ 0.2*. Hence, a newly proposed "z-less" mixing length-scale is (based on a two-term binomial expansion):

$$\Lambda_{1,2} \approx c_{1,2} \frac{(TKE)^{1/2}}{\$} (1 + \frac{Ri}{(3,2)\,\text{Pr}}) \,, \tag{6}$$

which appears as a modification of (1). For generality (5) was not plugged in (6), the latter only needs the asymptotic range of values for the ratio *Ri/Pr*. In fact, there is a whole class of the above parameterizations, between 1st and 2nd order closures, that yield to the same basic formulation: $\Lambda \sim (TKE)^{1/2}/\$$.

If K_m and K_h were parameterized in (3) as $K_m = a_3(TKE/N)$ and $K_h = a_3$ *TKE/(PrN)*, respectively (strictly $Ri > 0$), which also makes sense for the VSABL, one would end up, instead of (4), with

$$\Lambda_3 = c_3 \frac{(TKE)^{1/2}}{\$} \frac{Ri^{1/2}}{(1 - Ri/\text{Pr})} \,, \tag{7}$$

again, due to (5), this allows its binomial series for the denominator's second factor, similar to that in (6). Also note from (4) (or (6)) and (7) that Λ for the 1st order parameterization is somewhat less sensitive to the ratio of *Ri/Pr* than the higher-order closures. Furthermore, we conclude that most of sensible parameterizations, that are between 1st and 2nd order turbulence closures, for the SABL (above the immediate surface sub-layer) are best handled with a "z-less" mixing length scale of type

$$\Lambda = const \frac{(TKE)^{1/2}}{\$} f(Ri, \text{Pr}) \,, \tag{8}$$

where *0 < const < 1* and *f(Ri, Pr)* is a relatively simple function, or even a simpler series expansion, already shown for two overall cases to be $\approx 1 + Ri/(3Pr)$, or $1 + Ri/(2Pr)$; while in the third case discussed it is $\approx (Ri)^{1/2}(1+Ri/Pr)$. For both 1st order- and higher-order closure schemes in general, the respective single coefficient entering to the RHS of either (6), (7) or (8) is a priori known number from the respective definitions of eddy diffusivities (see between (3) and (4), or above (7)) in each particular NWP and climate models used. Mesoscale models with advanced higher-order turbulence closures, as e.g. MIUU model [2] [7] [8] [1] [23], usually possess a multiple combination/choices for obtaining eddy diffusivity and conductivity under stable conditions; note that a suitable set of options and entering coefficients is already accommodated implicitly with the proposed Λ. Namely, any combination of these parameterizations discussed end up with (8), i.e. $\Lambda \sim (TKE)^{1/2}/\$$. This generality of Λ is provided by the systematic reduction of the eqn. for *TKE*, (2) toward (3), which still secures a three-term balance used for the estimation of Λ. Aside from the constant in (8), it is indicated that the function multiplying the "z-less" length-scale $(TKE)^{1/2}/\$$, i.e. *f(Ri, Pr)*, is progressively more sensitive to *Ri* and *Pr* inter-relation for higher-order closure parameterizations than for the 1st order closure. This finding suggests that higher-order closures should be better in handling the VSABL structures and its turbulence than the 1st order closures because the former ones are more responsive to multiple-scale processes and their variations of *Ri* and *Pr*. Preliminary tests with MIUU model show that Λ_2 from (6) behaves in accordance with the expectation, i.e. there is no noticeable difference between the katabatic flow simulations already displayed using (1), and the one obtained with (6), in particular with $\Lambda = 0.2685$ $(2 \cdot TKE)^{1/2}/\$ \cdot (1 + 0.5 \cdot Ri/Pr)$, where all relevant coefficients are revealed now. Moreover, test runs are stable even after 30h of simulations. More testing is necessary before the newly proposed length-scale can be reliably used in all types of the SABL, but a further generalization of Λ is already being derived based on a renormalization procedure like that for (6) through (8).

An indirect advantage of the formulation (6) through (8) is that the explicit inclusion of the wind shear, followed by *Ri* and *Pr* local numbers, will be more sensible to minute flow variations, than former mixing length expressions for the SABL (e.g. the one above (1), not to mention the prescribed Blackadar scale, etc.). This enhanced Λ sensitivity to shear effects could be instrumental in sensing other, even non-local effects on turbulence, such as buoyancy waves, thus indirectly Λ being susceptible even to the transport/redistribution terms. Further testing is left for new studies and is beyond this analysis. Although (4), (6), (7) and/or (8) may have problems in handling turbulent mixing with the wind shear diminishing faster than $(TKE)^{1/2}$, which is possible in certain strongly-stratified weak-wind conditions, it remains to be seen if this length-scale proposed will bring some practical improvements in modeling VSABL flows. The latter type of VSABL is apparently determined most of its lifetime by unknown dynamics and physics [20] [3] [19] [10] [5]. Without suitable measurements there, we may not even know whether the relatively weak turbulence in the weak-wind VSABL is transported and/or redistributed from elsewhere and then only (partially) destroyed in this VSABL.

3. Concluding Remarks

Several aspects of the ABL are reviewed and discussed; the emphasis has been on modeling, in particular, on parameterizing turbulence in the SABL. While the "classical" SABL, that is always weakly stratified (i.e. $Ri << \infty$, typically, $0 < Ri \leq 1$), is modeled reasonably well during the last few decades or so, strongly stable cases, i.e. the VSABL (where typically $Ri >> 1$) is generally not understood well [21] [20] [3] [28] [11] [19] [14]. Various approaches have been envisioned [9] [10] [6] [5]. Here, a pragmatic approach is undertaken. In particular, excessively diffusive and too deep SABL flows, often appearing in numerical models, are discussed in the light of the recently proposed alleviation of this problem [11]. The latter demanded an explicit inclusion of the vertical shear of horizontal wind. A generalization of this proposal is given here in the view of a simplified *TKE* eqn. and a set of subsequent parameterizations for the mixing length-scale.

Long lived SABL over relatively long inclined surfaces often consists of persistent katabatic flows triggering the corresponding cross-slope wind component V due to the Coriolis effect [30] [17] [27] [29]; V diffuses upwards without a steady state. This type of the SABL flow has been modeled analytically, through the calibrated Prandtl model (with vertically varying prescribed eddy diffusivity and conductivity, solved either numerically, or via the *WKB* method), and simulated via MIUU mesoscale model. The results from these two very different models agreed only when the latter model used a more appropriate "z-less" mixing length scale (1).

The new generalized "z-less" mixing length-scale Λ is proposed (4), (7) or (8). It is derived from a couple of most recent studies [3] [11] [31] that indicated a few important shortcomings of the current turbulence parameterizations for the SABL and its turbulence as modeled in NWP, air-chemistry and climate models. It basically states that $\Lambda \sim (TKE)^{1/2}/\$$, almost regardless of the other parameterization details. This new length-scale remains yet to be checked in simulations against observations; a few preliminary tests for the katabatic flows already discussed, but now using the newly proposed Λ_2 from (6), show a promising behavior and agreement with (1).

Acknowledgement

The author is thankful to: Iva Kavčič for reproducing the left panel of Figure 3, and to Dragutin Mihailović and Boro Rajković for recommending the overall frame to this work and more. This study was inspired by collaboration with Danijel Belušić, Sergej Zilitinkevich, Leif Enger, Thorsten Mauritsen, Željko Večenaj and Larry Mahrt; it was supported by EMEP4HR project No. 175183/S30 provided by the Research Council of Norway and by the Croatian Ministry of Science, Education & Sports, projects BORA No. 119-1193086-1311.

Appendix - list of symbols

List of Symbols		
Symbol	**Definition**	**Dimensions or Units**
a, a_i, b	dimensionless coefficients in turbulence parameterizations	
ABL	atmospheric boundary layer	
α	terrain slope	rad
C	surface potential temperature deficit	^{0}C
CABL	convective ABL	
c_i	coefficients obtained from a_i and b	
ε	dissipation of TKE	$m^2 s^{-3}$
f	Coriolis parameter	s^{-1}
g	acceleration due to gravity	ms^{-2}
h	height of the maximum value of a prescribed eddy diffusivity or conductivity	m
K_h	eddy conductivity	$m^2 s^{-1}$
K_m	eddy diffusivity	$m^2 s^{-1}$
K_{max}	maximum value of a prescribed gradually varying eddy conductivity	$m^2 s^{-1}$
LLJ	low-level jet	
l_{STAB}	"z-less" turbulent mixing length-scale	m
Λ	"new generalized "z-less" mixing length-scale	m
MIUU	Meteorologiska Institutionen Uppsala Universitetet	
N	buoyancy frequency	s^{-1}

NWP	numerical weather prediction	
p'	turbulent fluctuation of pressure	Pa
Pr	Prandtl number	
Ri	gradient Richardson number	
ρ_0	mean density	kgm^{-3}
S	vertical shear of the horizontal wind	s^{-1}
$\mathcal{S}$	$=\|S\|$, absolute shear	s^{-1}
SABL	stable atmospheric boundary layer	
t	time	s (or h)
TKE	turbulent kinetic energy	$m^2 s^{-2}$
Θ	potential temperature	K
θ'	turbulent fluctuation of potential temperature	K
U,V	(or with a bar on its top) horizontal mean wind components	ms^{-1}
$u,'v,'w'$	turbulent fluctuations of the wind field	ms^{-1}
VSABL	very (strongly) SABL	
WKB	name of a singular perturbation method	
z	vertical coordinate	m
$\overline{(...)}$	suitable averaging	

References

1. B.J. Abiodun and L. Enger, The role of advection of fluxes in modeling dispersion in convective boundary layers. *Quart. J. Roy. Meteorol. Soc.* **128**, 1589-1608 (2002).
2. A. Andrén, Evaluation of a turbulence closure scheme suitable for air-pollution applications. *J. Appl. Meteor.* **29**, 224-239 (1990).
3. A. Baklanov and B. Grisogono (Eds.), Atmospheric Boundary Layers: Nature, Theory and Applications to Environmental Modelling and Security (Springer, 2007).
4. R.M. Banta, Stable-boundary-layer regimes from the perspective of the low-level jet. *Acta Geophysica*, **56**, 58-87 (2008).
5. D. Belušić and L. Mahrt, Estimation of length scale from mesoscale networks. *Tellus*, **60A**, 706-715 (2008).
6. J. Cuxart and M.A. Jiménez, Mixing processes in a nocturnal low-level jet, an LES study. *J. Atmos. Sci.*, **64**, 1666-1679 (2007).

7. L. Enger, Simulation of dispersion in moderately complex terrain -Part A. The fluid dynamics model. *Atmos. Environ.* **24A**, 2431-2446 (1990).

8. L. Enger and B. Grisogono, The response of bora-type flow to sea surface temperature. *Quart. J. Roy. Meteorol. Soc.* **124**, 1227-1244 (1998).

9. E. Fedorovich and R. Conzemius, Effects of wind shear on the atmospheric convective boundary layer structure and evolution. *Acta Geophysica*, **56**, 114-141 (2008).

10. H.J.S. Fernando, Turbulent patches in a stratified shear flow. *Phys. Fluids.* **15**, 3164-3169 (2003).

11. B. Grisogono and D. Belušić, Improving mixing length-scale for stable boundary layers. *Quart. J. Roy. Meteorol. Soc.* **134**, 2185-2192, (2008).

12. B. Grisogono and D. Belušić, A review of recent advances in understanding the meso- and micro-scale properties of the severe Bora wind. *Tellus* **A 61**, 1-16 (2009).

13. B. Grisogono and L. Enger, Boundary-layer variations due to orographic wave-breaking in the presence of rotation. *Quart. J. Roy. Meteorol. Soc.* **130**, 2991-3014 (2004).

14. B. Grisogono, A. Jeričević and L. Kraljević, The low-level katabatic jet height versus Monin-Obukhov height. *Quart. J. Roy. Meteorol. Soc.* **133**, 2133-2136 (2007).

15. A. Jeričević and B. Grisogono, The critical bulk Richardson number in urban areas: verification and application in a numerical weather prediction model. *Tellus.* **58A**; 19-27 (2006).

16. A. Jeričević and Ž. Večenaj, Improvement of vertical diffusion under stable atmospheric conditions. *Boundary-Layer Meteorol.* In press, (2009).

17. I. Kavčič, and B. Grisogono, Katabatic flow with Coriolis effect and gradually varying eddy diffusivity. *Boundary-Layer Meteorol.* **125**, 377-387 (2007).

18. J. Kondo, O. Kanechika and N. Yasuda, Heat and momentum transfers under strong stability in the atmospheric surface layer. *J. Atmos. Sci.* **35**, 1012-1021 (1978).

19. L. Mahrt, Bulk formulation of the surface fluxes extended to weak-wind stable conditions. *Quart. J. Roy. Meteorol. Soc.* **134**, 1-10 (2008).

20. L. Mahrt, The influence of small-scale nonstationarity on the turbulent flux for stable stratification. *Bound.-Layer Meteorol.* **125**, 245-264 (2007).

21. L. Mahrt, Stratified atmospheric boundary layers and breakdown of models. *Theoret. Comput. Fluid Dyn.* **11**, 263-279 (1998).

22. T. Mauritsen, G. Svensson, S. Zilitinkevich, I. Esau, L. Enger and B. Grisogono, A total turbulent energy closure model for neutrally and stably stratified atmospheric boundary layers. *J. Atmos. Sci.* **64**, 4113-4126 (2007).

23. M. Mohr, Problems with the mean sea level pressure field over the western United States. *Mon. Wea. Rev.* **132**, 1952-1965 (2004).

24. F.T.M. Nieuwstadt, Some aspects of the turbulent stable boundary layer. *Bound.-Layer Meteorol.* **30**, 31-38 (1984).

25. F.T.M. Nieuwstadt, The turbulent structure of the stable, nocturnal boundary layer. *J. Atmos. Sci.,* **41,** 2202–2216 (1984).

26. O. Parmhed, J. Oerlemans and B. Grisogono, Describing surface-fluxes in katabatic flow on Breidamerkurjökull, Iceland. *Quart. J. Roy. Meteorol. Soc.* **130**, 1137-1151 (2004).

27. A. Shapiro and E. Fedorovich, Coriolis effects in homogeneous and inhomogeneous katabatic flows. *Quart. J. Roy. Meteorol. Soc.* **134**, 353-370 (2008).

28. G.J. Steeneveld, Understanding and prediction of stable atmospheric boundary layers over land. *PhD thesis, Wageningen Univ.*, the Netherlands, 199 pp (2007).

29. I. Stiperski, I. Kavčič, B. Grisogono and D.R. Durran, Including Coriolis effects in the Prandtl model for katabatic flow. *Quart. J. Roy. Meteorol. Soc.* **133**, 101-106 (2007).

30. M. Van den Broeke, D. Van As, C. Reijmer and R. Van De Wal, Sensible heat exchange at the Antarctic snow surface: A study with automatic weather stations. *Int. J. Clim.* **25**, 1081-1101 (2005).

31. S. Zilitinkevich, T. Elperin, N. Kleeorin, I. Rogachevskii, I. Essau, T. Mauritsen and M. W. Miles, Turbulence energetics in stably stratified geophysical flows: strong and weak mixing regimes. *Quart. J. Roy. Meteorol. Soc.* **134**, 793-799 (2008).

Chapter 4

MODELLING OF STRATIFIED AND TURBULENT FLOW

VLADIMÍR FUKA, JOSEF BRECHLER and ALEŠ JIRK

*Department of Meteorology and Environment Protection, Faculty of
Mathematics and Physics, Charles University, Prague, Czech Republic*

When modelling a flow in the atmosphere and the processes strongly influenced by it (*e.g.*, the dispersion of air pollution), it is important to appreciate that the properties of both the flow itself and the dispersion are affected by the flow regime; *i.e.*, whether the flow is turbulent (as is almost always the case in the atmosphere) or laminar. A second factor that might complicate atmospheric flow is stability, which depends on the nature of vertical temperature stratification.

In the first part of this chapter, we demonstrate the impact of vertical temperature stratification on flow structure, modelled via the Boussinesq approximation and by varying the Froude number (Fr). The flow is assumed to be laminar and is modelled in 2D.

Next, we review several approaches to treating turbulence in modelling studies, with an emphasis on an implicit large-eddy simulation. The results of Taylor–Green vortex computations performed using this method are compared with the results of a direct numerical simulation at moderate Reynolds numbers. Several quantities are considered, including the kinetic energy dissipation rate, probability density functions of turbulent fluctuations, and 3D energy spectra.

Keywords: turbulence, large eddy simulation, Taylor-Green vortex, stratification, buoyancy, Boussinesq approximation, Navier-Stokes equations, continuity equation, square cylinder, kinetic energy spectrum

1. Introduction

The behaviour of fluids is a subject of interest across a broad field of fluid mechanics. There occur great differences in behaviour between fluids depending on, for example, whether they obey Newton's law (so-called Newtonian fluids). Fluids can also be classified based on the speed of fluid motion. In the case of a low fluid speed (*i.e.*, Mach number M (speed of fluid/speed of sound) $\ll 1$), the fluid can be treated as incompressible. This case applies to many problems regarding geophysical fluids, and explains why the atmosphere is commonly taken to be incompressible.

There are two main regimes of fluid flow: laminar and turbulent. It is important to bear in mind that whether a flow lies in the turbulent or laminar regime does not depend on the fluid itself, as it is not a material property. A given fluid can exist in the laminar flow regime under certain conditions, but become turbulent when the conditions change.

The parameter used to distinguish laminar from turbulent flow is the Reynolds number (Re), defined as the ratio of inertial forces to viscous forces. This is expressed mathematically as $Re = UL/v$, where L denotes the length scale of the problem, U is the velocity scale and v is the kinematic viscosity of the fluid. Flow becomes turbulent when inertial forces are dominant (Re $\gg$ 1), whereas laminar flow occurs when viscous forces are dominant.

Another important property that affects flow behaviour is vertical temperature stratification. This is mainly relevant to problems concerned with geophysical fluids, of which atmospheric flow is an important subject. The simplest case is an isothermal flow, for which there is no influence of thermal convection (either positive or negative), which is a consequence of the vertical temperature gradient. If a temperature gradient exists, its influence should be taken into account in the equations used to describe flow behaviour.

Many books and journal articles describe the physics of fluid flow and mathematical methods that can be used to solve model equations. Among such books, one must mention the classical work by Batchelor [1] and the work on fluid flow by Deville, Fischer and Mund [2]. The mathematical methods used for the computation and modelling of fluid flow represent a branch of science referred to as computational fluid dynamics (CFD), which has received much attention in the literature [3] [4] [5] [6].

The first part of this contribution deals with the modelling of stably stratified flow in the atmospheric planetary boundary layer (PBL). The second part considers a method of modelling the turbulence in the atmosphere. Most of the books that deal with CFD mention the Boussinesq approximation — a method of taking into account temperature stratification. This approach is relatively simple and is easily implemented in the Navier–Stokes equations that describe the conservation of momentum in the solved problem. This approach is employed in the modelling undertaken as part of the present study. For modelling turbulent flow, we used the Implicit Large Eddy Simulation (ILES) [7]. This recently proposed method appears to be highly promising. Here, we present preliminary results based on the use of ILES for a Green–Taylor vortex.

The remainder of this chapter is organised as follows. Section 2 contains the basic equations that describe flow, along with simplifications and modifications, and parameters such as the Reynolds and Froude numbers. Section 3 describes the stratified flow approach to modelling, and results are presented in Section 4. Section 5 deals with ILES and the numerical methods that enable its use, together with some results of ILES modelling. Finally, concluding remarks are presented in Section 6.

2. Basic Equations

This section introduces a basic set of equations that describe fluid flow, together with simplifications. As we are interested in the flow of air, we refer to air when we use the word "fluid". In the most general form, this set of formulas consists of equations that describe the conservation of momentum, mass and energy. Because these equations, together with their derivations, are well known and described in every textbook on fluid mechanics, only the dimensional forms of these equations are presented here.

The continuity equation in differential form reads as follows:

$$\frac{\partial \rho}{\partial t} + \nabla \cdot (\rho \boldsymbol{u}) = 0 \ , \tag{1}$$

where ρ is the density of air, t is time, $\boldsymbol{u}$ is a velocity vector with components (u, v and w) and ∇ is a Hamilton's (nabla) operator. This equation is valid to the limit at which the volume of the element for which it has been derived approaches zero.

In deriving the energy conservation equation, one can start directly from the 1st law of thermodynamics. When using the formulas of thermodynamics, the potential temperature Θ is defined as

$$\Theta = T \left(\frac{1000 \, \text{hPa}}{p} \right)^{\kappa} \ , \tag{2}$$

where $\kappa = R/c_p = 0.286$ and R is a gas constant for dry air. The thermodynamic energy equation reads

$$\frac{d\Theta}{dt} = \frac{\partial \Theta}{\partial t} + (\boldsymbol{u} \cdot \nabla)\Theta = \frac{\Theta}{c_p T} \frac{dQ}{dt} \ , \tag{3}$$

where T is temperature and dQ/dt is the diabatic heating rate, which consists of all the energy sources and sinks due to non-adiabatic processes (*e.g.*, phase changes or radiation). Given certain assumptions, it is possible to rearrange this equation to a more simplified form commonly used in atmospheric models.

The differential form of the momentum conservation equation is

$$\frac{\partial \rho \boldsymbol{u}}{\partial t} + \mathrm{div}\left(\rho \boldsymbol{u}\boldsymbol{u}\right) = -\mathrm{grad}\,p + \mathrm{grad}\left(\frac{\mu}{3}\,\mathrm{div}\boldsymbol{u}\right) + \mathrm{div}\left(\mu\,\mathrm{grad}\boldsymbol{u}\right) + \rho \boldsymbol{g}\ , \qquad (4)$$

where p is pressure, μ denotes dynamic viscosity, and $\boldsymbol{g}$ is the gravitational acceleration vector with components (0, 0 and $-g$). Equation (4) can be further simplified, as discussed below.

2.1. *Simplification of conservation equations*

The conservation equations stated above can be simplified in the case that certain assumptions are valid. Suppose that the Mach number of atmospheric flow is very low ($M \ll 1$). This assumption is valid for the majority of processes occurring in the atmosphere, especially in the boundary layer. It is then possible to treat the atmosphere as incompressible and to neglect local and temporal changes in density in the continuity equation. The continuity equation can then be written as a condition for the non-divergence of fluxes:

$$\nabla \cdot (\rho \boldsymbol{u}) = 0\ . \qquad (5)$$

The continuity equation for flow of an incompressible and constant-density fluid is

$$\nabla \cdot \boldsymbol{u} = 0\ . \qquad (6)$$

This equation describes a real situation, at least in the lower part of the atmospheric boundary layer.

Another simplification deals with the momentum conservation equation, which in the case of constant density can be written as

$$\frac{\partial \boldsymbol{u}}{\partial t} + \mathrm{div}\left(\boldsymbol{u}\boldsymbol{u}\right) = -\frac{1}{\rho}\,\mathrm{grad}\,p + \mathrm{div}\left(\nu\,\mathrm{grad}\boldsymbol{u}\right) + \boldsymbol{b}\ , \qquad (7)$$

where $\boldsymbol{b}$ is the buoyancy force, given as

$$b = \frac{\theta'}{\overline{\Theta}} k\, g \ , \tag{8}$$

where θ' is the perturbation of the potential temperature, $\overline{\Theta}$ is its spatial mean, and k is the vertical unit vector. This approach is referred to as the Boussinesq approximation [2] [3] [5].

If one models the incompressible and isothermal flow of a constant-density fluid, the basic set of equations is given by the continuity equation (Eq. (6)) and the momentum equation (Eq. (7)), with the buoyancy term b set equal to zero. If the model does not describe an isothermal situation, Eq. (3) must be added to the set of model equations and the buoyancy term b should be specified more precisely via Eq. (8).

2.2. *Dimensionless form of equations*

For any problem describing fluid flow, one can choose at least two parameters that characterise the problem. For constant-density and isothermal flow past an obstacle, these parameters are the length scale of the problem L (*e.g.*, for flow around a cylinder the length scale can be the diameter of the cylinder) and a velocity scale U (*e.g.*, the flow velocity far upstream of the cylinder). Using these scales (U and L), all the dimensionless variables can be defined as follows:

$$u^* = \frac{u}{U}, \qquad x^* = \frac{x}{L}, \qquad t^* = \frac{tU}{L}, \qquad p^* = \frac{p}{\rho U^2} \ . \tag{9}$$

When we express dimensional variables via the above scales and dimensionless variables and substitute them back into the equations, we obtain the model equations in the dimensionless form:

$$\nabla \cdot u^* = 0 \ , \tag{10}$$

and

$$\frac{\partial u^*}{\partial t} + \nabla \cdot \left(u^* u^* \right) = -\nabla p^* + \frac{1}{\mathrm{Re}} \nabla^2 u^* + \frac{b^*}{\mathrm{Fr}} \ . \tag{11}$$

The continuity equation remains unchanged, whereas the new parameters Re and Fr appear in the momentum equation. Re is the Reynolds number (as described above) and Fr is the Froude number, defined as the ratio of inertia forces to gravity forces:

$$\text{Fr} = \frac{U^2}{g\,L}\,.\tag{12}$$

For Re $\gg$ 1, inertia forces dominate and viscous forces, being negligible, can be omitted. The influence of turbulence must be taken into account in the momentum equations, as considered below in the section 5. The influence of the Froude number is discussed in greater detail in the section 3.

The method described above is the most frequently used in rendering the equation dimensionless, but alternative methods are available. The appropriate method depends on the nature of the problem and the employed basic scales. The Richardson number (Ri) is important in terms of stability analysis. There exist many different ways to express Ri [9]. Here, we use Ri $=$ Gr/Re^2, where Gr is the Grashof number [10]. In the following text, all variables are dimensionless (the use of an asterisk is omitted for simplicity). The use of dimensional variables is clearly stated where appropriate.

3. Stratified Flow

A dimensionless form of the momentum equation is given by Eq. (11). When b is expressed by the potential temperature and acceleration due to gravity g, the final form of the non-dimensional equation of motion with a Boussinesq approximation is

$$\frac{\partial u^*}{\partial t} + \nabla \cdot \left(u^* u^*\right) = -\nabla p^* + \frac{1}{\text{Re}}\nabla^2 u^* - \frac{\theta'^*}{\overline{\theta}^*}\,\frac{g^*}{\text{Fr}}\,,\tag{13}$$

where an asterisk denotes non-dimensional.

The system of equations must be closed by the equation for perturbations in potential temperature. Starting with the continuity equation (Eq. (1)), after some manipulation we have

$$\frac{\partial \rho'}{\partial t} + \left(\overline{\rho}+\rho'\right)\nabla\cdot u + u\cdot\left(\nabla\rho'\right) + u_3\frac{\partial \overline{\rho}}{\partial x_3} = 0\,.\tag{14}$$

Using Eq. (14), together with the application of the continuity equation in the form of Eq. (6), we obtain the desired equation for potential temperature. The dimensionless and dimensional forms of this equation have the same expression:

$$\frac{\partial \theta'}{\partial t} + \boldsymbol{u} \cdot (\nabla \theta') = -u_3 \frac{\partial \bar{\theta}}{\partial x_3} \boldsymbol{k} \ , \tag{15}$$

where $\boldsymbol{k}$ is the vertically pointing unit vector and all the variables are dimensionless.

3.1. *Stratified flow model*

In the previous section, we presented the equations that make up the model of stratified flow. The model, in which we assume a constant density except for density perturbations associated with temperature fluctuations, is formed using the continuity equation for incompressible flow (Eq. (6)), the momentum conservation equation with the Boussinesq approximation (Eq. (13)) and the equation that describes fluctuations in potential temperature (Eq. (15)). All of these equations are used in the dimensionless form, and, for simplicity, in the first trial computations a 2D approach was employed. Moreover, the value of the Reynolds number in all the treated problems ensured that the flow was laminar, meaning that it was unnecessary to use a turbulence model [15].

In the model, we used the fractional step method for incompressible flow [16]. In this method, the value of pressure for the next time step is computed by the solution to the Poisson equation. As a spatial discretization method, we used the finite volume method on a staggered grid [5]. The equations were solved using the total variation diminishing (TVD) Runge–Kutta method of the fourth order of accuracy. The non-linear advection (convective) terms were treated using the WENO scheme [17] [18] [19]. Additional details of this method can be found in the thesis by Jirk [15].

4. Results of Stratified Flow Modelling

We present the results of our modelling of stratified flow in two parts: first, the results of lid-driven cavity flows are presented and discussed; second, we consider flow past an obstacle placed in the stream of the flowing fluid. In both cases, the impact of stratification is clearly seen in the resulting flow patterns.

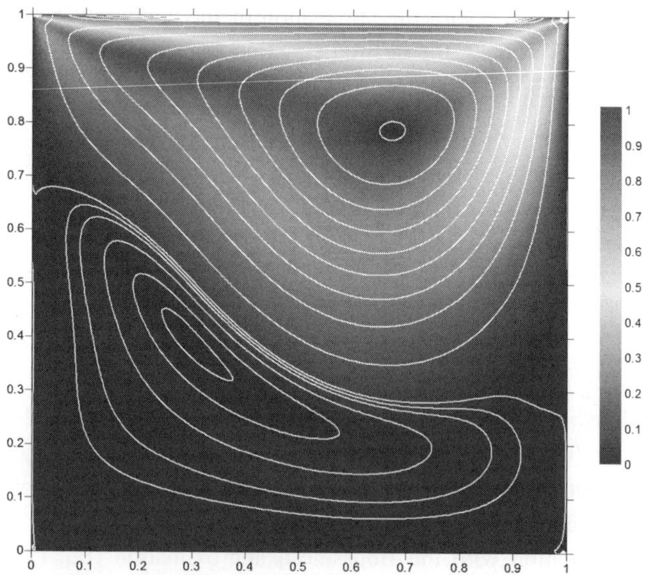

Figure 1. Lid-driven cavity flow problem. Moderately stable stratified flow, Fr = 1.

4.1. *Lid-driven cavity*

The lid-driven cavity flow problem was used as a testing scenario, in which neutral stratification was chosen (not shown here) to ensure that the results could be compared with those of previous studies [20-23]. The value of the Reynolds number was set to 100 (as in all lid-driven cavity flow problems), corresponding to laminar flow. The results obtained for neutrally stratified flow are in good agreement with the results of previous studies.

The influence of stratification on the flow can be controlled via the Froude number (Fr), which was set to 1 and 0.1. In general terms, the lower the value of Fr, the more stable the stratification.

The situation depicted in Figure 1 corresponds to a moderately stable situation. In comparing this result with the case of neutral stratification, we observe an important difference in the pattern of the circulation system. The main (primary) vortex is relatively confined to the upper lid and its axis is shifted to the upper-right (upwind) corner. In the case of stratification, only one secondary vortex exists, filling the entire bottom part of the cavity. As with the primary vortex, the secondary vortex is strongly asymmetric. This vortex is located in a region with a relatively low flow velocity.

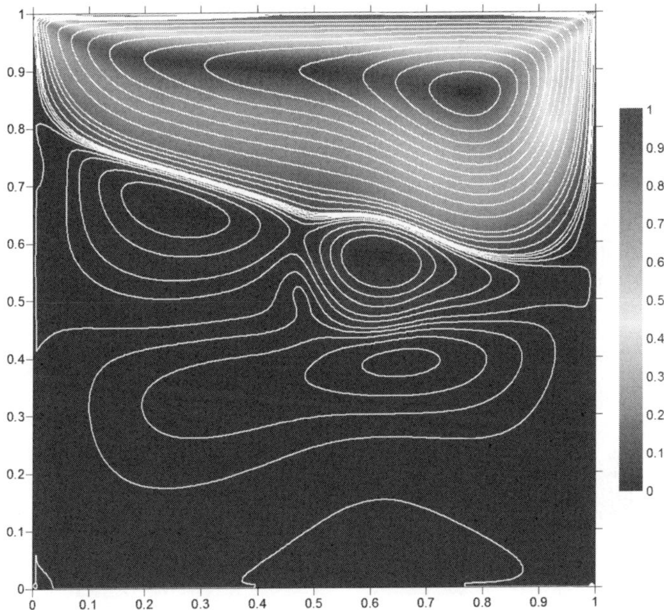

Figure 2. Lid-driven cavity flow problem. Stable stratified flow, Fr = 0.1.

With increasing stability (corresponding to decreasing Fr), the pattern of circulation changes. In this relatively stable situation, vertical exchange is strongly suppressed and the primary vortex is more strongly confined to the lid of the cavity. The majority of the bottom part of the cavity is filled with a chaotic system with very slow flow. The primary vortex is more asymmetric than in the previous case (Figure 2). The results depicted in Figs. 1 and 2 are in good qualitative agreement with those published by Iwatsu *et al.* [10] and Iwatsu and Hyun [24].

4.2. *Flow past a square cylinder*

The second set of results deals with the problem of flow past an obstacle immersed in the stream of fluid within a channel. The employed boundary conditions are as follows. At the inflow, we used the Dirichlet condition (with $v = 0$) and a parabolic profile of the u component of the inflow velocity, with a maximum dimensionless value equal to 1 at the centreline and values of zero on each edge of the parabola. For both side walls of the domain, we applied the conditions $u = 0$ and $v = 0$. At the outflow boundary, we used the Neumann condition ($\partial u/\partial x = 0$ and $\partial v/\partial x = 0$) together with a linear extrapolation to the boundaries for perturbations of potential temperature.

The parameters characterising the flow were Re = 200 and Fr = 1, 0.1 and 0.01. For the basic temperature field, we used a constant temperature gradient between the lower and upper walls. Decreasing Fr corresponds to an increase in the impact of stable stratification on the pattern of the flow field. The selected value of Re corresponds to a laminar flow. Figures 3 and 4 show the flow-field patterns, together with perturbations in the potential-temperature field, for the above values of Fr and Re.

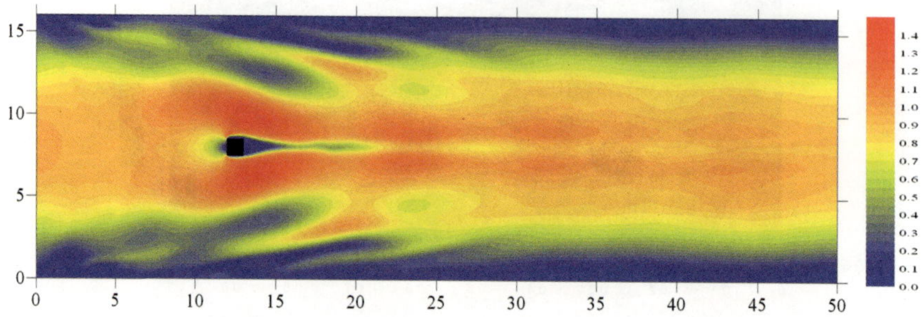

Figure 3(a). Field of velocity magnitude for Re = 200 and Fr =1.

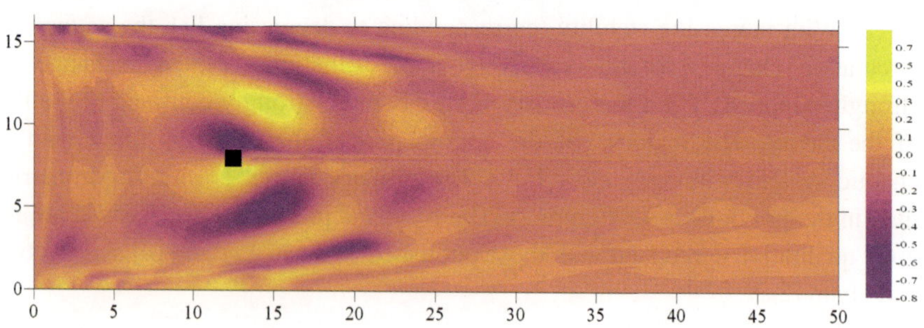

Figure 3(b). Field of perturbations in potential temperature for Re = 200 and Fr =1.

The relatively unstable stratification derived from Fr = 1 is shown in Figure 3a) and b). The magnitude of velocity is shown in Figure 3a) and the corresponding field of perturbations in potential temperature is shown in Figure 3b). Because the impact of stable stratification is relatively weak, wake structures behind the obstacle are clearly observed.

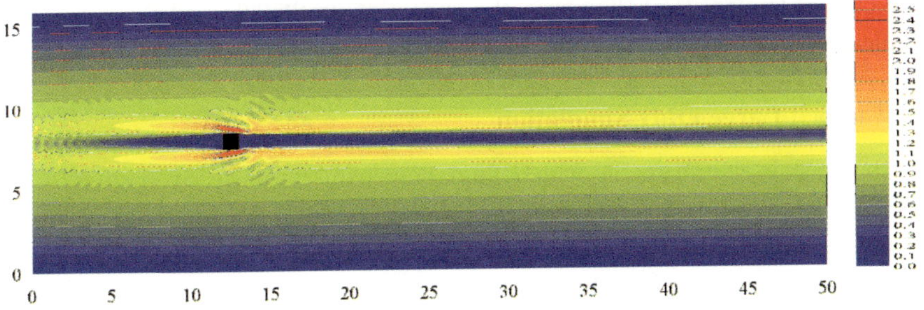

Figure 4(a). Field of velocity magnitude for Re = 200 and Fr = 0.01.

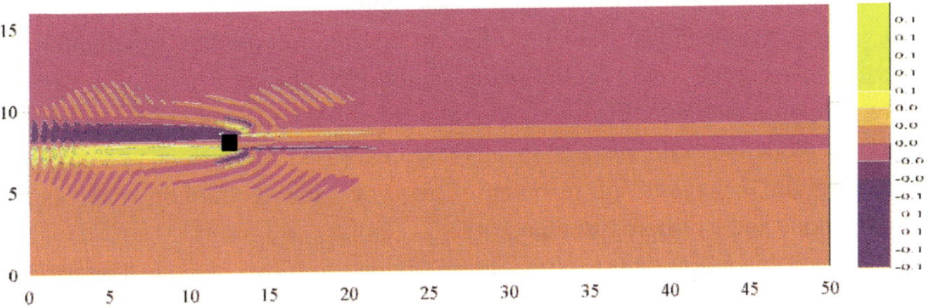

Figure 4(b). Field of perturbations in potential temperature for Re = 200 and Fr = 0.01.

In the next step, we computed the flow pattern and perturbations in potential temperature for the same value of Re but for relatively stable conditions (*i.e.*, Fr = 0.1; results not shown). In this case, wake effects observed behind the obstacle are less pronounced, but internal gravity waves are generated by the obstacle due to the increasing influence of stability.

In the third and final case, the Froude number is set to 0.01 and the Reynolds number remains unchanged (Figure 4a) and b)). In this case, the influence of stability is responsible for a rapid damping of the wake effects and the generated waves have a higher frequency and shorter wavelength compared with those generated in the previous situation. This result is expected because the restoring force is enhanced by the greater stability of the fluid flowing past the obstacle.

In both of the above cases (lid-driven cavity flow and flow past an obstacle), stratification plays an important role in determining the flow pattern. The flow was laminar, meaning that the waves generated by the obstacle were not dissipated as intensively as would occur if the flow had been turbulent. In the case of laminar flow, however, the impact of stratification on flow structure is clearly observed.

5. Implicit Large Eddy Simulation

Turbulence is a process in which fluid flow exhibits unsteady motions of a chaotic nature at various temporal and spatial scales. This wide range of scales explains why most turbulent flows cannot be computed using models that solve only the basic Navier–Stokes equations (*i.e.*, direct numerical simulations; DNS). When the Reynolds number is large enough, the computational power required for this direct approach is many orders of magnitude larger than that current available [25]. Because such cases represent the majority of industrial and environmental flows, it is important to be able to compute them using an alternative method that is less-computer-intensive, despite a loss in accuracy. Several approaches have been developed in this respect. Small scales of motion must be modelled with respect to large scales based on various assumptions regarding the behaviour of turbulent flow (*e.g.*, the Kolmogorov theory of homogenous and isotropic turbulence).

Such modelling approaches can be divided into two classes. The earliest approach consists of the Reynolds-averaged Navier–Stokes equations (RANS), in which all turbulent motions are modelled using various turbulence models [26]. Such methods are strongly dependent on the right choice of turbulence model, which must be selected with respect to the type of flow. For example, jets and free share layers can be successfully simulated using models that fail to accurately predict bounded flows.

The second approach, that of Large Eddy Simulation (LES), uses spatially filtered Navier–Stokes equations, and only models those scales that cannot be resolved explicitly on the computational grid. The sub-grid stress models in LES are similar to the turbulence models used in RANS. The earliest sub-grid models were based on the concept of turbulent viscosity, which depends on the filtered flow field (*e.g.*, the strain rate directly bound to turbulent viscosity in the Lilly–Smagorinsky model). Additional details regarding classical LES can be found in the books by Sagaut [27], Geurts [25] and Lesieur *et al.* [28].

A relatively new field of turbulence modelling (*i.e.*, ILES) [7] enables the computation of turbulent flow using methods most commonly employed for compressible fluid dynamics with shocks. These so-called shock-capturing schemes can describe flows with very strong gradients using convection terms with nonlinear numerical diffusion. This numerical dissipation is stronger in areas with higher gradients and weaker in areas with lower gradientsNonlinearity can be achieved using explicit numerical diffusion terms [29] (similar to traditional sub-grid stress models) or implicitly using flux or slope limiters [30] [31]. This latter approach has proved to be efficient in simulating compressible turbulent flows. Because of the monotonicity-preserving property of these schemes, this type of LES is also known as a monotonically integrated large eddy simulation (MILES). The most widely used of these types of methods are flux-corrected transport (FCT) [32], PPM [33] and MPDATA [34]. MPDATA is mainly employed for geophysical applications, which is also the aim of our work.

Most of the present ILES methods were developed in the framework of compressible flow solvers. Our aim was to develop and test a similar method using traditional incompressible code. Schemes for incompressible flows generally employ an artificial compressibility method or projection methods. In artificial compressibility methods, it is straightforward to use schemes for compressible fluid dynamics.

Projection methods, also known as fractional step methods [16], can be divided into two groups: exact and approximate projection methods. Approximate projection methods generally employ cell-centred grids and enable the ready use of high-resolution methods [35] [36] at the cost of difficult treatment of velocity–pressure coupling. Exact projection methods [37] provide excellent velocity–pressure coupling on staggered grids, but the usage of high-resolution methods is complicated.

The first example of an exact projection method on a staggered grid with a high-resolution advection scheme was proposed by Tau [38]. This scheme used the projection method developed by Bell, et al. [39] with the Godunov method for advective fluxes modified for a staggered grid in 2D. This unique combination was emphasised by Rider [40]. We chose this method as a base for our 3D model for an incompressible ILES. The main stages of each time step of the projection method are as follows:

$$\frac{\mathbf{u}^* - \mathbf{u}^n}{\Delta t} + \nabla p^{n-1/2} = -\left[(\mathbf{u}.\nabla)\mathbf{u}^n\right]^{n+1/2} + \frac{\nu}{2}\left(\Delta\mathbf{u}^n + \Delta\mathbf{u}^{n+1}\right), \tag{16}$$

$$\tilde{\mathbf{u}} = \mathbf{u}^* + \nabla p^{n-1/2}, \tag{17}$$

$$\Delta\varphi = \nabla.\tilde{\mathbf{u}}, \tag{18}$$

                    V. Fuka, J. Brechler and A. Jirk

$$\mathbf{u}^{n+1} = \tilde{\mathbf{u}} - \Delta t \nabla \varphi, \tag{19}$$

and

$$p^{n+1/2} = \varphi, \tag{20}$$

where u^* does not obey the continuity equation and in the next stage is projected onto a solenoidal field using the so-called pressure form of the exact projection. The scheme has a high spatial resolution because of the use of slope limiters, which limit the derivatives of the variables in computational cells, meaning that no new extrema are created [41].

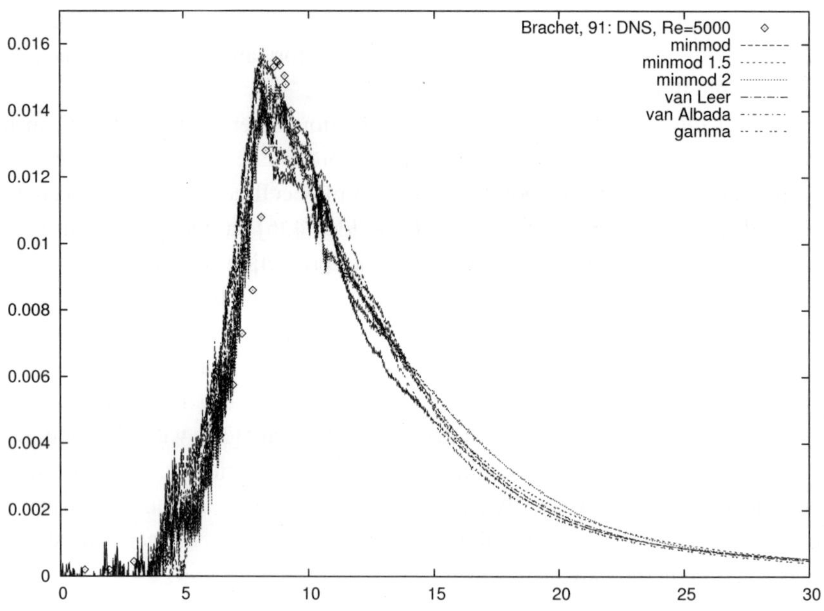

Figure 5. Dissipation of kinetic energy at a resolution of 256^3.

5.1. *Results of computations of the Taylor–Green vortex using ILES*

Here, we present the results of our computation of the Taylor–Green vortex using the ILES method. The Taylor–Green vortex serves as a simple example of free flow with a transition to turbulence and subsequent turbulent decay. This flow has previously been used for this purpose [42-44]. A comparison is usually made with the DNS data of Brachet [45-46]. This flow is defined in a box of $(0, 2\pi)^3$ by initial conditions for the velocity in the form

$$u = \sin(x)\cos(y)\cos(z), \tag{21}$$

$$v = \cos(x)\sin(y)\cos(z), \tag{22}$$

$$w = 0, \tag{23}$$

and by the pressure given by

$$p = p_0 + (2 + \cos(2z))(\cos(2x) + \cos(2y))/16, \tag{24}$$

where the value of p_0 is arbitrary (in the case of incompressible flow).

The Taylor–Green vortex shows complex behaviour. The flow is initially purely two-dimensional, but eventually forms vortex sheets that break into smaller eddies; after $t \approx 5$ the flow becomes turbulent. According to the results of DNS [46], $t \approx 9$ is the peak of enstrophy and kinetic energy dissipation. After this peak, the turbulent eddies dissipate in a self-similar energy cascade.

We performed our calculations on uniform grids with 128^3 and 256^3 cells. Molecular diffusion was set to zero. For comparison, we looked at DNS

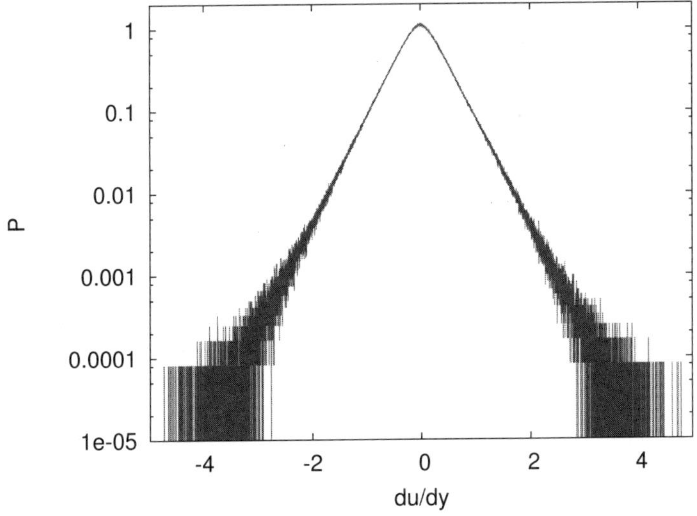

Figure 6. Probability density function of $\partial u/\partial y$ in $t = 40$.

results [46] at Re = 5000, which are very similar to the results at Re = 3000 and are therefore close to being independent of the Reynolds number. Figure 5 shows the time history of the dissipation of kinetic energy, defined as

$$\varepsilon = -\frac{\mathrm{d}K}{\mathrm{d}t} = -\frac{\mathrm{d}}{\mathrm{d}t}\frac{\left\langle \|\mathbf{u}\|^2 \right\rangle}{2},\tag{25}$$

and scaled to enable a direct comparison with results obtained using DNS. It is clear from the figure that the main features of the time dependence are captured by the calculations. The 128^3 runs produced broader and lower peaks than the 256^3 runs. In both cases, however, the top of the main peak was lower than that obtained using DNS, in good agreement with other ILES simulations [44]. The peaks also occurred a little later than that in DNS. The differences between individual limiters are clearly visible. The most distinct is the minmod limiter, which is overly diffusive. Among the other limiters, the extended minmod appears to produce results most similar to those obtained using DNS. The results presented below correspond to a resolution of 256^3 and the extended minmod limiter.

In the case of kinetic energy, for a short time interval around $t = 10$ there exists an area with scaling of approximately $K = t^{-1.2}$, but for $t > 12$ there is a

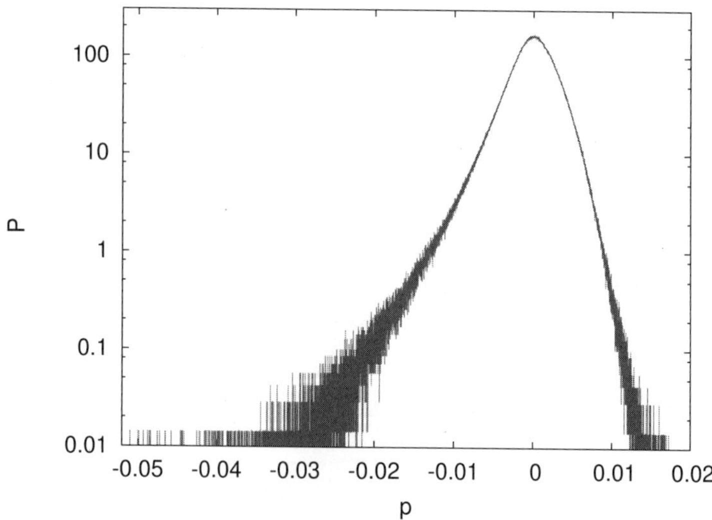

Figure 7. Probability density function of p in $t = 40$.

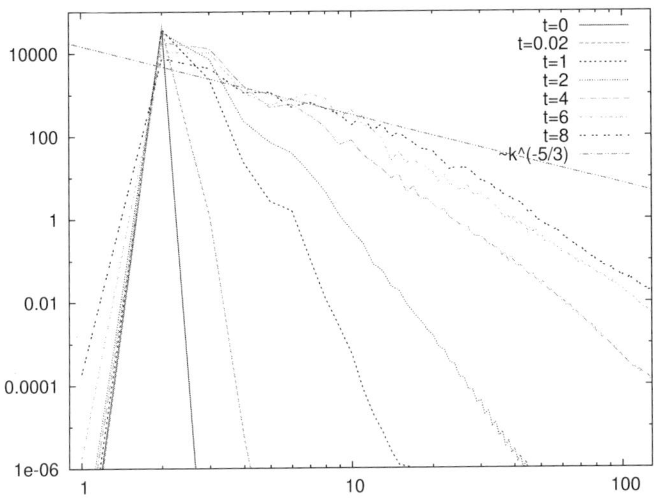

Figure 8. Temporal development of 3D energy spectra over the time interval (0,8).

clear scaling of $K \sim t^{-2}$. This finding is in agreement with previous results[44], but our $t^{-1.2}$ interval appears to be relatively short.

At time $t = 40$, we also computed the probability density functions (PDFs) of the velocity gradients (Figure 6) and pressure (Figure 7). For the velocity, the tails of the distributions show an almost exponential behaviour, as expected. For the pressure, the distribution is asymmetric. Figure 8 shows the temporal development of the 3D kinetic energy spectra, revealing the development of the energy cascade from the single initial wavenumber. Over a sufficient time period, the energy spectrum contains a clear inertial subrange with the Kolmogorov $k^{-5/3}$ power law, at least for moderate wavenumbers. The higher wavenumbers are probably affected by the numerical dissipation of the scheme. Similar results have been reported in a previous study [43].

7. Conclusions

In this chapter, we presented several options for treating two important features of environmental flow in CFD models: stratification causing buoyant forces to act on the fluid, and turbulence that strongly changes and complicates fluid behavior at high Reynolds numbers. Our analysis focussed on the Implicit Large Eddy simulation, and we presented numerical results for stratified and turbulent flows.

     V. Fuka, J. Brechler and A. Jirk

Acknowledgements

This research was supported by the Grant Agency of the Czech Academy of Sciences (grant no. 1ET400300414), by the Grant Agency of the Czech Republic (grant no. 205/06/0727) and by the Czech Ministry of Education, Youth and Sports under the framework of research plan MSM0021620860.

APPENDIX- LIST OF SYMBOLS

List of Symbols		
Symbol	**Definition**	**Dimensions or Units**
F	artificial momentum source term	$[L{\cdot}T^{-2}]$
Fr	Froude number	
Gr	Grasshof number	
K	kinetic energy density	$[L^2{\cdot}T^{-2}]$
L	length scale	
M	Mach number	
R	specific gas constant	$J{\cdot}kg^{-1}{\cdot}K^{-1}$
Re	Reynolds number	
Ri	Richardson number	
T	Temperature	K
U	velocity scale	
c_p	specific heat at constant pressure	$J{\cdot}kg^{-1}{\cdot}K^{-1}$
b	buoyancy	$[L{\cdot}T^{-2}]$
g	gravitational acceleration	$[L{\cdot}T^{-2}]$
k	unit vector in direction z	
p	pressure	$[L{\cdot}T^{-2}]$
t	time	$[T]$
u	velocity vector with components (u,v,w)	$[L{\cdot}T^{-1}]$
μ	dynamic viscosity	$[M{\cdot}L^{-1}{\cdot}T^{-1}]$
v	kinematic viscosity	$[L^2{\cdot}T^{-1}]$
ρ	density	$[M{\cdot}L^{-3}]$
θ	potential temperature	K

References

1. G. K. Batchelor, *An introduction to Fluid Dynamics*. (Cambridge University Press, Cambridge, 2000).
2. M. Deville, P. Fischer, and E. Mund, *High-Order Methods for Incompressible Fluid Flow*. (Cambridge University Press, Cambridge, 2002).
3. P. Wesseling, *Principles of Computational Fluid Dynamics*. (Springer Verlag, Berlin, 2001).
4. H. K. Versteeg and W. Malalasekera, An introduction to Computational Fluid Dynamics, The Finite Volume Method. (Longman, Harlow, 1995).
5. J. H. Ferziger and M. Perić, *Computational Methods for Fluid Dynamics*. (Springer Verlag, Berlin, 1997).
6. D. Drikakis and W. Rider, High-Resolution Methods for Incompressible and Low Speed Flow. (Springer, Berlin, 2005).
7. F. Grinstein, L. Margolin, and W. Rider, *Implicit Large Eddy Simulation. Computing Turbulent Fluid Dynamics*. (Cambridge University Press, Cambridge, 2007).
8. M. Z. Jakobson, *Fundamentals of Atmospheric Modeling*. (Cambridge University Press, Cambridge, 2000).
9. R. Stull. R. Iwatsu, J. M. Hyun, and K. Kuwahara, *Int. J. Heat Mass Transfer*. **35**, 1069–1076 (1992).
10. R. Iwatsu, J. M. Hyun, and K. Kuwahara, *Int. J. Heat Mass Transfer*. **35**, 1069–1076 (1992).
11. C. S. Peskin, *Annu. Rev. Fluid Mech.* **14**, 235–259, (1982).
12. R. Mittal and I. G., *Annu. Rev. Fluid Mech.* **37**, 239–261, (2005).
13. J. Kim, D. Kim, and H. Choi, *J. Comput. Phys.* **171**, 132– 150, (2001).
14. E. A. Fadlun, R. Verzicco, P. Orlandi, and J. Mohd-Yusof, *J. Comput. Phys.* **161**, 35–60, (2000).
15. A. Jirk. Thermally stratified atmospheric flow modelling (in czech). Master's thesis, Charles University in Prague, (2008).
16. D. L. Brown, R. Cortez, and M. L. Minion, *J. Comput. Phys.* **168**, 464–499, (2001).
17. G. Jiang, *J. Comput Phys.* **126**, 202–228 (1996).
18. C.-W. Shu and S. Osher, *J. Comput. Phys.* **77**, 439–471, (1988).
19. V. A. Titarev and E. F. Toro, *J. Comput. Phys.* **201**, 238–260, (2004).
20. E. Erturk, T. C. Corke, and C. Gk̦cl, *Internat. J. Numer. Methods Fluids*. **48**, 747–774, (2005).
21. U. Ghia, K. N. Ghia, and C. T. Shin, *J. Comput. Phys.* **48**, 387–411, (1982).
22. U. Ghia, K. N. Ghia, and C. T. Shin, *J. Comput Phys.* **48**, 387–411 (1982).
23. V. Fuka. A high resolution model of airflow (in czech). Master's thesis, Charles University in Prague, (2006).

24. R. Iwatsu and J. M. Hyun, *Int.J. Heat Mass Transfer.* **38**, 3319–3328, (1995).

25. B. Geurts, *Elements of direct and large-eddy simulation.* (R.T. Edwards, Philadelphia, 2004).

26. D. C. Wilcox, *Turbulence Modeling for CFD.* (DCW Industries, La Canada, 1993).

27. P. Sagaut, Large Eddy Simulation for Incompressible Flows. (Springer, Berlin, 2002).

28. M. Lesieur, O. Metais, and P. Comte, *Large-Eddy Simulations of Turbulence.* (Cambridge University Press, Cambridge, 2005).

29. A. Jameson, W. Schmidt, and E. Turkel. Numerical simulation of the euler equations by finite volume methods using Runge-Kutta time stepping schemes. In *AIAA 5th Computational Fluid Dynamics Conference (1981)*, pp. AIAA Paper 81–1259, (1981).

30. A. Harten, *J. Comput. Phys.* **49**, 357–393, (1983).

31. B, van Leer, *J. Comput. Phys.* **32**, 101–136, (1979).

32. J. P. Boris and D. L. Book, *J. Comput. Phys.* **135**, 172–186, (1997).

33. P. Colella and P. R. Woodward, *J. Comput. Phys.* **54**, 174–201, (1984).

34. P. K. Smolarkiewicz and L. G. Margolin, *J. Comput. Phys.* **140**(2), 459–480, (1998).

35. A. Almgren, J. Bell, and W. Szymczak, *SIAM J. Sci. Comput.* **17**, 2, (1996).

36. Almgren, J. Bell, P. Colella, L. Howell, and M. Welcome, *J. Comput. Phys.* **142**, 1–46, (1998).

37. J. Kim and P. Moin, *J. Comput. Phys.* **59**, 308–323, (1985).

38. E. Y. Tau, *J. Comput. Phys.* **115**, 147–152, (1994).

39. J. B. Bell, P. Colella, and H. M. Glaz, *J. Comput. Phys.* **85**, 257–283, (1989).

40. W. Rider, *Int. J. Numer. Methods Fluids.* **28**, 789–814, (1998).

41. V. Fuka and J. Brechler. In iEMSs 2008:International Congress on Environmental Modelling and Software, (2008).

42. R. E. Bensow, M. G. Larson, and P. Vesterlund, *Internat. J. Numer. Methods Fluids.* **54**, 745–756, (2007).

43. E. Garnier, M. Mossi, P. Sagaut, P. Comte, and M. Deville, *J. Comput. Phys.* **153**, 273–311, (1999).

44. F. Grinstein, L. Margolin, and W. Rider, In *Implicit Large Eddy Simulation. Computing Turbulent Fluid Dynamics*, chapter Numerics for Iles, Limiting Algorithms. Cambridge University Press, Cambridge, (2007).

45. M. E. Brachet, D. I. Meiron, S. A. Orszag, B. G. Nickel, R. H. Morf, and U. Frisch, *J. Fluid Mech.* **130**, 411–452, (1983).

46. M. E. Brachet, *Fluid dyn. Res.* **8**,1-8, (1991).

Chapter 5

THE ENVIRONMENTAL HYDRAULICS OF TURBULENT BOUNDARY LAYERS

ROBERT J. SCHINDLER and JOSEF DANIEL ACKERMAN

*Physical Ecology Laboratory, Department of Integrative Biology,
University of Guelph, Guelph, Ontario, Canada*

Turbulent flow over rough boundaries is a common occurrence in nature and the subject of much interest in a range of disciplines. It has long been recognized that the geometry of the boundary (or surface) dictates the flow and turbulence structure on a mean and instantaneous time scale. However, the mechanisms linking flow characteristics to roughness geometry remain poorly quantified, which has implications for our understanding of a variety of processes, particularly those occurring in the near-boundary region. It has been demonstrated that temporal and spatial variations in flow structure are sensitive to a range of geometric parameters describing the boundary geometry. We review the experimental evidence for rough boundary/flow interactions across different disciplines. A synthesis reveals that (1) different approaches have led to the adoption of a variety of parameters that are used to describe boundary roughness, and (2) that different criteria are used to evaluate the relative effects of boundary roughness. Moreover, (3) much of the experimental data relates to idealized surfaces that do not reflect the complexity of natural boundaries, or (4) is taken in low Reynolds number flows, and generally cannot be applied to aquatic flows in nature. The implications for our understanding of near-bed aquatic processes in turbulent boundary layers are discussed, and suggestions for future research approaches are presented.

Keywords: boundary layer; d-type roughness; flow separation; flow structure; K-type roughness; Relative roughness; Reynolds number; roughness geometry; roughness index; roughness layer; similarity hypothesis; turbulence; turbulent wake

1. Introduction

Unidirectional turbulent flow over rough boundaries (or surfaces) is a common occurrence in nature and the subject of much interest in a range of disciplines that can classified under the rubric of environmental hydraulics. One of the best models of this phenomenon is given by flow over a flat plate [40]. This model provides an ideal or theoretical condition on which to study hydraulic flows in the laboratory and field, and reasonable framework to contrast with nature.

87

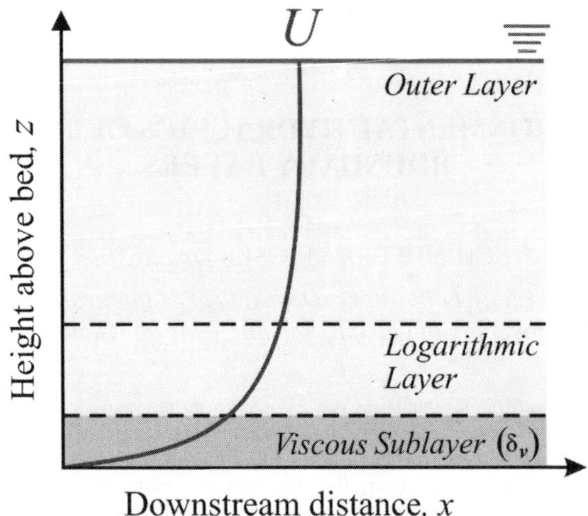

Figure 1. Boundary layer vertical structure with the $\partial u/\partial z$ indicated by the solid line.

The basic concept is that fluids cannot usually penetrate a boundary and this leads to a velocity gradient ($\partial u/\partial z$, where u is the velocity in the x, or downstream, direction and z is the height) perpendicular to the boundary, which increases outward to the free stream velocity (U_0). The region of reduced flow (i.e., $< 0.99\ U_0$) is arbitrarily referred to as the boundary layer, which has a thickness δ, and is a function of the Reynolds number, $Re= lu/v$ (where l is a length scale and v is the kinematic viscosity) and x, in the case of a flat plate. Close to the surface, the boundary layer will become turbulent when the local Re ($Re_x = ux/v$) approaches a critical value of 3 to 5 x 10^5, in the case of a flat plate oriented parallel to the flow. In nature this transition is accelerated by the presence of roughness or obstacles on the boundary [40]. The vertical structure is also important in a fully developed boundary layer: (1) The layer adjacent to the boundary is referred to as the viscous sublayer ($\delta_v \approx 10\,v/u_*$, where u_* is the friction velocity, a scale related to $\partial u/\partial z$ near the boundary) in which viscous forces dominate. The viscous sublayer includes a thin diffusional sublayer immediately adjacent to the boundary ($\delta_D \approx v/u_*(v/D)^{-1/3}$, where D is the molecular diffusivity) [33] where diffusive processes dominates. (2) The next layer is the inertial sublayer or log layer ($\delta_l \approx 0.15\ \delta$), which is a region of rapidly increasing velocity where inertial forces dominate. (3) The outer layer of the boundary layer represents a transition to the free stream flow (Figure 1).

Given that the boundary layers concept is based on spatial and velocity scales it should not be surprising to find that they can be embedded in one another, i.e., there may be a series of boundary layers of increasing sizes defined for (1) a larval mayfly on (2) a pebble in (3) a pebble cluster, on (4) the streambed [6]. Boundary layer flow is not isolated from upstream or downstream obstructions, nor is the flow necessarily uniform over them [19]. Lastly, boundary layers can also be generated by other types of water motion (e.g., wave current boundary layers).

The velocity within a boundary layer can be modeled mathematically using *the law of the wall*, which is a representation of the aforementioned vertical structure. Specifically, the velocity (u) is given as a function of the height (z) given by

$$ u = \frac{u_*}{\kappa} \ln\left(\frac{z}{z_0} \right) \tag{1} $$

where $\kappa = 0.4$ is the von Kármán constant, and z_0 is the roughness height. This model allows for the modeling of many flow conditions and also for the estimation of parameter values from data. For example, u_* is equal to κ multiplied by the slope of the linear regression of u on the natural logarithm of z in the log layer, and z_0 is equal to the base of the natural logarithm (e) raised to the value of the y intercept of the same regression [3] (Figure 2). Such estimates provide the ability to quantify the boundary, bed or wall shear stress (τ_w), which is the quotient of the shearing force and the area of the boundary, i.e., $\tau_w = \rho u_*^2$. Indeed, much of the challenge associated with environmental hydraulics has been to model flows and estimate parameters under non-ideal conditions that exist due to a number of factors especially bed roughness. These issues are the subject of this chapter, which focuses on flowing or lotic systems (i.e., rivers and streams).

 R.J. Schindler and J.D. Ackerman

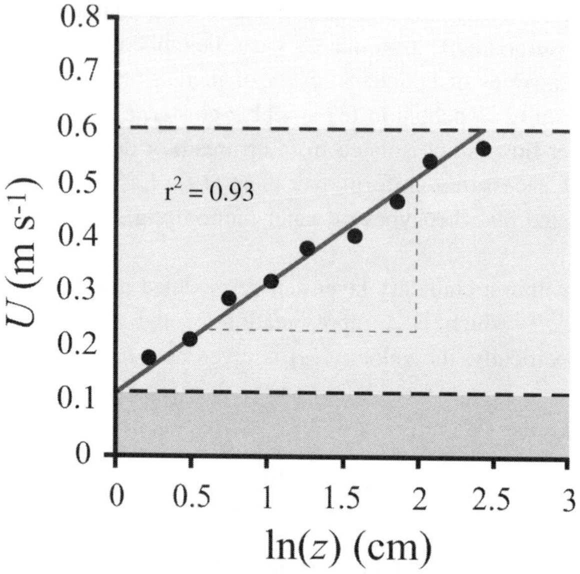

Figure 2. Law of the wall ($u_* = 0.068$ ms^{-1}, $z_o = 0.37\ 10^{-2}$ m, $\tau_w = 4.46$ Pa).

2. Understanding Boundary Roughness

In river channels, the boundary layer typically extends to the water surface due to relatively small ratio of flow depth (d) to roughness height (k) (i.e., d/k or relative roughness). The velocity characteristics of fluvial boundary layers have been extensively studied (e.g., [7] [9] [31] [2]). The shape of the downstream velocity profile, the velocity gradient, the turbulent flow characteristics and their links to sediment transport and bedforms have been common research themes in past decades [38].

A basic distinction can be made between the log-layer and outer region in a hydrodynamically rough turbulent boundary layer. The majority of boundary layer research has focused on uniform boundary roughness, or the identification of a common equivalent roughness height, k_s, which represent a spatial averaging of the roughness. The flow is said to be hydrodynamically rough when $k_s > 5\delta_v$ (i.e., the roughness element extends above the viscous sublayer; see below). This condition is met in most rivers, and other fully turbulent flows, where vortices generated around individual roughness elements or impinging from the log layer permanently disrupt the laminar sublayer [38]. These concepts can be integrated into the law of the wall for straight uniform channels with uniform bed material using the displacement height $(d_0,)$

$$u = \frac{u_*}{\kappa} \ln\left(\frac{z - d_0}{z_0} \right) \tag{2a}$$

or by using an equivalent roughness height (k_s) [38] given by

$$u = \frac{u_*}{\kappa} \ln\left(\frac{d}{k_s} \right) + 2.40 \tag{2b}$$

when the roughness height is uniform, e.g., $k_s = 30.1 z_0$ (see Figure 3).

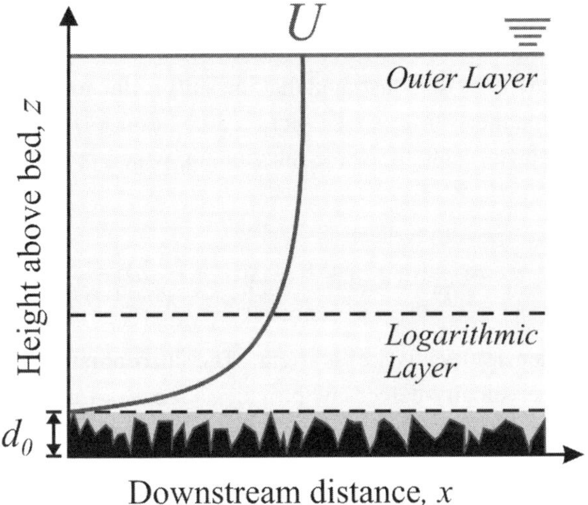

Figure 3. Law of the wall over boundary with uniform roughness.

However, where the roughness is not of uniform height, significant spatial variations in velocity profiles and associated parameters exist. Despite this, eqn. (1) and (2) have been applied to channels with heterogeneous bed material and/or bedforms. k_s can be considered a spatial average of effective roughness height. Numerous studies have shown that k_s is much greater than the median grain size, D_{50} (where the subscript represents the percentile of the grain size range), due to the disproportional effects of larger roughness elements on retarding the flow (e.g., [7]). Typical coefficients are $k_s = 6.8\ D_{50}$ and $k_s = 3.5\ D_{84}$ [9].

Engineering and physical sciences approached tend to model roughness using additional parameters to modify the law of the wall over a smooth bed. A non-dimensional form of the law of the wall is given by

$$u^+ = \frac{1}{\kappa}\ln(z^+) + B \tag{3}$$

where the subscript $^+$ denotes a dimensionless parameter and B is the smooth-wall log-law intercept equal to 5.60 [28]. According to [8] [18] the effects of surface roughness in the roughness layer (see section 3.1) cause a downward shift in the log-law relationship, which they described using a roughness function, Δu^+, given by

$$u^+ = \frac{1}{\kappa}\ln(z^+) + B - \Delta u^+ \tag{4}$$

More recent research has demonstrated that the shift is due to greater momentum absorption over rougher surfaces (see section 3.0). Eqn. (4) can be extended to include the outer layer using the wake function, ω, which provides a 'law of the wake' defined as

$$u^+ = \frac{1}{\kappa}\ln(z^+) + B - \Delta u^+ + \frac{\Pi}{\kappa}\omega\left(\frac{z}{\delta}\right) \tag{5}$$

where Π is dimensionless wake strength. The incorporation of the surface velocity (u_s) in the velocity-defect law provides a means of unifying the overlap between inner and outer regions

$$\left(\frac{u_s}{u_*}\right) - u^+ = -\frac{1}{\kappa}\ln(z/\delta) + B - \Delta u^+ + \frac{\Pi}{\kappa}\omega\left(\frac{z}{\delta}\right). \tag{6}$$

Some studies have reported that Π is greater for rough-wall flows [22]. This implies that accounting for roughness effects on the mean flow through Δu^+ may be limited. Conversely, there is evidence for a universal velocity-defect profile for smooth and rough walls [8] [18].

2.1. *Defining boundary roughness*

In the simplest terms, roughness can be thought of as any deviation from an idealized form such as a flat plate described above. In nature, roughness is caused by biotic (e.g., periphyton, aquatic larvae) and abiotic features (e.g., substrate, bedforms) on the boundaries. The physical characteristics of natural roughness include the roughness height (k), density, and spacing (wavelength λ, groove width, j, downstream length, l), and their effects on the flow, which are critical issues to consider in environmental hydraulics (Figure 4). For example, the flow over an array of roughness types (sand, variously sized and spaced hemispheres, spheres, and fences) indicated that different boundary roughness geometries may have similar effects on velocity profiles, even though the mechanisms of turbulence generation, mixing and dissipation may differ [40]. This was based on a surface density parameter (total projected frontal area of roughness per unit wall-parallel projected area), which demonstrated that the effect of roughness increased with the density parameter to a critical value after which the mutual sheltering of roughness elements led to a decrease in the effect

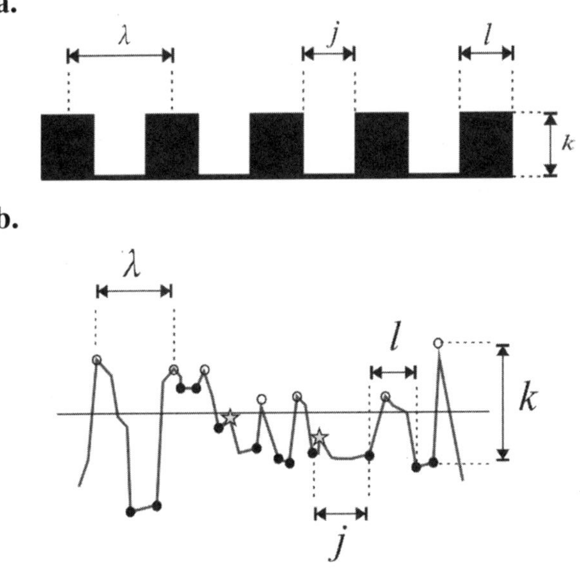

Figure 4. Physical descriptions of (a) idealized roughness elements characterized by blocks and (b) natural roughness elements in a bed profile. Each roughness element can be described by its height (k), length in the downstream direction (l), spacing with respect to other elements (λ), and groove width between elements (j). Legend: solid dots = troughs within a profile; hollow dots = peaks within a profile; and stars = insignificant peaks (i.e., below threshold indicated by solid line) within a profile.

of roughness. That analysis was somewhat limited because it did not fully characterize the surface, and correlations may be to limited to particular experiments. Consequently, most attempts to examine the relationship between roughness and flow characteristics are restricted to surfaces whose geometry is easily described [21]. Studies of flow over these idealized objects have been successful in elucidating the effects of relative size, spacing, arrangement and two-dimensional (2D) or three-dimensional (3D) roughness configuration upon flow resistance, turbulence generation and dissipation (see below).

Three types of flow-roughness interactions were defined by [29] based on the premise that resistance was primarily a function of energy extraction due to the formation of turbulent wakes around individual roughness elements. Using elements of equal height, length and width, but differently spaced in x, [29] identified: (i) *Isolated roughness flow*, where the coherent structure generated by the wake of an upstream element is dissipated before the next element downstream; (ii) *Skimming flow*, where the elements are sufficiently close to prevent the spaces between elements being filled with low-velocity water; and (iii) *Wake interference flow* marking the transition between (i) and (ii), where the wake structures of successive elements impinge on one another to generate additional turbulence. A fourth type was included by [10] (iv) *Chaotic flow* in shallow rough flows where $d < 3k$ (Figure 5). This conceptual framework was an attempt to categorize flow conditions via roughness parameters such as k, λ, l, and j.

The importance of the connectivity of the fluid between roughness elements and the fluid above the roughness elements underlies the distinction that has been made between (1) k-type roughness, where k_s is proportional to k and (2) d-type roughness, where k_s is proportional to δ rather than k [35]. The usual explanation is that d-type roughness sustains stable recirculation (or vortices) that isolate the fluid between the roughness from the outer flow. The effective length scale determining k_s is not k, but $k - d$ (or the 'error-in-origin'; [35]). Where the distance between the edges of elements, or groove width (j), is wider than $3 - 4k$, the recirculation zone reattaches before the next roughness element, exposing it to the outer flow. The thresholds between each flow type are dictated by k, and the downstream distance between roughness elements given by λ, j, and l (e.g., [29] [34] [10] [44] [21]). Ratios of these length scales have been used to develop a number of dimensionless parameters that are used to describe bed roughness geometry (Table 1).

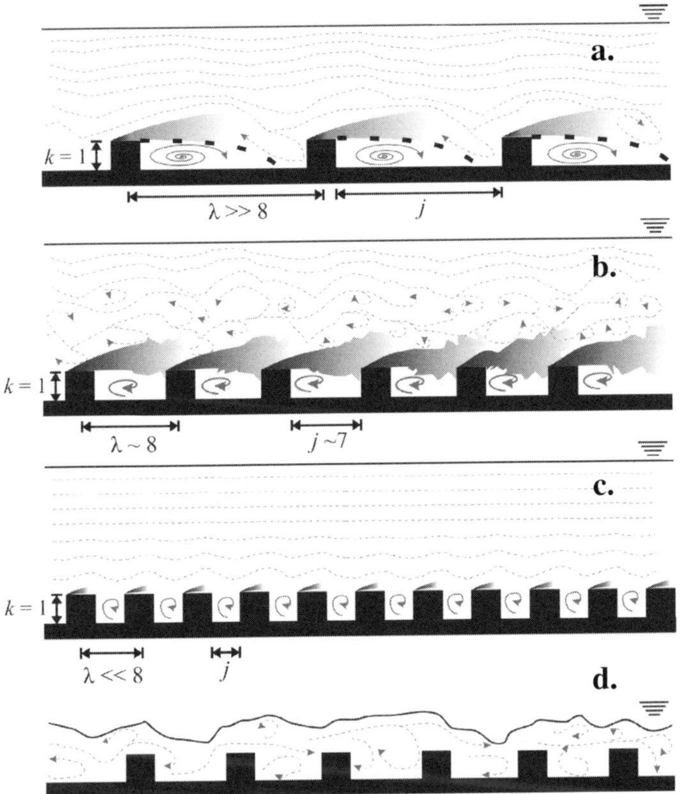

Figure 5. Hydrodynamic regimes defined by flow-roughness interactions include (a) isolated roughness flow, (b) wake interference flow (c) skimming flow, and (d) chaotic flow. Regimes are defined by water depth k, λ, and j (after [29] [10]).

For high Reynolds numbers ($Re = Ud/\nu$), the threshold roughness spacing, $\lambda_{Crit,}$ between isolated roughness flow and wake interference flow is given by

$$\frac{\lambda / k}{C_D(1 - nj / P)} = 67.2 / \left(\frac{100}{(2\log(d / \lambda) + 1.75)^2} \right) \qquad (7)$$

where C_D is drag coefficient, n is the number of elements in a cross-section, P is wetted perimeter and d is flow depth [29]. The term nj/P represents the proportion of the flat surface between the roughness elements, and $C_D (1 - nj/P)$ varied with roughness density. When roughness density was constant, the critical value λ_{Crit} varied with k and d, but [44] showed λ_{Crit} increased with d according to

$$\lambda_{Crit} = 2.46k\left(\frac{d}{k}\right)^{0.42} \tag{8}$$

Note that eqn. (8) is limited because it is based on a single value of C_D (1- ns/P) and assumes long, flat roughness elements.

Table 1. Summary of bed geometry parameters derived from experiments over idealized objects.

Parameter	Ratio	Reference
Roughness Index	λ/k	[29]
Relative Roughness Spacing	k/λ	[29]
Roughness Spacing Index	λ/j	[29]
Dimensionless Groove Width	j/k	[44]
Blockage Ratio	δ/k, $\delta = d$ in most natural channels	[21]

2.2. *Parametrizing boundary roughness*

The threshold between skimming flow and wake interference flow is relevant as it represents a transition from a bed characterized by sheltered zones of separated flow between elements to that perturbed by turbulent structures. Physically, the threshold is defined as the point where $j > k$. The turbulent wake occupies the entire grove width and is stable, i.e., skimming flow. When $j > k$, the wake will is increasingly unstable in a manner common to flow separation in the lee of obstacles, leading to wake interference flow [44].

The exact ratio j/k that defines the threshold between each state is a function of the shape of the roughness elements (see section 3.3.5). More importantly, for a coarse clastic river bed the variables λ, j and k are not easily determined. Initially, a "significant roughness element" must be chosen arbitrarily. For instance, a small element below the average roughness height may be considered significant if exposed, and may exert more flow resistance than a much larger element sheltered by a close neighbour. A visual interpretation of the roughness profile in cobble-bed rivers was used in [44] to determine λ. It was estimated that k was equivalent to two times the average deviation away from the height, and the groove width (j) approximated the roughness spacing minus the D_{50} of the long axis on the basis that cobbles generally lie with their short axes vertical and their long axes aligned to the flow. This provided a value of $j/k = 1.0$ for 'stream bed roughness'.

The 'average' flow condition in nature appears to be wake interference [44]. Using high spatial resolution profiles of downstream velocity and visual observations in modelled cobble beds in flume channels, [44] isolated all three flow states, plus chaotic flow around a relatively large obstacle ($d < 3k$). The term C_D $(1 - ns/P)$ ranged from 0.4 - 0.8 and the dominant condition was wake interference flow, validating the approach. It is evident that the choice of criteria for the selection of significant roughness element can have a profound affect on the outcome. For example, Figure 6a presents a bed profile taken at 2.5 cm increments with a 1 mm vertical resolution in mid channel in a cobble bed stream in southern Ontario. The selection of significant roughness elements using objective criteria such as relative depths (k/d) of 10, 30 or 50 led to dramatically different outcomes in terms of the interpretation of the parameters used to describe the roughness (e.g., k) from the same dataset (Figure 6b; Schindler and Ackerman, unpublished). Whereas this approach is somewhat subjective, the results are intriguing and warrant further study for application across a range of bed morphologies (see section 3.3.3).

2.3. *Determining the spatial complexity of boundary roughness*

Given the complexity of channel bed topography, it is not surprising that a single, universal parameter to describe the effects of boundary roughness on flow and turbulence structure has been sought [42]. Because of the simplicity of relating downstream velocity profile characteristics to k_s or D_x, this approach is often used in environmental hydraulics. However, this is a somewhat of a 'black box' approach because it neglects variations in bed structure [7] [9] [31] [2].

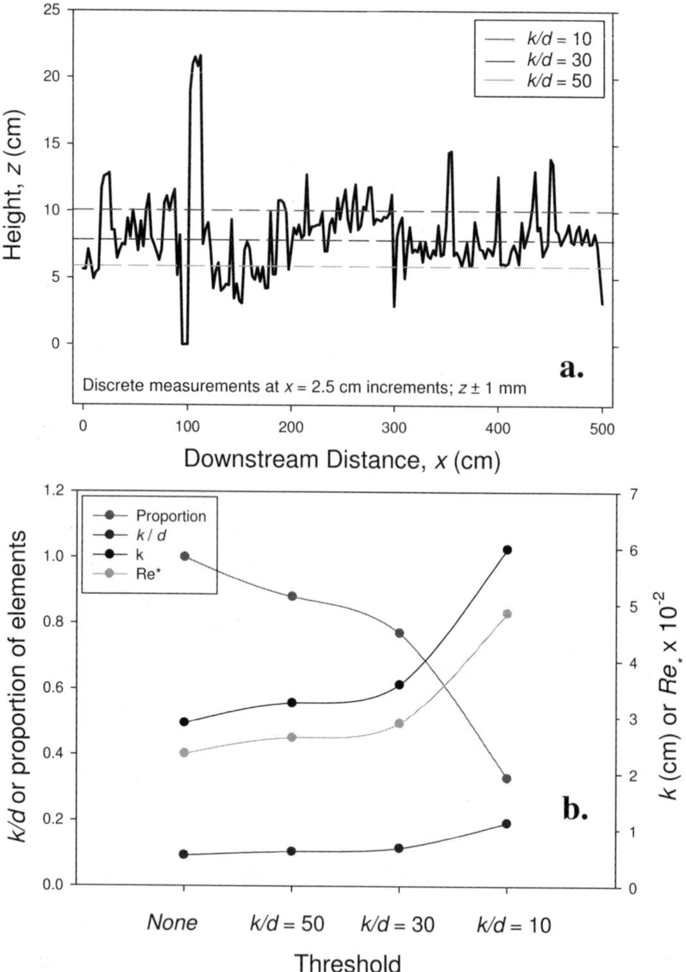

Figure 6. The effect of the threshold selection on roughness parameters. (a) a bed profile in mid channel in a cobble bed stream, and (b) the effect of different thresholds on roughness parameters (Schindler and Ackerman, unpublished).

Apparently, bed structure (particle shape, orientation, height variation, degree of packing, structural arrangement) may significantly affect the overall hydraulic resistance because turbulent structures generated at the boundary are ultimately responsible for the extraction of mean flow energy, i.e., resistance [31]. In other words, the use of D_x as a roughness parameter requires that independently formed roughness surfaces with the same value of D_x show exactly the same resistance to flow for identical hydraulic conditions [1]. The size, intensity and longevity of turbulent structures governing the mean flow field are, therefore, a product of (1) the geometry of the roughness obstacle and (2) flow conditions (assuming the absence of bedforms).

2.3.1. *Statistical measures of boundary roughness*

Various statistical measures of bed roughness have been suggested. For example, the difference between the highest point on the particle and the average elevation of the points of contact with adjacent particles could be used for the effective roughness height. The evaluation given by [11] of percentile grain sizes supported the use of K_3, which is the maximum difference between three adjacent points on a bed profile. This statistical approach is limited, as it depends on the sampling interval used [31], and [43] showed that k_s was dependent on the density of roughness elements, and $k_s = 3D_{84}$ even when the flow deviated significantly from a vertical logarithmic profile.

Another method uses the statistical moments of the distribution of discrete measurements of z to describe the roughness surface. It was shown by [2] that gravel-bed channels exhibiting a step-pool morphology or exhibiting high gradients were better described by the standard deviation of bed elevations (σ^2_z) than by D_x. Conversely, [1] found that σ^2_z was closely related to changes in D_{50} and D_{84}. The skewness of the distribution also increased with the level of armoring, which reflects the settling of finer particles, ultimately reducing the magnitude of surface elevations below mean bed height.

2.3.2. *Random field approach to characterizing boundary roughness*

An alternative way of accounting for bed surface geometry is to describe the roughness as a random field of surface elevations in the downstream and transverse directions $z(x,y)$ (e.g., [31] [2]). This method is popular due to technological advances (e.g., laser scanning) that have facilitated the detailed measurement of bed surfaces.

Random roughness has also been characterized using second-order structure functions that investigate the scaling properties of various rough surfaces [31]. This approach is equivalent to semivariograms used by [9] to describe gravel bed surfaces. The second-order structure function $D(\Delta x, \Delta y)$ of bed elevation $z(x,y)$ is defined as an average square increment

$$\{z(x+\Delta x, y+\Delta y) - z(x, y)\}^2, \tag{9}$$

which describes the 'average variance' of the z surface in downstream $(D(\Delta x, \Delta y = 0))$ and transverse $(D(\Delta x = 0, \Delta y))$ directions based on different spatial sampling lags. The resulting curves of D vs. lag show that at sufficiently large lags, D becomes constant and approximates $2\sigma^2_z$, known as the region of saturation. Small lags corresponds to a scaling region, which can be approximated by a power function i.e., $D(\Delta x) \propto \Delta x^{2H_x}$, where H_x is the scaling exponent. Whereas D provides information on the 'average' surface properties, H_x or H_y (in the transverse direction) can be considered a measure of the complexity of the surface topography (a larger H indicates a smoother profile [5]).

These concepts suggest that statistical moments and the 'random field' approach provide deeper insight into the geometry and surface-forming processes of gravel-bed rivers than a simple characteristic grain size. However, it should be noted that: (1) the concepts have yet to be examined in channels exhibiting smaller or larger ranges in grain size; (2) the application of statistical moments assumes that the distribution of z is Gaussian or near-Gaussian; and (3) the concept is descriptive rather than mechanistic. Detailed comparison of the turbulence field and statistical parameters is required before this approach can be used to understand the role of bed geometry in modifying boundary layer structure and associated turbulent dissipation.

3. Hydrodynamics of Rough Boundaries

The aims of this section are: (1) to highlight the differences between near-bed turbulence over smooth- and rough-beds; (2) to elucidate the key geometrical or dimensional parameters describing bed geometry that ultimately dictate near-bed flow and turbulence modification; and (3) to examine the effect of different roughness types (e.g., 2D and 3D surfaces) on the modification of near-bed flow structure.

3.1. *Outer flow region vs the roughness layer*

It is apparent that the ability to describe the flow region at and immediately above roughness elements should provide important insight into their interaction with flow. It would then be possible to expand the description of the vertical structure of a boundary layer over an impermeable rough bed with sufficient depth to include: (1) the interfacial sublayer, δ_t, within the groove space (j) between roughness crests and troughs; (2) the form-induced sublayer, δ_f, in the immediate vicinity of the roughness elements; (3) the logarithmic layer in the flow above the bed; and (4) the outer region [32]. The interfacial sublayer and the form-induced sublayer comprise the roughnesss layer, δ_r, which is subject to turbulent motions that are directly influenced by the roughness characteristics. Its upper-boundary is defined by z_r (Figure 7).

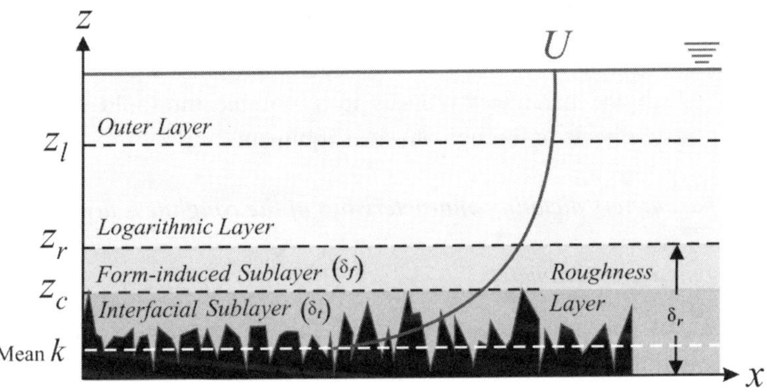

Figure 7. Rough turbulent boundary layers defined using the roughness layer, which is composed of the form-induced and interfacial sublayers.

An important driver of the recent interest in rough boundary flows is the 'similarity hypothesis', which states that at high Re, turbulent motions become independent of wall roughness and viscosity at a certain height away from the boundary. This notion can be considered an extension of the Reynolds-number similarity hypothesis for turbulent flows. The theory dictates that smooth and rough walls will exhibit the same turbulence structure in the outer layer, which is a region of weak shear and, therefore, not the site of the dominant instability mechanisms that generate the turbulence, nor of the strongest production of turbulent kinetic energy. Typically, profiles of second-order velocity moments will collapse to common curves in the outer-layer, regardless of wall roughness characteristics [21]. Several experiments have provided substantial support for the similarity hypothesis (e.g., [4] [17] [37]). Other experiments have shown that the outer flow can exhibit significant structural differences due to the influence of large-scale (e.g., [12] [22] [23]) and small-scale ([42]) coherent turbulent structures specific to individual roughness geometries.

The emphasis on the roughness layer and the validity of the similarity hypothesis also provides an opportunity to elucidate the effects of a large range of bed roughness types and flow conditions. The height to which the roughness layer extends, z_r, is generally considered to be 2 - 5k [23] [37] [21], although [4] suggest that it may be as high as 8k. The range varied with each type of boundary because the turbulent structures that dictate z_r are controlled by different length scales across different surfaces [37]. z_r can be determined in several ways and its thickness is somewhat dependent on the methodology used for parameter selection [4]: (1) the position of the maximum Reynolds shear stress, used to define the upper boundary in shallow flow over rough beds, which is often ill-defined, leading to some uncertainty [36]; and (2) profiles of the variance or standard deviation of a flow parameter can be used to isolate the height at which the parameter reduces to a suitable threshold defined from previous measurements or through physical arguments.

3.2. *Key parameters dictating characteristics of the roughness layer*

3.2.1. *Geometrical parameters*

The height of the roughness layer, and the characteristics of turbulence generated within it, varies with changes to the boundary roughness characteristics. The most important roughness parameters are roughness height, k, and spacing, λ, both of which should be considered in terms of relative roughness height (k/d) and roughness index (λ/d). As discussed above, the roughness index dictates the degree to which turbulent structures formed around individual elements interact. The relative roughness height dictates the degree to which elements interact with the entire water depth. Consequently, the shape of the roughness elements, in particular the degree of angularity with respect to the flow (i.e., degree of bluntness), must also be considered (see section 3.3.5).

A distinction can be made between 2D and 3D surfaces, although the bias has been towards 2D geometries due to the ease of measurement in the roughness layer, and in particular the interfacial sublayer. 2D surfaces are almost exclusively comprised of bars or rods oriented in the transverse direction across channels, and are generally termed *2D transverse bars*. Compared with 2D transverse bars, much less is known about the influence of length scales for 3D roughness [37], due primarily to the inherent complexity of 3D surfaces. The variety of 3D roughness surfaces studied is considerable, and each will exhibit different turbulent characteristics in the roughness layer.

3.2.2. *Key flow parameters*

One effect of wall roughness is the injection of turbulent energy into the flow field. The roughness Reynolds number (Re_*) provides an indication of the influence of the roughness surface on the buffer zone between the inner logarithmic region and the outer flow region in terms of the balance of viscous and form drag forces. It incorporates the height of the roughness surface, k, to describe the influence of the roughness in the boundary layer flow [10]

$$Re_* = u_* k / v \tag{10}$$

Values typically range from 1×10^2 to 1×10^3 for zero-pressure gradient boundary layer flows. There is little consensus on how to calculate the threshold between hydrodynamically smooth and hydrodynamically rough flows, principally because the use of k neglects the effects of other geometrical parameters, although as mentioned above, $k_s > 5\delta_v$ has been used. For example, [27] found that the roughness types and uniformity dictated the threshold values. Their rough surface, consisting of uniform close-packed spheres, displayed a narrow transitionally-rough regime with $Re_{*smooth} \approx 15$ and $Re_{*rough} \approx 55$. Less uniform close-packed sand grains had a wider transitionally rough regime where $Re_{*smooth} \approx 5$ and $Re_{*rough} \approx 70$. [21] reviewed the data for a variety of roughness types and concluded that k_s/k was independent of Re_* when $Re_* \geq 80$, although 70 has been cited [24] (Figure 8).

When a surface is hydrodynamically smooth the bed shear stress is entirely viscous and $\Delta u^+ = 0$ (see eqn. (4)). Flow near a rough boundary may generate coherent flow structures similar to those in smooth boundaries. However, their shape and size are altered during the enhanced momentum exchange between the near-wall and outer flow regions [36]. This, in turn, depends on the roughness geometry and configuration [25]. Sufficiently large roughness elements destroy the autonomous burst-sweep cycle characteristic of smooth boundaries, often generating more disrupted or disorganized structures that take

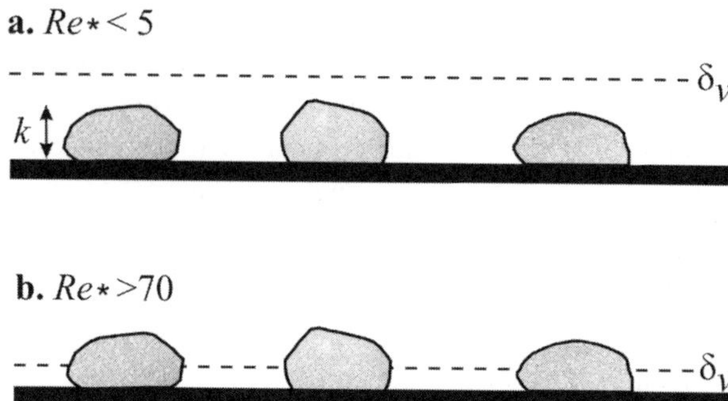

Figure 8. (a) Hydrodynamically smooth flow where $Re_* < 5$ and roughness elements are contained within δ_v, and (b) hydrodynamically rough flow where $Re_* > 70$ and roughness elements extend beyond δ_v.

over the role of generating turbulence at the wall [4] [13]. Turbulent structures generated by individual roughness elements provide an additional source of turbulent kinetic energy ($TKE = 1/2\left(\overline{(u')^2} + \overline{(v')^2} + \overline{(w')^2}\right)$, where $u = \overline{u} + u'$, i.e., instantaneous u = mean u + fluctuation from the mean, similarly for the other velocity components), which enhance turbulent mixing in the near wall region. Therefore, hydraulically rough flows are dominated by form drag on the roughness elements, i.e., velocity is independent of viscosity [41].

During the transition from hydrodynamically smooth to hydrodynamically rough regimes both skin and form drag are significant. Initially extra form drag is generated, which weakens the burst-sweep cycle, thus decreasing skin friction. As Re_* increases, the burst-sweep cycle becomes increasingly disrupted until it is destroyed, and the reduction in skin friction is lost, leading to an overall increase in drag. Different surfaces have different balances of both effects. [21] stated that for surfaces with sparsely distributed elements, the form drag increases before the burst-sweep cycle is modified and the reduction in skin drag is never realized. On this basis, uniformly rough surfaces are likely to be most effective for drag-reducing roughness. In the fully rough regime, Re_* is given by

$$\Delta u^+ = \frac{1}{\kappa} \ln Re_* + B - 8.5 \tag{11}$$

and B is the smooth-wall log-law intercept equal to 5.60 [38] (see eqn. (3)). The relationship among Δu^+, Re_*, and k has been obtained experimentally for a variety of rough surfaces in atmospheric research (review in [14]), however, no equivalent data exist for aquatic systems.

3.3. *2D transverse bar roughness*

3.3.1. *Spatial flow variation*

Whereas the precise length scales and flow conditions vary among studies, a general pattern of the spatial variation in flow and turbulence structure over 2D transverse bars has emerged. Flow separation occurs in the lee of the roughness elements and generates a well-defined recirculation zone. Downstream vorticity is small within the separation zone in the lee of each bar. Conversely, intense vorticity is created at the height of the roughness crests due to the generation of a shear layer between the outer flow and the recirculation cell. Shear layer vortices are intensified by vortex stretching due to the high strain generated at the crest of the bars. Consequently, this region is responsible for most of the production and dissipation of TKE [4]. Under isolated roughness flow (see Figure 5a) reattachment occurs within the cavity (or jaw space, j) between adjacent elements before flow streamlines are oriented upwards as the next element is approached. In addition to the shear layer extending from each roughness crest, secondary vortices of opposing directions can form at the corners of the element, although this is dependent upon the angularity of the roughness elements.

Comparisons of the TKE production and dissipation revealed that the roughness layer over 2D transverse bars is far from an equilibrium state. Dissipation is maximized at the position of the crests due to the intensity of the vorticity associated with shear layer formation, although the maximum vorticity is smaller compared with a smooth bed. Conversely, the roughness induces an increase in the maximum production rate.

3.3.2. *The role of roughness geometry*

The manner in which the shear layers generated over each roughness crest interacts with the roughness elements themselves is not trivial [13]. The vertical length scale of the vortices generated at the shear layers approximates k, and consequently it is reasonable to assume they dominate flow in this region. The importance of roughness index (λ/k, see Table 1) on near bed flow structure was first noted by [29] primarily due to its impact on shear layer interaction. In addition, z_r increased with the transverse dimension of roughness elements [37]. A consideration of other parameters (e.g., shape) is also relevant (see sections 3.3.5 and 3.4).

The thickness of the roughness layer was a product of both k and d in measurements of downstream turbulence intensity, TI, over k-type 2D transverse square bars performed by [36]. Planar laser-induced fluorescence was used by [13] to evaluate the differences in near-bed mean flow over 2D square bar roughness where $\lambda/k = 8$ and 16. Both cases showed flow separation from the downstream edge of the roughness elements and a well-defined recirculation in its lee. The extent of this region varied between cases as flow reattachment occurred at ~ 3.5k and 4.2k for $\lambda/k = 8$ and 16, respectively. In general, the two cases exhibited similar mean flow characteristics despite the difference in λ/k, and matched the direct numerical simulation (DNS) results of [20] [25] derived using similar λ/k ratios. An examination of TI and Reynolds shear stress suggested that when $\lambda/k = 16$ higher downstream and vertical TI and Reynolds shear stress were generated in the upper 90% of flow compared to $\lambda/k = 8$, but the values remained the same near the bed.

The roughness function, Δu^+ (see eqn. (11)), can be considered a measure of the capacity of the rough surface to absorb momentum [36]. There is sufficient evidence to suggest that Δu^+ is a function of λ/k. For example, [13] showed that for $\lambda/k = 8$ and 16, Δu^+ diminished from 13.3 to 12.9, respectively and [25] used DNS to determine equivalent values of 12.0 and 11.3. A wider array of λ/k was used by [16] who found similar values at the same roughness indices. More importantly, they found that Δu^+ was maximized at $\lambda/k = 8$ and that beyond this threshold λ/k led to a decline in Δu^+. [13] demonstrated that peak turbulence production at $\lambda/k = 16$ was 14% higher than at $\lambda/k = 8$, and suggested that this represented the reduction of viscous effects in the former case. The precise mechanisms that result in deviations from the mean flow properties on either side of $\lambda/k = 8$ can be described as wake interference flow (see Figure 5b). Put simply, viscous drag and form drag combine to maximize resistance over 2D transverse roughness in turbulent flows when $\lambda/k = 8$ [24] [25].

A comparison of flow data taken over 2D transverse bars revealed that skimming flow ($\lambda/k = 2$; [26]) generated far less form drag, due to the limited connectivity between the fluid between the bars and the outer flow, than isolated roughness flow ($\lambda/k = 11$; [13]) (see Figure 5a and c). Recirculation existed across the cavity at low λ/k and was only partly destroyed by intermittent ejections into the outer flow (see also [35]). At higher λ/k, the recirculation zone was found immediately next to each bar and was intermittent due to repeated disruption from shear layers generated at the upstream crest, which increased the connectivity between the interfacial cavities and the flow above, in particular the wake-dominated roughness layer. This suggests that the mechanism(s) of near-bed turbulence production differ between d-type and k-type roughness. In fact, low speed vortical streaks characteristic of smooth walls were observed when $\lambda/k = 2$, but were absent for $\lambda/k > 11$ [26]. d-type roughness and smooth beds may be more similar hydrodynamically than d-type and k-type roughness.

The effect of λ/k over 2D square bars was examined by [39] using two-equation Reynolds-Averaged Navier-Stokes (RANS) numerical modelling. They maintained a constant $k/d = 0.1$, characteristic of individual roughness elements [21] (see section 3.3.3). The roughness index varied between d-type and k-type roughness ($5 \leq \lambda/k \geq 30$) and the resistance was measured by the Darcy-Weisbach friction factor, f,

$$f = \frac{8\tau_w}{\rho u^2},$$

(12)

which rose and fell around $\lambda/k \sim 8.8$ at a range of turbulent Re. They used profiles of the spatially-averaged downstream velocity to determine the 'virtual origin', or height to which the space-averaged downstream velocity vanished. The virtual origin was inversely related to λ/k, and maintained an approximately logarithmic decrease until $\lambda/k \sim 15$, whereupon it declined asymptotically. More importantly, they showed that Δu^+ was related to λ/k in a similar manner to f. This may explain the apparent contradiction between studies that have shown that Δu^+ both declined [13] [25] [16] and increased [26] with increased λ/k.

3.3.3. *The role of relative depth*

Whereas there is compelling evidence that λ/k is an important parameter dictating near-bed flow structure, the difference in scale between roughness length scale and the largest turbulence scales may explain the differences between a range of studies, regardless of roughness configuration [21]. The relative height, δ/k (or blockage ratio, k/δ), provide an alternative parameter that can be used to examine whether boundary roughness affects the near bed flow. A recent review suggested that, on average, the transition for individual roughness elements occurs when $k/\delta \leq 50$ (or $\delta/k \leq 0.02$) [21].

                    R.J. Schindler and J.D. Ackerman

The relative height of the roughness elements may also influence the evaluation of the similarity hypothesis over 2D transverse bars. The results of studies from channel flows bounded on two sides [23] with regular, square bar k-type roughness [25] [26] and boundary layer flows with 2D transverse bars of uneven height [20], and square bars [36] (details below) showed that the outer region was significantly affected by the rough boundary. In each case, k/d was between $0.1d$ and $0.2d$, well above the threshold of $0.02d$ for transition to individual roughness elements. The effect of variable relative roughness of 2D square bars on resistance, f, and Δu^+, in fully turbulent and hydrodynamically rough flows was examined by [39] where the magnitude of resistance, f, increased with k/d across all roughness index values. Moreover, resistance became independent of Re earlier as height increased, but the flow depth had no influence on Δu^+ or the virtual origin at all Re and λ/k. In other words, flow depth had no influence on the height to which the roughness elements influenced the flow, which was in the range $0.025< k/\delta > 0.1$, i.e., above the threshold specified above. Note that this pertains only to fully turbulent, hydrodynamically rough conditions.

 DNS simulated fully-turbulent flows in a bounded channel flow with smooth and rough boundaries comprised of 2D transverse bars were examined by [4] at a much lower relative depth. The relevant geometrical parameters in the rough bed case were $k/d = 0.07$, $\lambda/k = 7$ and $Re_* = 395$. They showed that higher vorticity was exhibited within the roughness layer, that there was less Reynolds stress anisotrophy, and that structures typical of smooth boundaries became disrupted [4]. The transverse vortices were disorganized at the boundary but became increasingly coherent and organized with height until they equalled the size and intensity of those over the smooth bed, although the outer layer was unaffected. This implies that, providing mean shear is sufficient, similar structures to smooth beds will be formed, albeit away from the boundary, when relative roughness is sufficiently low. Third-order velocity moments, which are sensitive to variations in turbulent transport processes due to their high degree of non-linearity, showed that the rough bed substantially increased the transport of kinetic energy towards the boundary, but reduced it at the edge of the roughness layer. In the outer region, the smooth and rough bed cases exhibited similar characteristics. The results provide strong support for the similarity hypothesis.

3.3.4. *The importance of Reynolds number*

Experimental and modelled data from flows over 2D transverse bars at Re_* spanning transitional ($Re_* = 63$) and fully rough ($Re_* = 121$) hydrodynamic conditions were examined by [24]. Three key differences were noted: (1) the total flow resistance for transitional flow was influenced by viscous and form drag, but the viscous forces were negligible for hydrodynamically rough conditions; (2) this leads to greater coherence in near-wall structures in transitional flows; and (3) the influence of roughness on Δu^+ was less pronounced for transitional cases. Similarly, [39] determined that the effect of Re on f decreased with increasing Re, and Re independence occurred above a threshold value dictated by λ/k. The roughness function increased with Re_* across d-type and k-type roughness, ultimately exhibiting a logarithmic relationship above a threshold Re. At this point, the slope of the relationship recovered to κ^{-1} at sufficiently high Re_*. This facilitated the calculation of equivalent roughness height, k_s, for each bar arrangement because the relationship paralleled the correlation for sand-grain roughness (i.e., [40]). Overall, the results show that once $Re_* > 500$ the flow becomes independent of Re.

The precise mechanisms to account for relationships between f, Δu^+ and flow and roughness parameters remain uncertain. However, some insight is provided by [36] who examined turbulence structure in some detail across a range of hydrodynamically rough conditions ($30 \leq d \geq 95$ mm; $6{,}000 \leq Re \geq 48{,}450$; $355 \leq Re_* \geq 515$) over 2D transverse bars of fixed geometry (6.4 mm height and width spaced at $\lambda/k = 8$). They used quadrant decomposition to provide information regarding the turbulence structure in the near-bed region (i.e., turbulent events were sorted into quadrants of the u', w' *plane* (where u' and w' are fluctuations from the mean u and w, respectively) associated with ejections ($Q2$; $u'w' < 0$ with $u' < 0$) and sweeps ($Q4$; $u'w' < 0$ with $w' < 0$)) (Figure 9). The influence of the roughness elements on the spatial variability in time-averaged velocities was similar across all flow conditions. Upstream of each element u decreased and w increased relative to the mean values, leading to dominant $Q2$ activity. $Q1$ events dominated above the roughness element because of the acceleration in u, and w became negative downstream of the element resulting in $Q4$ dominance. As the subsequent bar is approached, the flow decelerated, leading to $Q3$ dominance. The sequence of quadrant

                    R.J. Schindler and J.D. Ackerman

dominance – clockwise through $Q2$-$Q1$-$Q4$-$Q3$ – is typical of surfaces where roughness elements behave individually and flow reattachment occurs (e.g., 'k-type' roughness; isolated roughness flow; where $\lambda/k \geq 8$). Through the use of two heights within and two heights above the roughness layer, they showed that the influence of the roughness elements on the spatial variability in time-averaged velocities declined with distance from the boundary. However, the sequence remained the same at all heights.

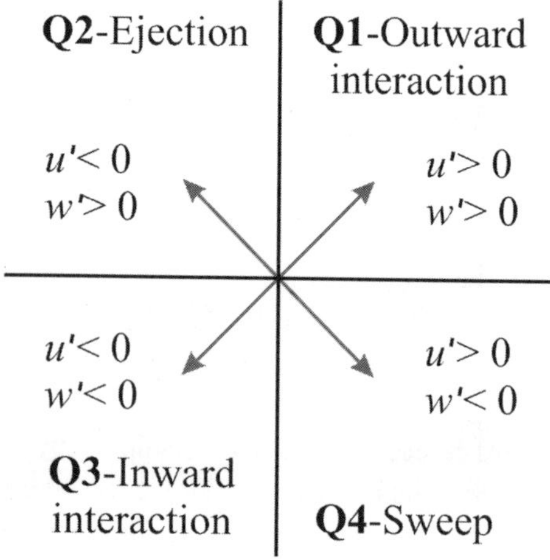

Figure 9. Quadrants are defined by the joint distribution of the velocity fluctuations from the mean downstream (u') and vertical (w') velocity components.

The form-induced stress, defined as the covariance of the spatial variation in u and w, peaked close to $z/k = 1$ and constituted 15% of τ_w. The stress rapidly declined above this height as the flow adjusted via local momentum transfer due to local imbalances in momentum and stress components [36]. This indicates that the form-induced variations in velocity disturbances were far greater in the vertical extent than the form-induced variations in stress, and that velocity disturbances may be found where stress disturbances do not exist. This results in alternating regions of upwards and downwards time-averaged form-induced momentum flux, which is particularly relevant to scalar transport and dispersion in the roughness layer [36].

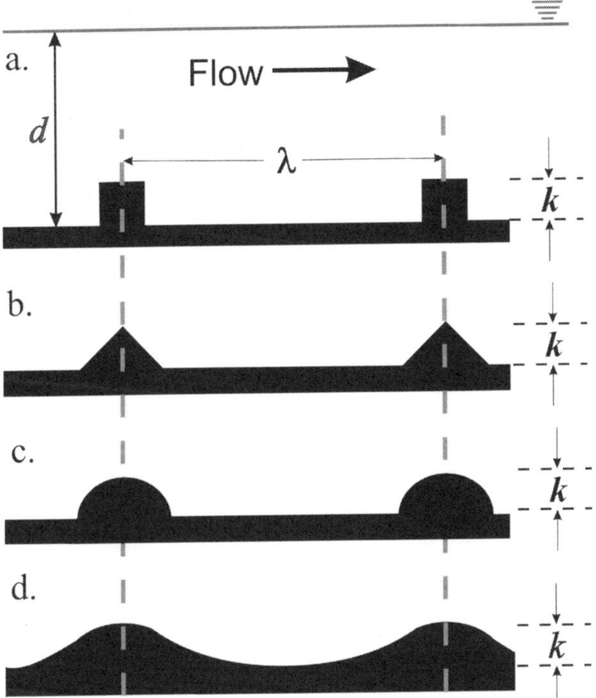

Figure 10. Schematic of channels roughened using 2D transverse bars: (a) square bars; (b) triangular bars; (c) semicircular bars; and (d) wavy wall (after [39]).

3.3.5. *The importance of 2D transverse bar shape*

The majority of studies of flow over 2D transverse bars have concentrated on square bars, and to a lesser extent on circular or semi-circular 'rod' roughness. However, as mentioned above, [39] used RANS modelling to examine the effect of differently shaped transverse bars (square, triangular and semicircular transverse bars, and a wavy wall) on near-bed flow structure (Figure 10). They maintained a constant relative roughness, $k/d = 0.1$, varied the roughness index to span d- and k-type roughness ($5 \leq \lambda/k \geq 30$) and used a range of Re_* and boundary layer Re ($Re_\delta = U\delta/v$). The bottom of the velocity profile (i.e., virtual origin) was displaced by ~$0.8d$ ($0.83d$ square bar, $0.81d$ triangular bar, and $0.79d$ semicircular bar and less so by the wavy wall condition $0.68d$). The profiles exhibited a logarithmic region, with the exception of the wavy wall at high *Re*, and the effective roughness height, k_s, extended higher into the flow as *Re* was increased in all cases.

All bar shapes generated an initial increase *f* with λ/k (Figure 11). In general, the resistance was positively related to the degree of angularity, e.g., when *Re* = 20,000 the maximum resistance was at $\lambda/k \sim 8.8$ for the square bar, 7.5 for the triangular bar and 7.0 for the semicircular bar, close to the threshold of $\lambda/k = 8$ (see section 3.3.2). This was followed by a decline toward the smooth wall limit, well beyond the maximum value of $\lambda/k = 30$ examined. Interestingly, the resistance offered by the square bars dropped below that of the triangular bars where $\lambda/k < 7$. This was likely due to the onset of *d*-type or wake interference behaviour in the square bars. *d*-type behaviour has not been reported for other bar types, but may be attainable using sufficiently small values of the roughness index. In contrast, *f* in the wavy wall rose and fell around $\lambda/k \sim 8$, and exhibited a decline over $5 < \lambda/k > 30$.

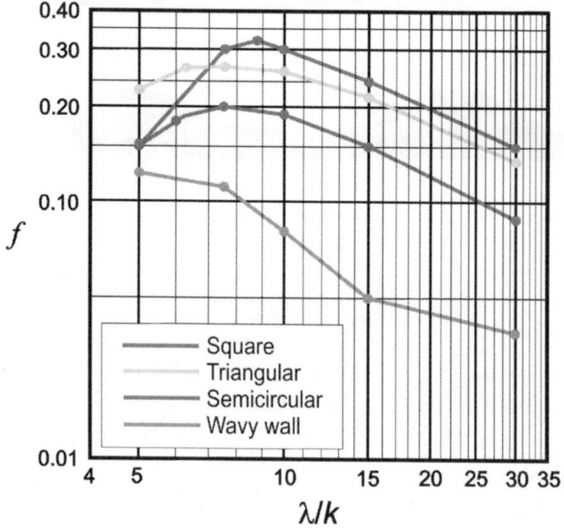

Figure 11. Mean resistance, *f*, for various bar shapes presented in Figure 10 at *k/d* = 0.2 across a range of λ/k at *Re* = 10^4 (after [39]).

The roughness function, Δu^+, was related to λ/k in a similar manner to *f*. Whereas all surfaces imparted some resistance, Δu^+ can equal zero under hydrodynamically smooth conditions, i.e., at sufficiently large λ/k. The square, triangular and semicircular bars did not exhibit $\Delta u^+ \sim 0$ across the range of λ/k

examined. However, Δu^+ of the wavy wall dropped steeply to zero (i.e., the smooth wall condition) at $\lambda/k = 30$.

The resistance over square and triangular bars increased with Re_δ before becoming independent of Re_δ. The wavy wall showed a continual decline, similar to a smooth wall but at higher values due to form drag. The semicircular bars exhibited both a rise and fall around a peak. The behaviours in the latter two cases resulted from the movement of separation and reattachment points on the curved surface [39].

As λ/k increased, form-induced stresses declined and the near-bed velocity values decreased. Whereas the shape of the bar was unimportant, the height of the bottom of the profile was inversely related to Re_*, until it declined asymptotically to a constant value. As would be expected for the profile, Δu^+ was a logarithmic function of Re_* with a slope of κ^{-1}. The correlation in each case differed by a constant that varied with bar shape and λ/k. In contrast, the slope for the wavy wall deviated considerably from κ^{-1}.

Overall, the changes in f and Δu^+ with λ/d and Re_* suggest that the hydrodynamics of the wavy wall were fundamentally different to the 'true bar' roughness types. In general, the angularity of the 'true bar' roughness elements was positively related to boundary resistance. The exception occurred for square bars at low λ/k due to the transition to d-type or wake interference flow, when triangular bars became hydrodynamically rougher. This suggests that square bars also behave differently from other bar shapes at low λ/d. Similar experiments at lower λ/d are required to determine if d-type roughness is induced for roughness exhibiting less acute angles. There were pronounced differences in the behaviour of flow over the 'wavy wall' compared to the other 2D roughness. This suggests that our understanding of 2D transverse bars may be useful for some applications, but more effort is needed in nature where bed roughness varies in angularity and regularity.

3.4. *Three-dimensional surfaces*

It is useful to categorize 3D roughness into (1) 'Idealized' 3D roughness surfaces, as those surfaces comprising idealized blocks of specific dimensions, and (2) 'Complex' 3D roughness surfaces, where the specific geometry of individual roughness elements cannot be determined. In the latter, a single length scale, k, can be used for characterization and a qualitative description of the surface is provided in lieu of other geometrical parameters.

3.4.1. *Idealized 3D roughness*

Many of the relationships between the geometry and flow conditions addressed for 2D transverse bars have also been used for idealized 3D roughness elements. For example, [34] described the effects of 3D multi-scale roughness in boundary layers. They used discrete, regular blocks organized to establish surfaces of different roughness density, nA_e/A_t, which defined the ratio of the plan area of n roughness elements of individual area, A_e, to the total bed area, A_t. Roughness density values ranged from $0.008 - 0.125$, which included the range common to coarse-grained rivers [34]. They used the equation

$$\frac{u(z)}{u_*} = \left(\frac{z}{d}\right)^n \tag{13}$$

in conjunction with downstream velocity profiles to determine n, which peaked at a roughness density of 0.1 (range = 0.005-0.23) before declining rapidly. Profiles of normalised downstream turbulence intensity ($TI = RMS_u/u$, where RMS_u is the root mean square velocity), revealed 3 distinct regions: (1) outer flow ($z/d > 0.35$) where TI decreased linearly towards the surface across all roughness densities, and behaved as a conventional outer boundary layer [34]; (2) an intermediate region $0.35 < z/d > 0.2$ that exhibited approximately constant TI, conforming to a zone of wake generation and dissipation; and (3) an inner region ($z/d < 0.2$) where TI diverged according to roughness density, and where the height of maximum TI was inversely related to the roughness density. Peak TI at the maximum roughness density occurred at 0.25 z/d and declined towards the bed. This height was considered a zone of transition between wake-dominated flow and the near-bed region.

Profiles of dimensionless turbulent energy dissipation, $\varepsilon d / u_*^3$, where ε is the turbulent energy dissipation rate given by

$$\varepsilon = 15\nu \left(\frac{du}{dx}\right)^2 \tag{14}$$

and ν is kinematic viscosity, increased from 10^{-2} W m^{-3} at the surface to 5 W m^{-3} at the bed. The changes in boundary layer structure in the inner and wake regions are related to roughness density and can be explained in terms of the original classification in [29] (see Figure 5). Skimming flow occurred at the highest roughness densities where the flow shifted above the roughness elements as did the maximum ε (to $z/d = 0.1$, close to the top of the elements at $z/d = 0.125$). Intermediate roughness densities provided constant values of TI from the bed to $z/d = 0.35$, indicative of interacting wakes. A notable result in [34] was the absence of roughness effects in the outer flow region despite the high relative depth ($k/d = 0.125$) characteristic of individual roughness elements [21]. This suggests that the transverse movement of fluid between individual 3D elements led to differences in flows over 2D vs. 3D surfaces.

[39] modelled the effects of flow over uniformly distributed square and rectangular blocks in boundary layer flow with constant relative roughness, $k/d = 0.1$ and where the roughness index was varied to span both d-type and k-type roughness ($5 \leq \lambda/k \geq 30$). Note that no attempt was made to account for the variation in block size, and that the effective frontal blockage for the rectangular blocks was double that for square blocks. Flow was examined around (1) in-line roughness elements and (2) staggered roughness elements, where every second row was offset so that the transverse cavity between elements was blocked by elements in the subsequent row. In general, flow in the in-line arrangement remained well organized, and flow disturbance was limited to the 'immediate neighbourhood' of the elements. Conversely, the transverse movement induced by the staggered arrangement led to intense vortical motion. Comparison of spatially-averaged downstream velocity profiles showed that the in-line blocks generated logarithmic profiles, which were better defined as Re increased and k/d decreased. The flow disturbance over staggered blocks meant the log-layer was only definable when $k/d \leq 0.05$. This described a threshold where the blocks could be considered as individual element rather than components of a roughness surface.

A comparison of the resistance, f, at $k/d = 0.05$ (i.e., within the roughness layer) indicated that the resistance of the square blocks was lower than rectangular blocks across $6 < \lambda/k > 18$. The results were the same for both in-line and staggered arrays, but they were opposite in the case of rectangular blocks where the in-line arrangement generated a smaller resistance than the staggered case with peak values of f approximately 2.5 times smaller. The peak in f occurred over the range $10 < \lambda/k > 15$, which was similar to 2D transverse bars (section 3.3). However, the square block and the in-line rectangular arrays showed a

       R.J. Schindler and J.D. Ackerman

lower resistance than the 2D transverse bars across all cases of λ/k. This was likely due to the reduced frontal blockage compared to the 2D transverse bars. The staggered, rectangular case generated higher resistance, which suggests that the degree of transverse deviation required for flow continuity, and the resistance imparted by vorticity in this case outweighed the reduction in blockage. The Δu^+ (eqn. 11), varied logarithmically with a slope of κ^{-1} as Re_* increased, which was also the case for 2D transverse bars (see section 3.3).

3.4.2. *Complex 3D roughness*

The surface characteristics of complex 3D roughness appear to explain the divergence of turbulent statistics both within and outside of the roughness layer. It is difficult to synthesize the data even when surfaces are similar, as experiments involved different: (1) measurement methods; (2) parameter calculation; (3) emphasis on different turbulent quantities; and (4) flow conditions. For example, closely packed spheres (e.g., [27]), sandpaper and sanded surfaces [15] had little effect on the outer flow. Other roughness types, such as 3D woven mesh, have been used to both support [15] and dispute [22] the concept of wall similarity.

Unfortunately, complex 3D surfaces do not lend themselves to the measurement of element spacing, not least because parameterization would need to incorporate downstream and transverse effects. Consequently, the parameterization of an irregular 3D roughness is not trivial. However, [41] examined the effect of changing Re_* on turbulence within a boundary layer over a surface with bi-directional scratches in a diamond shaped pattern for $2{,}175 < Re < 27{,}080$. The flow was in the hydrodynamically smooth and transitionally rough range ($2.3 \leq Re_* \geq 26$) given the small roughness height ($k = 193$ μm; $k/d = 0.0025$). Measurements were also taken over a smooth surface at smooth and transitional Re ($Re = 3{,}110$ and $13{,}140$, respectively), facilitating comparison with the rough surface.

A range of turbulence statistics presented for each value of Re_* revealed the relative importance of boundary roughness on the roughness layer and outer flow region. For example, profiles of the downstream Reynolds normal stress, $\overline{u^2}$, for $Re_* \leq 9.2$ showed no discrepancy with the smooth bed condition. As Re_* increased, however, there was a rise in $\overline{u'^2}$ in the roughness layer, and the rate at which it increased declined until $Re_* = 26$. The absence of a near-wall rise in

$\overline{u'^2}$ provided a sensitive indicator of the boundary layer reaching the fully-rough regime [27]. It represented the eventual break-up of downstream vortices generated through the burst-sweep cycle over smooth boundaries. When this occurred, the skin friction was dominated by form drag and viscous effects were negligible close to the wall. Hence, their data showed a much lower threshold value for the transition to hydrodynamically rough conditions than cited by [21] ($Re_{*rough} \geq 80$) and [24] ($Re_{*rough} \geq 70$). Moreover, [27] showed that the roughness types and uniformity dictated the threshold values. Their rough surface, consisting of uniform closely packed spheres, displayed a narrow transitionally-rough regime with $Re_{*smooth} \approx 15$ and $Re_{*rough} \approx 55$. Less uniform close packed sand grains had a wider transitionally rough regime where $Re_{*smooth} \approx 5$ and $Re_{*rough} \approx 70$. This indicates the importance of the type of roughness surface.

The effect of roughness on vertical or 'active' turbulent motions has been the topic of considerable debate. For example, vertical active turbulence is more sensitive to roughness geometry than the downstream component [13]. Profiles presented by [41] of vertical Reynolds normal stress, $\overline{w'^2}$, exhibited no difference between rough and smooth-bed cases at the full range of Re_* cited above, and maximum values were found in the log-layer. Similarly, their profiles of Reynolds shear stress, $-\overline{\rho u'w'}$, exhibited no significant changes between cases. Similarly, [37] concluded that there was similarity in $\overline{w'^2}$ outside the roughness (or viscous) layer for rough and smooth walls. Results over complex 3D surfaces do not support the similarity hypothesis. For example, [22] and [37] observed distinct differences in mean velocity and turbulent stresses in the outer region over wire mesh, and [42] stated that surface roughness had an unambiguous effect in enhancing *TI* and Reynolds shear stress over most of the boundary layer over wire mesh of a similar height. These differences are indicative of changes in the turbulent structures generated by each boundary.

Quadrant analysis (see Figure 9) was used by [41] to establish the relative contribution of individual quadrants to $-\overline{\rho u'w'}$ over smooth and rough surfaces. The contribution of *Q2* (ejection) and *Q4* (sweep) events to $-\overline{\rho u'w'}$ was consistent between smooth and rough-bed cases. Similar findings were made by [15] over sandpaper and woven mesh roughness, and [17] showed that sweeps and ejections occurred over rounded pebbles ($k = 9$ mm) regardless of boundary condition. Ejections were often coherent and identifiable within the outer flow but sweeps were confined closer to the bed, and the source of fluid for ejections in the rough case was the low-momentum fluid within cavities between roughness elements [17]. Strong contributions to Reynolds shear stress by each quadrant were isolated by [41] who revealed that, whereas the outer flow exhibited similar contributions between smooth and rough bed cases, there was an increase in $Q4$ and a decrease in $Q2$ in the inner flow region ($y/\delta \leq 0.025$) as Re_* increased. This indicated that near the rough boundary, strong sweeps dominate over ejections in terms of their contribution to $-\overline{\rho u' w'}$. In flow over glass beads, the sweep to ejection ratio increased with roughness and high-magnitude, low-frequency events, dominating momentum transfer, increased with roughness [30]. In contrast, [22] and [37] observed a significant increase in $Q2$ contributions, and a smaller increase in $Q4$ contributions, across much of the boundary layer. The time between successive $Q2$ or $Q4$ events also increased, contradicting earlier assertions that roughness does not influence timing between quadrant events.

Ultimately, broad conclusions regarding the effects of roughness geometry and flow condition over complex 3D boundaries are limited, because of differences among studies. For example, in terms of Re_*, [22] and [37] studied flow at Re_* = 340 in the hydrodynamically rough range, which was well above the maximum $Re_* = 26$ used by [41] who used relative roughness, $k/\delta = 0.0025$ (δ was substituted for d to account for differences in flow conditions), which was well below the threshold of 0.02 for the transition to 'individual roughness elements' cited by [21]. Both [22] and [37] used $k/\delta = 0.066$ (woven mesh), and suggested that each wire behaved independently, whereas [15] used $k/\delta = 0.016$ (sandpaper) and $k/\delta = 0.022$ (woven mesh), which would be in the 'transitional' range (i.e., $k/\delta > 0.0125$ or 0.025) for outer-region similarity [21]. In [42] $k/\delta = 0.016$ (wire mesh) and flow was below the hydrodynamic rough threshold ($Re_* = 17$). Flow over an array of roughness geometries (sand, variously sized and spaced spheres, hemispheres and fences) may provide the same effect on the mean downstream velocity profile, although mechanisms of turbulence generation, mixing and dissipation may differ [40]. Unfortunately, few direct comparisons can be made between studies to isolate the specific effects of 3D roughness type. However, [42] were able to isolate the effects of a smooth surface, a grain roughness ($k = D_{50} = 1.2$ mm) and a wire mesh ($k = 0.6$ mm) on turbulence structure by maintaining an approximately constant Re (u and d constant). Their data can be augmented by [37] who used wire mesh of k = 0.69 mm under similar flow conditions. In each case, the rough surfaces increased Δu^+ and downstream *TI* increased at all heights relative to a smooth surface. Profiles of the vertical *TI* on rough surfaces were higher than the smooth-wall data, likely because the surface roughness was substantial for the wire mesh data, but only marginal for sand grains. As mentioned above, surface roughness has a stronger effect on the vertical than downstream *TI* [23], and [42] demonstrated that the effects of surface roughness on $\overline{-\rho u'w'}/U$ were more distinct in the roughness layer.

The grain roughness increased peak *TI* values, which were located at $z/d = 0.08$, by ~20% compared to the smooth boundary. The mesh roughness caused an increase of ~40%, and the peak was located at $z/d = 0.09$, which was similar to peak values at $z/d = 0.04$ found by [37]. Although the wire diameters were only about 50% of the nominal height of the sand grains, the wire mesh data had significantly higher *TI* and Reynolds shear stress values. This further indicates that the effects of 3D roughness geometry and spacing are more than simply a function of k.

4. Summary and Key Findings

Whereas much has been learned about flow over rough boundaries, much of the environmental hydraulics of natural systems remains to be examined. This is due to several factors including: (1) the emphasis on mean effects rather than detailed structure of the flow; (2) the use of different parameters to examine turbulence; (3) the difficulty in finding a standard parameterization for a wide variety of roughness surfaces, particularly complex 3D roughness; and (4) the difficulty in comparing data from different measurement techniques. This has led to conflicting results in some cases and lack of clearly defined trends in others. Despite these issues, the following conclusions can be drawn:

- **The height to which the roughness layer extends, z_r,** is generally found within the range $2 < k > 5$. However, some examples indicate it may be as high as $8k$ due in part to the precise geometry of the surface, and the way in which z_r is determined (e.g., location of the maximum Reynolds shear stress, third-order velocity statistics).
- **The roughness Reynolds number, $Re_* = u_*k/v$,** is a fundamental parameter that describes the balance between viscous and form drag forces imparted by boundary roughness. *Hydrodynamically rough flows* occur at high values of Re_* and represents the dominance of form drag induced by the generation

of TKE around roughness elements, where velocity is independent of viscosity. Under *hydrodynamically smooth conditions* the boundary shear stress (τ_w) is largely viscous and velocity is governed by viscous forces. Under *transitionally rough flows* both viscous and form drag dictate near-bed velocity and turbulence production. An increase in Re_* will disrupt the burst-sweep cycle inherent to viscous forces and promote form drag.

- There is little consensus on how to determine the **transition threshold between hydrodynamically smooth and rough flows** because the use of simple measures such as k neglect the effects of other geometrical parameters such as spacing, shape and 2D vs. 3D dimensionality. The transition has been reported to occur between $Re_{*smooth} \approx 15$ and $Re_{*rough} \approx 55$ for uniform close-packed spheres, and between $Re_{*smooth} \approx 5$ and $Re_{*rough} \approx 70$ for less uniform close packed sand grains. These latter values have been supported by an evaluation of flow over a variety of roughness types, which found that k_s/k becomes independent of Re_* when $Re_* \geq 70 - 80$.

- **The ratio of the spacing between roughness elements and roughness height, λ/k,** is a geometrical parameter that dictates the influence of boundary roughness on the roughness layer. A number of non-dimensional indices have been used to evaluate roughness (e.g., roughness index, λ/k; relative roughness spacing, k/λ; and dimensionless groove width, j/k; Table 1) over 2D transverse bars.

- **The effect of λ/k** is well understood due to the focus on the effects of spacing between roughness elements for 2D transverse bars, in part due to the ease of measuring and changing geometrical parameters in 2D cases. For flows with $\lambda/k < 8$ the connectivity between the fluid in the space between roughness elements and the flow above becomes increasingly limited (i.e., *skimming flow*; *d*-type roughness). For flows with $\lambda/k \sim 8$ the turbulent wakes generated around individual roughness elements interact the most (i.e., *wake interference flow*), minimizing viscous drag but maximizing form drag. For flows with $\lambda/k > 8$ the interaction of wakes decreases towards *isolated roughness flow*, where the recirculation within the interfacial sublayer is intermittent and limited to the zone in the immediate lee of individual roughness elements.

- **The roughness function, Δu^+,** is a measure of the capacity of the rough surface to absorb momentum (i.e., a surrogate measure of resistance) and is related to k. It can be calculated from the dimensionless law-of-the-wall (eqn. 3). Where $\lambda/k < 8$, Δu^+ increases with λ/k, because turbulence production is lowered by the viscous forces in the near-bed region. Where $\lambda/k \geq 8$, Δu^+ decreases with λ/k, because viscous forces are reduced and turbulence production is higher, generating a larger sink for momentum.

- **The relative height, δ/k, or blockage ratio, k/δ,** can be used to determine whether boundary roughness can be considered as a single roughness surface or as one containing multiple, individual roughness elements. Variations in these parameters may be related to the contradictory statements regarding the similarity hypothesis. However, there is no threshold that can be applied to all surfaces given the influence of other geometrical parameters. A recent review suggested that, on average, individual roughness elements exist when $\delta/k \leq 50$ (or $k/\delta \geq 0.02$).

- **Block-like 3D surfaces** that can be parameterized in a similar manner to 2D transverse bars are better understood than those that cannot. Height is the only definable parameter, which is often insufficient for predicting turbulence in the roughness sublayer.

Whereas our knowledge of 2D systems has improved substantially, the application of this knowledge to natural systems is limited given the apparent (but as yet unclear) differences in the hydrodynamics of 3D systems. Ultimately, systematic investigations of flow over complex, 3D surfaces are required, and should incorporate: (1) hydrodynamically smooth, transitional and rough flows; (2) a range of relative depths, k/d, to establish the thresholds at which roughness elements start to behave 'individually' over different surfaces; and (3) a greater emphasis on the transverse flow component. Such a framework requires that a universal method of parameterizing 3D roughness geometry be developed, possibly from existing statistical or numerical modelling methods.

APPENDIX- LIST OF SYMBOLS

List of Symbols		
Symbol	**Definition**	**Dimensions**
A_e, A_t	Area of individual roughness elements, bed area	$[L^2]$
B	Smooth-wall log-law intercept (= 5.6)	-
C_D	Drag coefficient	-
d	Flow depth	$[L]$
d_0	Zero plane displacement	$[L]$
d/k	Blockage ratio	-

D	Molecular diffusivity	$[L^2 T^{-1}]$
D_x	Percent of cumulative grain size distribution	-
f	Darcy-Weisbach friction factor	-
$H_x \, or \, H_y$	Scaling exponent	-
j	Groove width	$[L]$
k, k_s	Roughness height, Equivalent roughness height	$[L]$
k/d or k/δ	Relative roughness	-
K_3	Max difference between 3 adjacent points on a bed	$[L]$
l	Downstream length of roughness element; length scale	$[L]$
n	Number of elements in a cross-section	-
P	Wetted perimeter	$[L]$
$Re, \, Re_*, \, Re_x, \, Re_\delta$	Reynolds number, Roughness Re, Local Re, Boundary Layer Re	-
TI	Turbulence intensity	-
TKE	Turbulent kinetic energy	$[L^2 T^{-2}]$
$U, \, U_0$	Mean downstream velocity, free stream velocity	$[L\,T^{-1}]$
u_s	Instantaneous surface velocity	$[L\,T^{-1}]$
$u, \, v, \, w$	Instantaneous velocity in x, y, and z directions	$[L\,T^{-1}]$
$u', \, v', \, w'$	Deviation from mean velocity in x, y, and z	$[L\,T^{-1}]$
u_*	Friction velocity	$[L\,T^{-1}]$
u^+	Dimensionless downstream velocity	-
$\overline{u'^2}, \overline{v'^2}, \overline{w'^2}$	Reynolds normal stress in x, y, and z directions	$[L^2 T^{-2}]$
$x, \, y, \, z$	Downstream, transverse, and vertical distance	$[L]$
z^+	Dimensionless height	-
$z_0, \, z_R, \, z_L, \, z_c$	Roughness height, height of roughness layer, height of logarithmic layer, and height of roughness crests	$[L]$
Δu^+	Roughness function	-
$\partial u / \partial z$	Vertical velocity gradient	$[T^{-1}]$

$\delta_D,\ \delta_v,\ \delta_I,\ \delta_F,\ \delta_R,\ \delta_I$	Diffusional sublayer thickness, Viscous sublayer thickness, Inertial sublayer thickness, Form-induced sublayer thickness, Roughnesss layer thickness, and Interfacial sublayer thickness	$[L]$
ε	Turbulent energy dissipation	$[L^2\,T^{-3}]$
σ_z	Standard deviation of bed elevations	$[L]$
ω	Wake function	-
Π	Wake strength	-
κ	von Kármán constant (equal to 0.4)	-
ν	Kinematic viscosity	$[L^2\,T^{-1}]$
λ	Wavelength	$[L]$
λ_{Crit}	Critical wavelength for wake interference flow	$[L]$
ρ	Density of water	$[M\,L^{-3}]$
$-\rho\overline{u'w'}$	Reynolds shear stress	$[M\,L^{-1}\,T^{-2}]$
τ_w	Bed or wall shear stress	$[M\,L^{-1}\,T^{-2}]$

Acknowledgements

This work was supported in part by funding from the Natural Sciences and Engineering Research Councils of Canada through a Discovery Grant and a Strategic Project Grant (STP 322317-05) to JDA.

References

1. J. Aberle and V. Nikora, *Water Resour. Res.* W11414 (2006).
2. J. Aberle and G.M. Smart, *J. Hydraul. Res.* **41**, 259– 269 (2003).
3. J.D. Ackerman and T.M. Hoover, *Limnol. Oceanogr.* **46**, 2080-2087 (2001).
4. A. Ashrafian and H.I. Andersson. *Int. J. Heat Fluid Fl.* **25**, 65-79 (2006).
5. N.E. Bergeron, *Math. Geol.* **28**, 537-1603 (1996).
6. B.P. Boudreau and B.P. Jørgensen, The benthic boundary layer (Oxford, 2001).
7. D.I. Bray, Flow resistance in gravel-bed rivers. R.D. Hey, J.C. Bathurst, and C.R. Thorne Eds., Wiley, 109 (1982).
8. F.H. Clauser, *J. Aeronaut. Sci.* **21**, 91–108 (1954).
9. N.J. Clifford, A.R. Robert and K.S Richards, *Earth Surf. Proc. Land.* **17**, 111-126 (1992).

10. J.A. Davis and L.A. Barmuta, *Freshwat. Biol.* **21**, 271-282 (1989).

11. C. de Jong, *Berlin. Geogr. Abh.* **59**, 1–229 (1995).

12. L. Djenidi, R. Elavarasan and R.A. Antonia, *J. Fluid Mech.* **395**, 271–294 (1999).

13. L. Djenidi, R.A. Antonia, M. Amielh and F. Anselmet, *Exp Fluids* **44**, 37-47 (2008).

14. J. Finnigan, *Annu. Rev. Fluid Mech.* **32**, 519–71 (2000).

15. K.A Flack, M.P. Schultz and T.A. Shapiro, *Phys. Fluids* **17**, 035102 (2005).

16. Y. Furuya, M. Miyata and H. Fujita, *J. Fluid. Eng.* **98**, 635–644 (1976).

17. A.J. Grass, *J. Fluid Mech.* **50**, 233–255 (1971).

18. F.R. Hama, Trans. Soc. Naval Archict. Marine Eng. **62**, 333–358 (1954).

19. T.M. Hoover and J.D. Ackerman. *J. Environ. Eng. Sci.* **3**, 365-378 (2004).

20. T. Ikeda and P.A. Durbin, *J. Fluid Mech.* **571**, 235–263 (2007).

21. J. Jimenez, *Ann. Rev. Fluid Mech.* **36**, 173–196 (2004).

22. P.A. Krogstad, R.A. Antonia and L.W.B. Browne, *J. Fluid Mech.* **245**, 599–617 (1992).

23. P.A. Krogstad and R.A. Antonia, *Exp. Fluids* **27**, 450–460 (1999).

24. P.A. Krogstad, H.I. Andersson, O.M. Bakken and A. Ashrafian, *J. Fluid Mech.* **530**, 327–352 (2005).

25. S. Leonardi, P., Orlandi, R.J. Smalley, L. Djenidi and R.A. Antonia, *J. Fluid Mech.* **491**, 229–238 (2003).

26. S. Leonardi, P. Orlandi, R.J. Smalley, L. Djenidi and R.A. *Int. J. Heat Fluid Fl.* **77**, 41–57 (2004).

27. P.M. Ligrani and R.J. Moffat *J. Fluid Mech.* **162**, 69–98 (1986).

28. B.J. McKeon, C.J. Swanson, M.V. Zagarola, R.J. Donnelly and A.J. Smits, *J. Fluid Mech.* **511**, 41-44 (2004).

29. H.M. Morris, *Trans. Am. Soc. Civil Eng.* **120**, 373-398 (1955).

30. S. Nakagawa and I. Nezu, *J. Fluid Mech.* **80**, 99-128 (1977).

31. V. Nikora, D. Goring and B.J.F. Biggs, *Water Resour. Res.* **34**, 517– 527 (1998).

32. V. Nikora, D. Goring, I. McEwan and G. Griffiths, *J. Fluid. Eng.* **127**, 123-133 (2001).

33. G.N. Nishihara and J.D. Ackerman. Mass transport in aquatic environments. C. Gualtieri and D.T. Mihailovic Eds. Taylor & Francis, 299 (2008).

34. R. M. Nowell and M.A. Church, *J. Geophys. Res.* **84**, 4816-4824 (1979).

35. A.E. Perry, W.H. Scholfield and P.N. Joubert, *J. Fluid Mech.* **165**, 383-413 (1969).

36. D. Pokrajac, L.J. Campbell, V. Nikora, C. Manes and I. McEwan, *Exp. Fluids* **42**, 413-423 (2007).

37. M.R. Raupach, R.A. Antonia and S. Rajagopalan, *Appl. Mech. Rev.* **44**, 1–25 (1991).
38. A.R. Robert, River processes (Arnold, 2003).
39. D.N. Ryu, D.H. Choi and V.C. Patel, *Int. J. Heat Fluid Fl.* **28**, 1098–1111 (2007).
40. H. Schlichting, Boundary layer theory 7th Ed (McGraw Hill 1979).
41. M.P. Schultz and K.A. Flack, *J. Fluid Mech.* **580**, 381–405 (2007).
42. M.F. Tachie, D.J. Bergstrom and R. Balachander, *Exp. Fluids* **35**, 338-346 (2003).
43. P.L. Wiberg and J.D. Smith, *Water Resour. Res.* **27**, 825–838 (1991).
44. W.J. Young, *Freshwat. Biol.* **28**, 383-391 (1992).

Chapter 6

REYNOLDS NUMBER EFFECT ON SPATIAL DEVELOPMENT OF VISCOUS FLOW INDUCED BY WAVE PROPAGATION OVER BED RIPPLES

ATHANASSIOS A. DIMAS and GERASIMOS A. KOLOKYTHAS

Department of Civil Engineering, University of Patras, Greece

Numerical simulations of the free-surface flow, developing by the propagation of nonlinear water waves over a rippled bottom, are performed assuming that the corresponding flow is two-dimensional, incompressible and viscous. The simulations are based on the numerical solution of the Navier-Stokes equations subject to the fully-nonlinear free-surface boundary conditions and appropriate bottom, inflow and outflow boundary conditions. The equations are properly transformed so that the computational domain becomes time-independent. For the spatial discretization, a hybrid scheme is used where central finite-differences, in the horizontal direction, and a pseudo-spectral approximation method with Chebyshev polynomials, in the vertical direction, are applied. A fractional time-step scheme is used for the temporal discretization. Over the rippled bed, the wave boundary layer thickness increases significantly, in comparison to the one over flat bed, due to flow separation at the ripple crests, which generates alternating circulation regions. The amplitude of the wall shear stress over the ripples increases with increasing ripple height or decreasing Reynolds number, while the corresponding friction force is insensitive to the ripple height change. The amplitude of the form drag forces due to dynamic and hydrostatic pressures increase with increasing ripple height but is insensitive to the Reynolds number change, therefore, the percentage of friction in the total drag force decreases with increasing ripple height or increasing Reynolds number.

1. Introduction

Sediment transport in the coastal zone results into the formation of sand ripples along the beach bottom. The presence of a rippled bed modifies the propagation of water waves and the development of the wave boundary layer. Wiberg and Nelson [10] studied experimentally the unidirectional flow over rippled bed and developed a flow model for the prediction of the mean velocity profiles and drag coefficients on each ripple. Ranasoma and Sleath [9] studied experimentally the combined current and oscillatory flow over ripples, considering a rigid lid approximation for the free surface, and concluded that the velocity profiles do not collapse to a single curve. Fredsøe et al. [5] studied experimentally and numerically the combined current and oscillatory flow over rippled bed and concluded that the velocity profiles comprise of two layers; one associated to

127

the ripple height and the other to the wave motion. Huang and Dong [6] studied numerically the wave propagation over rippled bed, utilizing the finite-analytic (FA) method, and concluded that the periodically averaged flow exhibits a current in the opposite direction of the wave propagation. Barr et al. [1] studied numerically and experimentally the oscillatory flow over sand ripples, considering a rigid lid approximation for the free surface, and concluded that the wave boundary layer over rippled bed increases, while the average shear stress on the wall decreases, with ripple steepness. Chang and Scotti [2] studied numerically the unidirectional and the oscillatory flow over sand ripples, considering a rigid lid approximation for the free surface, and concluded that the large-eddy simulation (LES) performs better than the Reynolds-Averaged Navier-Stokes (RANS) k-ω model, which underestimates the Reynolds stress and overestimates the amplitude of the vertical velocity oscillations. Malarkey and Davies [7] studied numerically the oscillatory flow over sand ripples, utilizing a discrete vortex method, and concluded that the flow in the lower part of the oscillating boundary layer is dominated by the process of vortex formation and shedding due to flow separation.

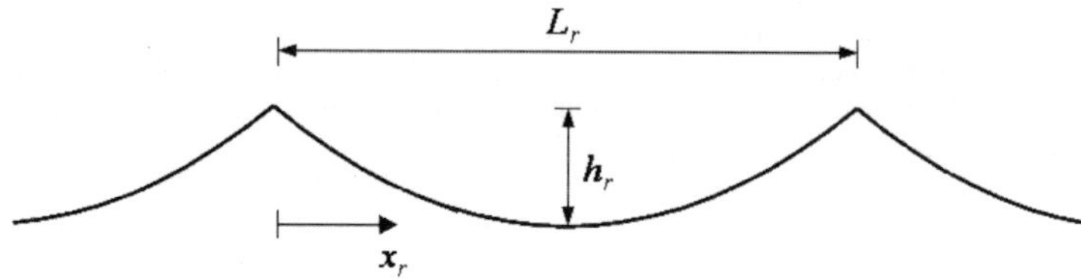

Figure 1. Sketch of bed ripple of parabolic shape.

Typical sand ripples have parabolic shape (Figure 1), while their dimensions, length L_r and height h_r, depend on wave period, wave height and water depth according to laboratory and field data [4]. In the present study, the case of wavelength to water depth ratio $\lambda/d_0 = 6$ and wave height to wavelength ratio $H_0/\lambda = 0.05$ is considered, which is consistent with $L_r/d_0 \leq 0.25$ and $0.02 \leq h_r/d_0 \leq 0.05$ according to Nielsen [8], at two Reynolds numbers. The objective is to simulate the corresponding free-surface flow, considering a rigid rippled bed, and study the effect of Reynolds number and ripple height on the flow behavior in the vicinity of the ripples, the distribution of wall shear stress and pressure along the bed, as well as the corresponding friction and form drag forces. In the following sections, the formulation, the numerical method and results are presented for the propagation of the above mentioned wave at two Reynolds numbers considering the two extreme values of ripple height $h_r/d_0 = 0.02$ and 0.05, which correspond to ripple steepness $h_r/L_r = 0.08$ and 0.2, respectively.

2. Formulation

The incompressible, viscous, two-dimensional, free-surface flow is governed by the continuity

$$\frac{\partial u_1}{\partial x_1} + \frac{\partial u_2}{\partial x_2} = 0 \tag{1}$$

and the Navier-Stokes equations

$$\frac{\partial u_1}{\partial t} + u_1 \frac{\partial u_1}{\partial x_1} + u_2 \frac{\partial u_1}{\partial x_2} = -\frac{\partial p}{\partial x_1} + \frac{1}{\mathrm{Re}}\left(\frac{\partial^2 u_1}{\partial x_1^2} + \frac{\partial^2 u_1}{\partial x_2^2} \right) \tag{2}$$

$$\frac{\partial u_2}{\partial t} + u_1 \frac{\partial u_2}{\partial x_1} + u_2 \frac{\partial u_2}{\partial x_2} = -\frac{\partial p}{\partial x_2} + \frac{1}{\mathrm{Re}}\left(\frac{\partial^2 u_2}{\partial x_1^2} + \frac{\partial^2 u_2}{\partial x_2^2} \right) \tag{3}$$

where t is time, x_1 and x_2 are the horizontal and vertical coordinates, respectively; u_1 and u_2 are the horizontal and vertical velocity components, respectively; p is the dynamic pressure and Re is the Reynolds number. The equations (1)-(3) are in dimensionless form utilizing the inflow depth d_0, the gravity acceleration g and the fluid density ρ, therefore, $\mathrm{Re} = d_0(gd_0)^{1/2}/\nu$ where ν is the fluid kinematic viscosity.

The kinematic and dynamic (normal and shear stress) boundary conditions at the free surface ($x_2 = \eta$), respectively, are

$$u_2 = \frac{d\eta}{dt} = \frac{\partial \eta}{\partial t} + u_1 \frac{\partial \eta}{\partial x_1}, \tag{4}$$

$$p = \frac{\eta}{\mathrm{Fr}^2} \quad \text{and} \quad \left(\frac{\partial u_1}{\partial x_2} + \frac{\partial u_2}{\partial x_1} \right)\left(1 - \left(\frac{\partial \eta}{\partial x_1} \right)^2 \right) + 4\frac{\partial u_2}{\partial x_2}\frac{\partial \eta}{\partial x_1} = 0 \tag{5}$$

where η is the free surface elevation and Fr is the Froude number, which under the present dimensionless formulation is equal to one. In addition, the no-slip and non-penetration boundary conditions at the bottom, respectively, are

$$u_1 - u_2 \frac{\partial d}{\partial x_1} = 0 \quad \text{and} \quad u_2 + u_1 \frac{\partial d}{\partial x_1} = 0 \tag{6}$$

where $d = d(x_1)$ is the bottom depth measured from the still free surface level.

Given that the free surface is time-dependent, the Cartesian coordinates are transformed, in order for the computational domain to become time-independent, according to the expressions

$$s_1 = x_1 \quad \text{and} \quad s_2 = \frac{2x_2 + d - \eta}{d + \eta} \tag{7}$$

According to (7), in the transformed domain, the free surface corresponds to $s_2 = 1$, while the bottom to $s_2 = -1$. The velocity components are also transformed in the following way

$$u_1 = v_1 \quad \text{and} \quad u_2 = v_2 + \frac{r}{2}v_1 = v_2 + v_\eta \tag{8}$$

where

$$r = \frac{1 + s_2}{2}\frac{\partial \eta}{\partial s_1} + \frac{1 - s_2}{2}\frac{\partial h}{\partial s_1} \tag{9}$$

and $\partial h/\partial s_1 = -\partial d/\partial s_1$ is the bottom slope.

Taking into account (7) and (8), the transformed continuity equation and the Navier-Stokes equations in rotational form, respectively, are

$$\frac{\partial v_1}{\partial s_1} + \frac{2}{d+\eta}\left[\frac{\partial v_2}{\partial s_2} + \frac{v_1}{2}\left(\frac{\partial \eta}{\partial s_1} - \frac{\partial h}{\partial s_1}\right)\right] = 0 \tag{10}$$

$$\frac{\partial v_1}{\partial t} = v_2 \zeta + \frac{1 + s_2}{d + \eta}\frac{\partial \eta}{\partial t}\frac{\partial v_1}{\partial s_2} + \frac{r}{d+\eta}\frac{\partial p}{\partial s_2} - \frac{\partial \Pi}{\partial s_1} +$$

$$\frac{1}{\mathrm{Re}}\left(\frac{\partial^2 v_1}{\partial s_1^2} + \left(r^2 + 1\right)\left(\frac{2}{d+\eta}\right)^2\frac{\partial^2 v_1}{\partial s_2^2} + A_1\right) \tag{11}$$

$$\frac{\partial v_2}{\partial t} = -v_1 \zeta + \frac{1+s_2}{d+\eta}\frac{\partial \eta}{\partial t}\frac{\partial \left(v_2 + v_\eta\right)}{\partial s_2} - a_\eta - \frac{2}{d+\eta}\frac{\partial \Pi}{\partial s_2} +$$

$$\frac{1}{\mathrm{Re}}\left(\frac{\partial^2 \left(v_2 + v_\eta\right)}{\partial s_1^{\,2}} + \left(r^2 + 1\right)\left(\frac{2}{d+\eta}\right)^2 \frac{\partial^2 \left(v_2 + v_\eta\right)}{\partial s_2^{\,2}} + A_2\right) \quad (12)$$

where

$$\zeta = \frac{\partial v_2}{\partial s_1} - \frac{2}{d+\eta}\frac{\partial v_1}{\partial s_2} \quad (13)$$

is the transformed vorticity,

$$a_\eta = \frac{\partial v_\eta}{\partial t} + v_1 \frac{\partial v_\eta}{\partial s_1} + v_2 \frac{2}{d+\eta}\frac{\partial v_\eta}{\partial s_2} \quad (14)$$

is the transformed vertical acceleration,

$$\Pi = p + \frac{1}{2}\left(v_1^{\,2} + v_2^{\,2}\right) \quad (15)$$

is the transformed dynamic pressure head and

$$A_i = -\frac{2}{d+\eta}\left(2r\frac{\partial^2 u_i}{\partial s_1 \partial s_2} + \left(\frac{\partial r}{\partial s_1} - r\frac{2}{d+\eta}\left(\frac{\partial \eta}{\partial s_1} - \frac{\partial h}{\partial s_1}\right)\right)\frac{\partial u_i}{\partial s_2}\right). \quad (16)$$

Finally, the free surface boundary conditions, obtained by the transformation of (4) and (5), are

$$v_2 = \frac{\partial \eta}{\partial t} \quad (17)$$

$$\Pi = \frac{\eta}{\mathrm{Fr}^2} + \frac{1}{2}\left(v_1^{\,2} + v_2^{\,2}\right) \quad \text{and} \quad \frac{2}{d+\eta}\left(1 + \left(\frac{\partial \eta}{\partial s_1}\right)^2\right)^2 \frac{\partial v_1}{\partial s_2} +$$

$$\left(1 - \left(\frac{\partial \eta}{\partial s_1}\right)^2\right)\frac{\partial v_2}{\partial s_1} - 2\frac{\partial \eta}{\partial s_1}\left(1 + \left(\frac{\partial \eta}{\partial s_1}\right)^2\right)\frac{\partial v_1}{\partial s_1} + \frac{\partial^2 \eta}{\partial s_1^{\,2}}\left(1 - \left(\frac{\partial \eta}{\partial s_1}\right)^2\right)v_1 = 0 \quad (18)$$

while the bottom boundary conditions, obtained by the transformation of (6), are

$$v_1 = 0 \quad \text{and} \quad v_2 = 0. \tag{19}$$

3. Numerical Method

The numerical solution of the unsteady, two-dimensional, Navier-Stokes equations is based on a fractional time-step scheme, which consists of three stages, for the temporal discretization and a hybrid scheme for the spatial discretization. In the latter, central finite-differences are applied along s_1 and a pseudo-spectral approximation with Chebyshev polynomials along s_2. Along s_1, the spatial interval Δs is constant, while along s_2, the number of collocation points N is equal to the highest order of the Chebyshev polynomials. At the first stage of each time step, the nonlinear terms and the A_i viscous terms of the equations of motion (11) and (12), are treated explicitly using a third-order Adams-Bashforth scheme without imposing any boundary conditions. At the second stage, an implicit Euler scheme is utilized for the treatment of the pressure head terms of (11) and (12). At this stage, the generalized Poisson's equation for the transformed pressure head, Π, is derived by satisfying the continuity equation (10) as well. The normal stress dynamic boundary condition (18), at the free surface, and the non-penetration condition (19), at the bottom, are imposed at this stage. The remaining viscous terms are treated implicitly, at the third stage of the time-step, using again an Euler scheme. The velocity components are computed by numerical solution of the corresponding generalized Poisson's equations subject to the shear stress dynamic boundary condition (18), at the free surface, and the boundary conditions (19), at the bottom. Finally, the free-surface elevation is computed by satisfying the kinematic boundary condition (17).

A sketch of the computational domain is shown in Figure 2. At the inflow boundary, velocity field, dynamic pressure and free-surface elevation are defined according to second order Stokes wave theory. The inflow region, L_I, of constant depth ($d = 1$) is placed just after the inflow plane to allow the incoming wave development, followed by the ripples region, L_R, which consists of 12 ripples, and the outflow region, L_E, of constant depth ($d = 1$). A damping zone of length $L_A < L_E$ is included in the outflow region just before the outflow plane to efficiently minimize wave reflection by the outflow boundary [3].

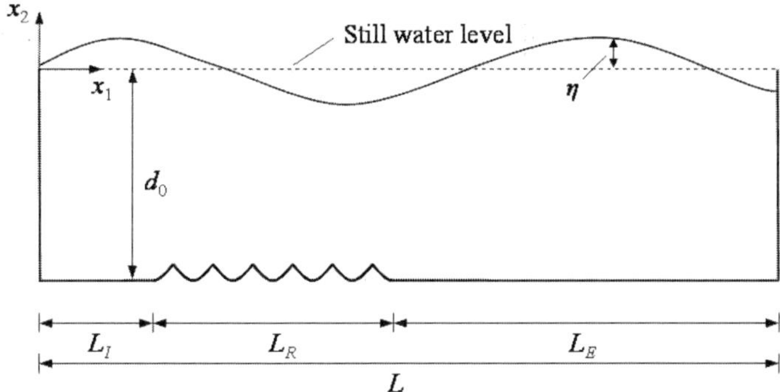

Figure 2. Sketch of computational domain for simulation of wave propagation over rippled bed.

4. Results

The numerical model is validated by simulating the laminar, oscillatory, current flow, which develops a uniform boundary layer over horizontal bottom of constant depth d_0 since the free surface is modeled as a rigid symmetry plane. In dimensional terms, the driving velocity is $u = U \sin(\omega t^*)$ where U is the velocity magnitude, $\omega = 2\pi/T$ is the radial frequency, T is the period and t^* is the time. In this case, flow variables are rendered dimensionless by d_0, U and ρ, while $x_2 = 0$ corresponds to the bottom and $x_2 = 1$ to the free surface. The computational domain length is equal to 4, in the x_1 direction, while the numerical parameters are $\Delta t = 0.002$ (in time units of d_0/U), $\Delta s = 0.0125$ and $N = 128$. The Reynolds and Strouhal numbers defined with respect to the Stokes length $\delta = (2v/\omega)^{1/2}$ are $\mathrm{Re}_\delta = U\delta/v = 80$ and $\mathrm{St}_\delta = \omega\delta/U = 0.025$, and correspond to $\delta/d_0 = 0.003$. The streamwise velocity, obtained by the analytical solution

$$u_1 = \frac{u}{U} = \sin\left(\omega t^*\right) - \exp\left(-\frac{x_2}{\delta/d_0}\right)\sin\left(\omega t^* - \frac{x_2}{\delta/d_0}\right), \qquad (20)$$

is in excellent agreement to the numerical solution (Figure 3).

Next, the simulation of finite-amplitude wave propagation over a rigid rippled bed is considered. The wavelength of the incoming wave is $\lambda/d_0 = 6$ and its steepness is $H_0/\lambda = 0.05$. Two cases of Reynolds numbers are considered at $\mathrm{Re} = 22{,}120$ ($\mathrm{Re}_\delta = 8$, $\delta/d_0 = 0.03$) and $\mathrm{Re} = 250{,}000$ ($\mathrm{Re}_\delta = 80$, $\delta/d_0 = 0.003$).

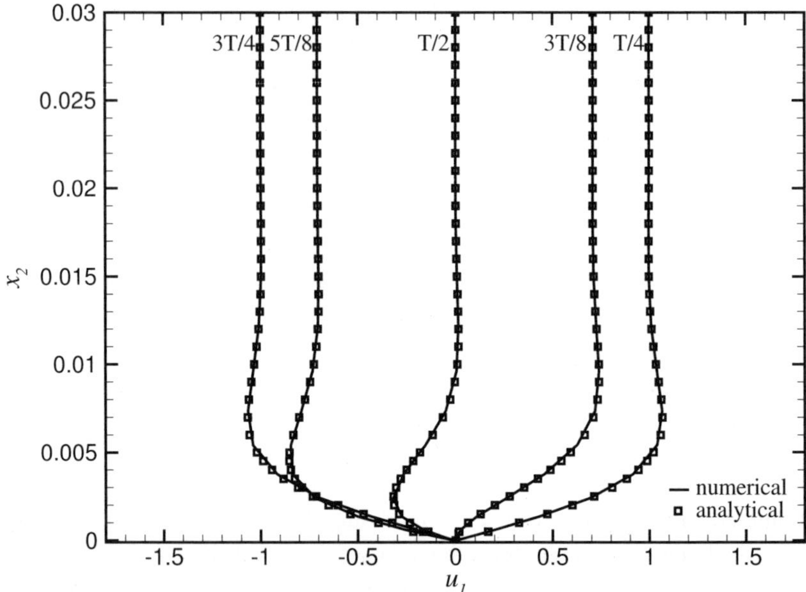

Figure 3. Profiles of streamwise velocity during a period, T, of oscillatory flow over flat bed for $Re_\delta = 80$. Solid lines correspond to the numerical simulation and symbols to the analytical solution.

The shape of the ripples (Figure 1) is parabolic

$$h = \frac{h_r}{d_0}\left(2\frac{x_r}{L_r}-1\right)^2 \tag{21}$$

and we consider two cases with ripple length $L_r/d_0 = 0.25$ and heights $h_r/d_0 = 0.02$ and 0.05. The domain length is set equal to $36.5 = 6.5\cdot\lambda/d_0$, the inflow region is equal to 6, the ripple train spans half a wavelength (from $x_1 = 6.125$ to $x_1 = 9.125$) and consists of 12 ripples, while the damping zone is equal to 24. The numerical parameters are $\Delta t = 0.002$, $\Delta s = 0.0125$ and $N = 128$.

Typical snapshots of the pressure field for the case $h_r/d_0 = 0.05$ at both Reynolds numbers are shown in Figure 4. The presence of the ripples does not alter the wave propagation, while the effect of the Reynolds number on the flow dynamics is minimal. The operation of the damping zone is also clearly demonstrated in Figure 4. The velocity field in the vicinity of a ripple for the case $h_r/d_0 = 0.05$ during one period of wave propagation is shown in Figure 5

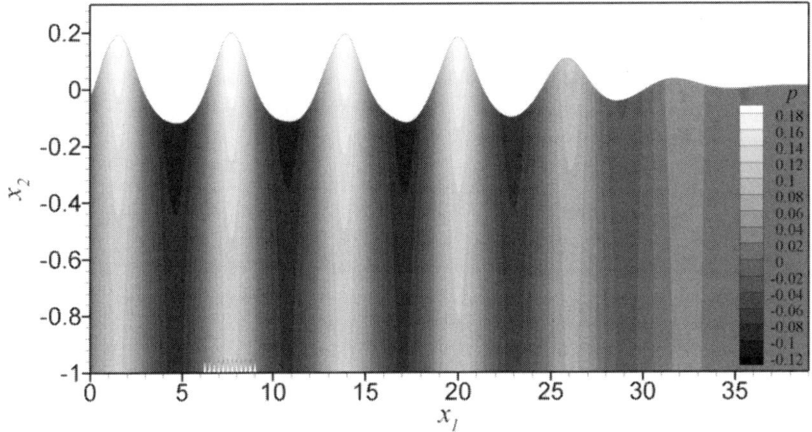

Figure 4. Contours of dynamics pressure at a time instant of the wave propagation over bed with ripples of $h_r/d_0 = 0.05$ at Re = 22,120 (top) and Re = 250,000 (bottom).

for Re = 22,120 and Figure 7 for Re = 250,000. The thickness of the boundary layer at Re = 250,000 is only about half the one at Re = 22,120, while the corresponding ratio of Stokes lengths is 1/10. This is due to the flow separation at the ripple crests which dominates over viscous diffusion. The flow separates along the wave propagation direction at the ripple crest, and forms a recirculation region with strong negative vorticity on the downslope side of the ripple, when a wave crest propagates above the ripple crest, while it separates opposite to the wave propagation direction, and forms a recirculation region with strong positive vorticity on the upslope side of the ripple, when a wave trough propagates above the ripple crest. The corresponding vorticity field in the vicinity of a ripple for the case $h_r/d_0 = 0.05$ during one period of wave propagation is shown in Figure 6 for Re = 22,120 and Figure 8 for Re = 250,000.

The wall shear stress distributions at Re = 22,120 for ripples with $h_r/d_0 = 0.02$ and 0.05 are shown in Figure 9, while the corresponding distributions at Re = 250,000 are shown in Figure 10. Note that wall shear stress is non-dimensionalized by $\rho g d_0$. The spatial development of the wall shear stress is modulated by the wavelength in the flat region and by the ripple length in the ripples region. The amplitude of the wall shear stress over the ripples is substantially larger than over the flat bed, and it increases according to the increase of the ripple height. The increase of the Reynolds number, on the other hand, decreases the amplitude of the wall shear stress.

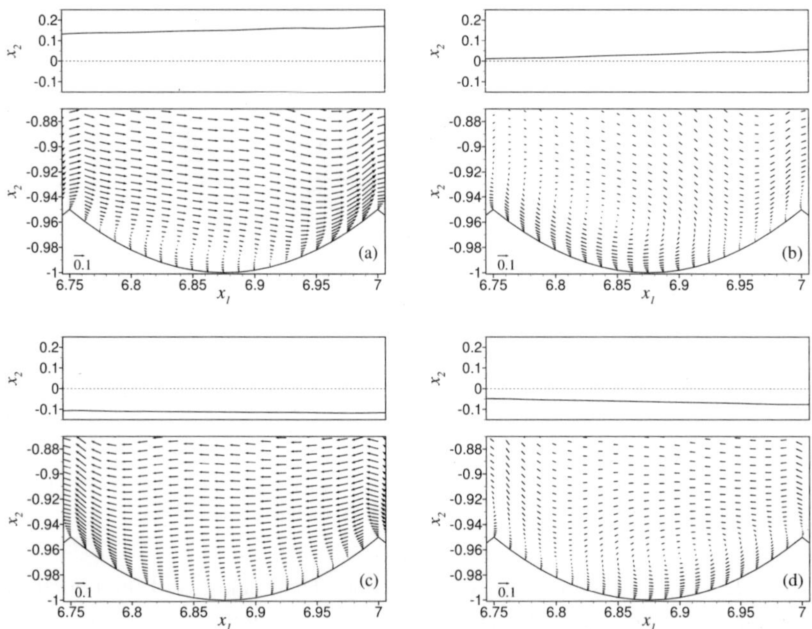

Figure 5. Free surface elevation and velocity field in the vicinity of a ripple during one period of wave propagation at Re = 22,120.

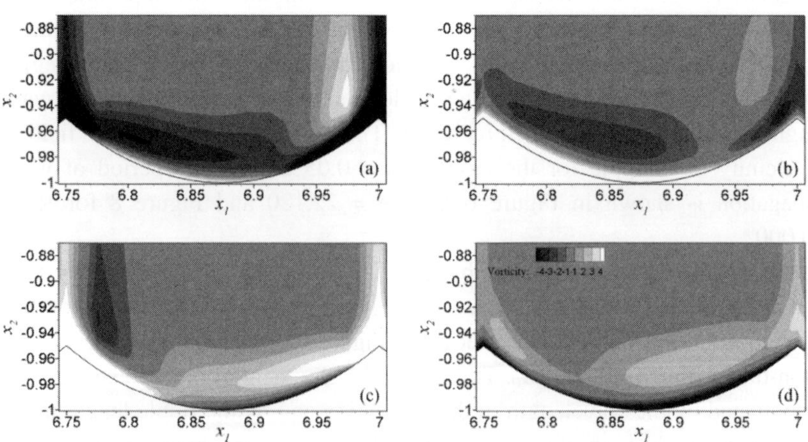

Figure 6. Contours of vorticity in the vicinity of a ripple during one period of wave propagation at Re = 22,120. The time instants are identical to the ones in Figure 5.

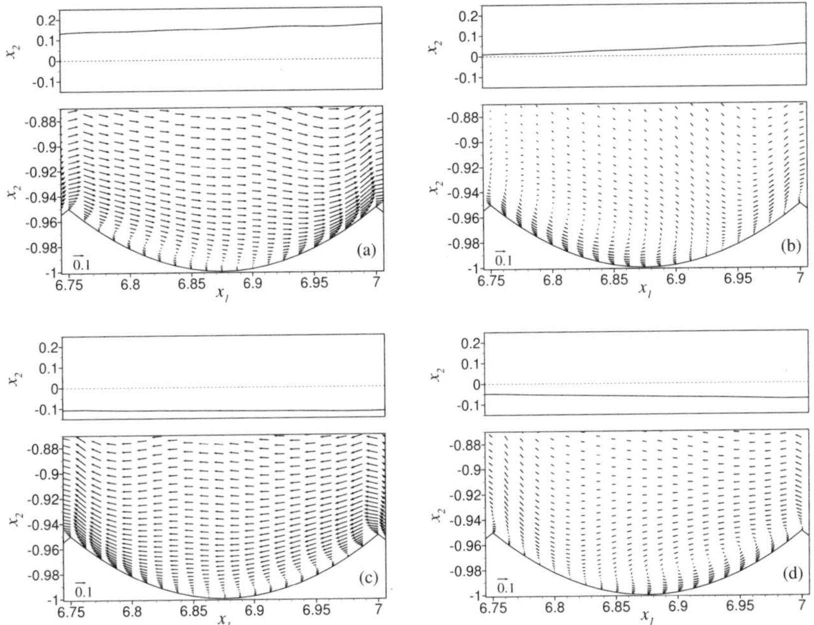

Figure 7. Free surface elevation and velocity field in the vicinity of a ripple during one period of wave propagation at Re = 250,000.

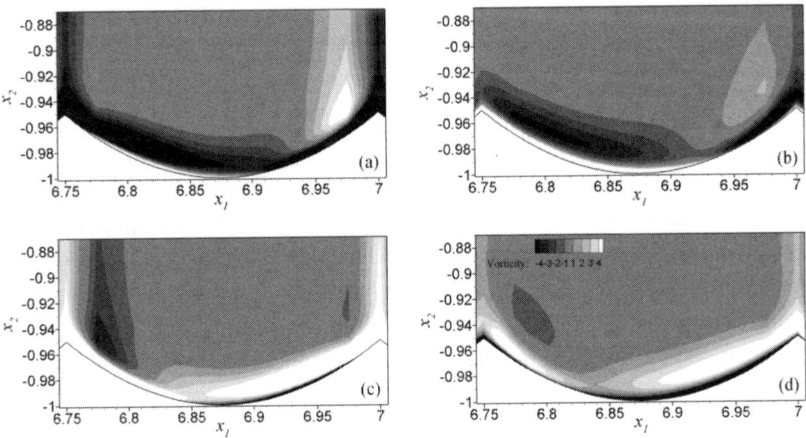

Figure 8. Contours of vorticity in the vicinity of a ripple during one period of wave propagation at Re = 250,000. The time instants are identical to the ones in Figure 7.

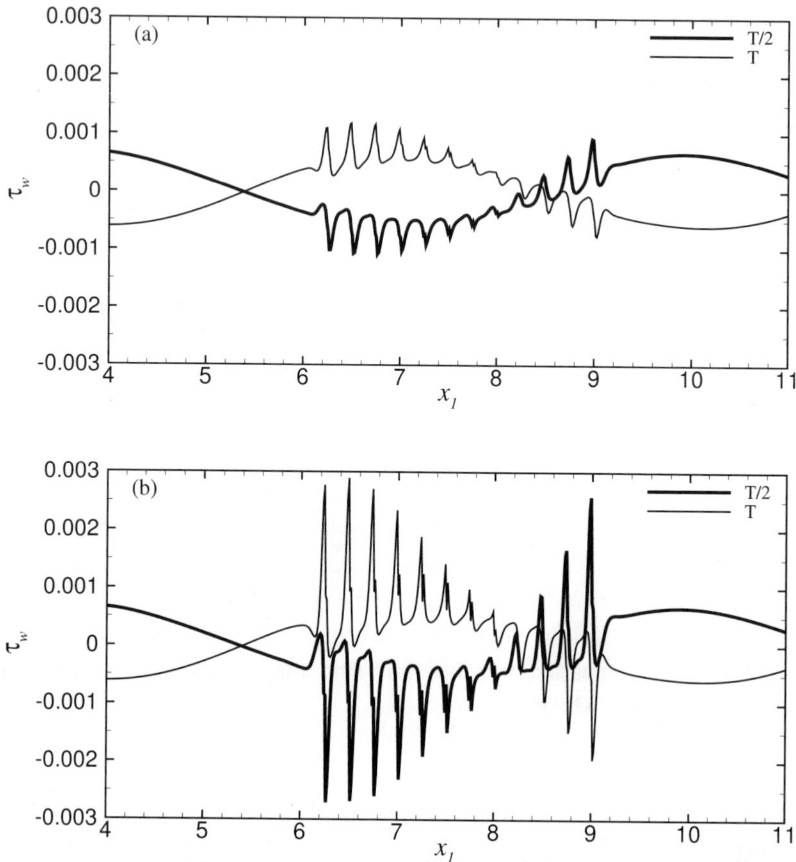

Figure 9. Distribution of wall shear stress during one period of wave propagation at Re = 22,120 for ripples of (a) $h_r/d_0 = 0.02$ and (b) $h_r/d_0 = 0.05$.

Finally, shear stress, dynamic pressure and hydrostatic pressure are integrated along the surface of one ripple to compute the horizontal components of the friction force F_F, the dynamic pressure drag force F_P and hydrostatic pressure drag force F_S, respectively, applied on the ripples. The sum of these forces is the total horizontal drag force, F_D, on a ripple. The corresponding drag coefficient is $C_D = F_D/\rho g L_r^2$, while the component drag coefficients C_F, C_P and C_S are defined accordingly. The time evolution of these drag coefficients on a ripple at Re = 22,120 is shown in Figure 11 for both $hr/d_0 = 0.02$ and 0.05, which correspond to ripple steepnesss $h_r/L_r = 0.08$ and 0.2, while the corresponding evolution at Re = 250,000 is shown in Figure 12. The amplitude of the hydrostatic pressure drag force is the highest, while the one of the friction force is the lowest. The amplitude of both pressure drag forces increases with increasing ripple height, while the amplitude of the friction force is insensitive to the increase of ripple height. On the other hand, the amplitude of both pressure drag forces is insensitive to the increase of the Reynolds number, while the amplitude of the friction force decreases with increasing Reynolds number. Therefore, the percentage of friction in the total drag force decreases with increasing ripple height or increasing Reynolds number.

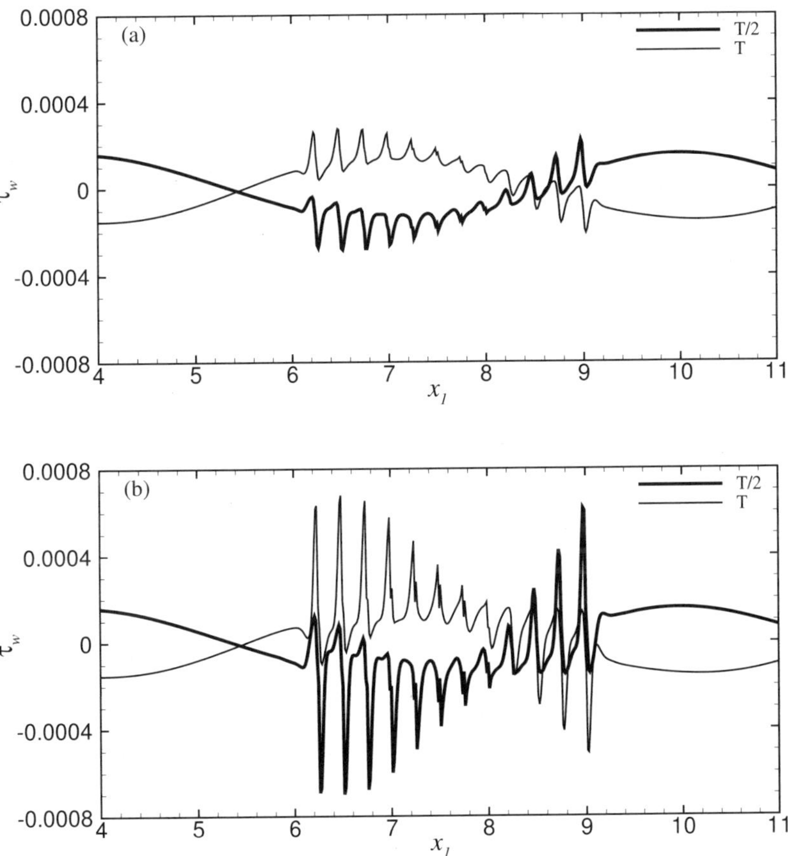

Figure 10. Distribution of wall shear stress during one period of wave propagation at Re = 250,000 for ripples of (a) $h_r/d_0 = 0.02$ and (b) $h_r/d_0 = 0.05$.

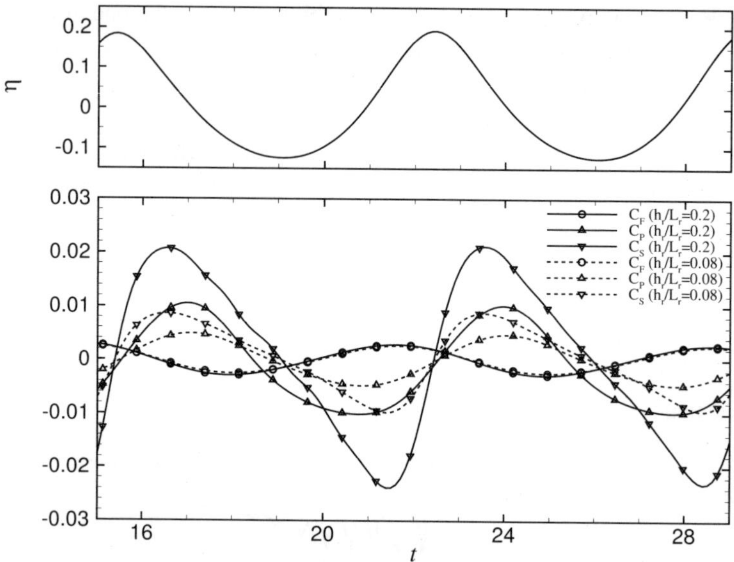

Figure 11. Time evolution of free surface elevation above and component drag coefficients on a ripple during two periods of wave propagation at Re = 22,120 for two values of ripple steepness.

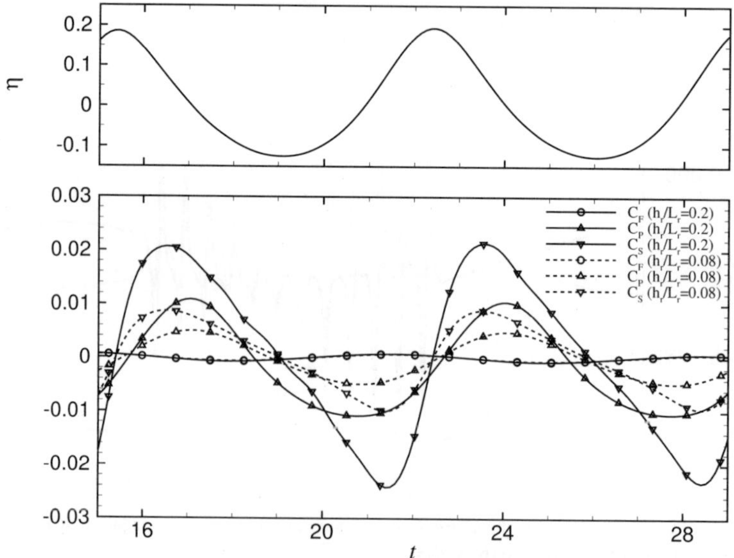

Figure 12. Time evolution of free surface elevation above and component drag coefficients on a ripple during two periods of wave propagation at Re = 250,000 for two values of ripple steepness.

5. Conclusion

The propagation of finite-amplitude waves over a rippled bed was simulated to deduce the effect of Reynolds number and ripple height on the flow dynamics. Flow separation at the ripple crests generates a circulation region between crests, which results into increased boundary layer thickness and wall shear stress. The amplitude of the wall shear stress over the ripples increases with increasing ripple height or decreasing Reynolds number, while the corresponding friction force is insensitive to the ripple height change. The amplitude of the form drag forces due to dynamic and hydrostatic pressures increase with increasing ripple height but is insensitive to the Reynolds number change, hence, the percentage of friction in the total drag force decreases with increasing ripple height or increasing Reynolds number. Therefore, ripple migration in the coastal environment is driven by form rather than friction drag and depends strongly on the ripple steepness.

Acknowledgements

We thank the European Social Fund (ESF), Operational Program for Educational and Vocational Training II (EPEAEK II), and particularly the Program PYTHAGORAS II, for funding the above work.

Appendix

<table>
<tr><th colspan="3">List of Symbols</th></tr>
<tr><th>Symbol</th><th>Definition</th><th>Dimensions</th></tr>
<tr><td>d</td><td>dimensionless bottom depth</td><td></td></tr>
<tr><td>d_0</td><td>inflow depth</td><td>[L]</td></tr>
<tr><td>Fr</td><td>Froude number</td><td></td></tr>
<tr><td>g</td><td>gravity acceleration</td><td>$[L\,T^{-2}]$</td></tr>
<tr><td>H_0</td><td>incoming wave height</td><td>[L]</td></tr>
<tr><td>h</td><td>dimensionless bed-ripple elevation</td><td></td></tr>
<tr><td>h_r</td><td>ripple height</td><td>[L]</td></tr>
<tr><td>L</td><td>dimensionless computational length</td><td></td></tr>
<tr><td>L_A</td><td>dimensionless absorption region length</td><td></td></tr>
<tr><td>L_E</td><td>dimensionless outflow region length</td><td></td></tr>
<tr><td>L_I</td><td>dimensionless inflow region length</td><td></td></tr>
<tr><td>L_R</td><td>dimensionless rippled region length</td><td></td></tr>
</table>

L_r	ripple length	[L]
p	dimensionless dynamic pressure	
r	linear interpolation of free surface and bed slope	
Re	Reynolds number	
Re_δ	Reynolds number based on Stokes length	
s_1	transformed horizontal coordinate	
s_2	transformed vertical coordinate	
St_δ	Strouhal number based on Stokes length	
T	dimensionless time	
t^*	dimensional time	[T]
T	period	[T]
u	oscillatory flow driving velocity	[L T^{-1}]
U	oscillatory flow velocity magnitude	[L T^{-1}]
u_1	dimensionless horizontal velocity	
u_2	dimensionless vertical velocity	
v_1	transformed horizontal velocity	
v_2	transformed vertical velocity	
x_1	dimensionless horizontal coordinate	
x_2	dimensionless vertical coordinate	
x_r	ripple horizontal coordinate	[L]
α_η	transformed vertical acceleration	
δ	Stokes length	[L]
Δs	dimensionless horizontal grid size	
Δt	dimensionless time step	
ζ	transformed vorticity	
η	dimensionless free surface elevation	
λ	wavelength	[L]
v	fluid kinematic viscosity	[L^2 T^{-1}]
N	Chebyshev modes number	
Π	transformed pressure head	
ρ	fluid density	[M L^{-3}]
ω	radial frequency	[T^{-1}]

References

1. B.C. Barr, D.N. Sinn, T. Pierro and K.B. Winters, *J. Geophys. Res.* **109**, C09009 (2004).
2. Y.S. Chang and A. Scotti, *J. Geophys. Res.* **109**, C09012 (2004).
3. A.A. Dimas and A.S. Dimakopoulos, *J. Wtrwy, Port, Coast., and Oc. Engrg.*, in press (2009).
4. J. Fredsøe and R. Deigaard, *Mechanics of coastal sediment transport* (World Scientific 1992).
5. J. Fredsøe, K.H. Anderson and B.M. Sumer, *Coastal Engrg.* **38**, 177 (1999).
6. C.J. Huang and C.M. Dong, *J. Wtrwy, Port, Coast., and Oc. Engrg.* **128**, 190 (2002).
7. J. Malarkey and A.G. Davies, *J. Geophys. Res.* **109**, C09016 (2004).
8. P. Nielsen, *J. Geophys. Res.* **86**(C7), 6467 (1981).
9. K.I.M. Ranasoma and F.A. Sleath, *J. Wtrwy, Port, Coast., and Oc. Engrg.* **128**, 331 (1994).
10. P.L. Wiberg and J.M. Nelson, *J. Geophys. Res.* **97**(C8), 12745 (1992).

Chapter 7

CALCULATION OF AGGREGATED ALBEDO IN RECTANGULAR SOLID GEOMETRY ON ENVIRONMENTAL INTERFACES

DARKO KAPOR[1], ANA CIRISAN[2], DRAGUTIN T. MIHAILOVIC[3]

[1]*Faculty of Sciences, University of Novi Sad, Novi Sad, Serbia*
[2]*Scientific Computing Laboratory, Institute of Physics, Belgrade, Serbia*
[3]*Faculty of Agriculture, University of Novi Sad, Novi Sad, Serbia*

The albedo of the interface has always been an important parameter for the evaluation of the radiation fluxes in environmental studies. The practical problem arises if the surface is a heterogeneous one. Various approaches to calculate the aggregated albedo have been developed. Our previous research demonstrated the existence of a geometrical effect when different parts of the interface have different heights. We have offered the general approach for the calculation of the flux that is lost due to the absorption on the vertical lateral boundaries. The multiple scattering effect and the dependence of the albedo on the zenithal angle of the incident radiation were disregarded. In the case of simple geometries we derived analytically the expressions for this loss coefficient, which, for some ideal urban geometries, coincides with Oke's sky-view factor. The aim of this chapter is an elaboration of this effect for more complex geometry, which does not allow analytic solutions. Therefore, it was necessary to develop an efficient numerical procedure, in this case the so-called ray-tracing Monte Carlo approach. It was first tested for known analytical solutions. As the next step, it was incorporated into one land surface scheme (LAPS), and then an example of central geometry was considered for: (i) a two-patch grid cell with a square geometrical distribution and (ii) different heights of its parts. Simulations were done for several patch areas, with different heights ("propagating building"). Various surface types were considered. The derived value for the albedo was compared with the result of the conventional approach. Changes in albedo lead to a significant change in partitioning of the energy at the environmental interface. The most remarkable changes were in values of sensible and latent heat fluxes, as well as the surface temperature. The values of effective surface temperature were calculated using the LAPS parameterisation scheme and then compared to the values obtained with a conventional parameterisation of the albedo.

Keywords: albedo; environmental interface; parameter aggregation; urban modelling.

1. Introduction

Albedo became one of the key variables in the parameterization of the land-surface radiative transfer over the grid cell in numerical modelling. Through its influence on the energy balance of the Earth, and thus indirectly on many other factors (fluxes of momentum, energy, carbon-dioxide, moisture *etc.*), surface albedo belongs to the class of fundamentally important meteorological parameters.

Nowadays, in order to obtain more precise parameterization of physical processes, one of the basic tasks is the determination of albedo over heterogeneous areas. Various manners to calculate the aggregated albedo have been developed [1] [2] [7]. The most common approach is to determine the albedo as the grid cell-average albedo using a simple arithmetical averaging. Our physics based analysis indicates that situations occur where the observed albedo significantly deviates from the simple averaged albedo.

We have proposed in our previous work [3] [8] [11] a new method for aggregating the albedo over a very heterogeneous surface in land surface schemes for use in environmental modelling. The basic assumption of our research is that the geometrical factor can play an important role in albedo calculation; even the same material at different heights might produce interesting effects.

In this chapter, we start with a brief summary of our previous work and continue with the extension of our approach to more complex geometries.

2. Physical Basis of Aggregation Short Review

2.1. *Introduction to basic assumptions*

In our aforementioned studies we have proposed a new approach for aggregating the albedo over a very heterogeneous surface. For the sake of simplicity, a two-patch grid cell was analysed with a simple geometrical distribution and different heights of its components. The basic constituent of the albedo coming from the grid cell in the present analysis describes the diffuse, homogeneous single scattering of incoming radiation from a given surface. The multiple scattering effect and the dependence of the albedo on the zenithal angle of the incident radiation were disregarded.

Under such conditions, the part of radiation reflected from the lower surface is completely absorbed by the lateral sides of the surface lying on a higher level. Obviously, this manifests itself in the decrease of reflected radiation and the aggregated flux is smaller. In order to describe it, we introduced a "loss coefficient", i.e. the relative flux that is completely lost due to the absorption on the vertical lateral boundaries (its definition is conceptually analogous to Oke's sky-view factor [12], particularly for some ideal urban geometries). The derived expression for the albedo of a particularly designed grid cell is compared with the conventional approach using a common parameterization of albedo over the same grid cell as used by Delage *et al.* [1].

2.2. *Analytical treatment*

We consider a grid cell divided into two subregions with different albedos; α_i is the corresponding albedo of a particular patch; σ_i is fractional cover, calculated as a ratio of patch's area S_i and total grid cell area $S\left(\sigma_i = S_i / S, i = 1, 2\right)$.

The average albedo over the grid cell according to conventional approach is given as

$$\overline{\alpha}_c = \alpha_1 \sigma_1 + \alpha_2 \sigma_2 \tag{1}$$

while our analytic treatment for calculating the average albedo over this grid is

$$\overline{\alpha}_n = \left(1 - k\right) \alpha_1 \sigma_1 + \alpha_2 \sigma_2 \tag{2}$$

k is the already mentioned loss coefficient, which can be derived from the following relation

$$k = \frac{\left(\dfrac{dE}{dt}\right)_l}{\left(\dfrac{dE}{dt}\right)_h} . \tag{3}$$

The denominator is the amount of flux emitted from the lower surface into the upper half space. The numerator is the total energy coming from the lower horizontal surface towards the lateral surface. The expression to calculate the radiant energy flux dE/dt was taken following Liou [5].

The amount of emitted flux reaching the vertical lateral boundary is determined by summing all infinitesimal amounts of radiant flux emitted from the infinitesimal surface element $dxdy$ (centred around the point with the position vector $\vec{r}$) confined in the solid angle $d\Omega$ under which the element $dxdy$ "sees" the lateral surface, shown shaded in Figures 1 and 2. According to our basic assumptions, the flux of radiation that reaches the vertical boundary surface is completely lost, so the contribution of the radiation reflected from the lateral surface is not taken into account in the total reflected flux of radiation.

The boundaries of the integration for a given point are determined over the local azimuthal $\left(\phi_l, \phi_u\right)$ and zenithal $\left(\theta_l, \theta_u\right)$ angles in terms of the x, y coordinates, as it is shown in Figures 1 and 2. The subscripts l and u denote

  D. Kapor, A. Cirisan and D.T. Mihailovic

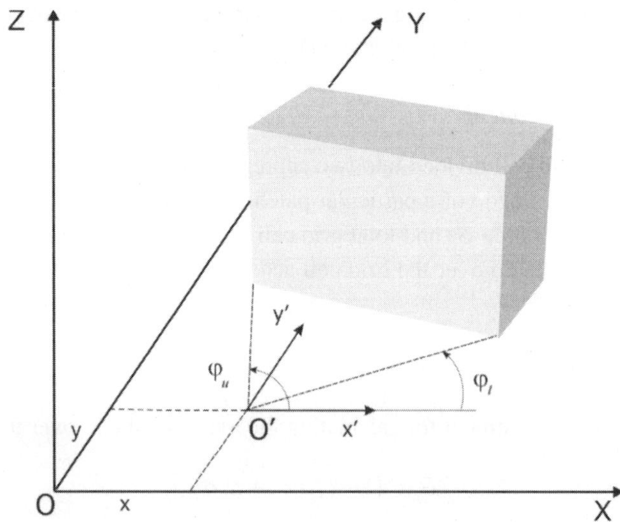

Figure 1. Definition of the boundaries for the integration over the local azimuthal angle.

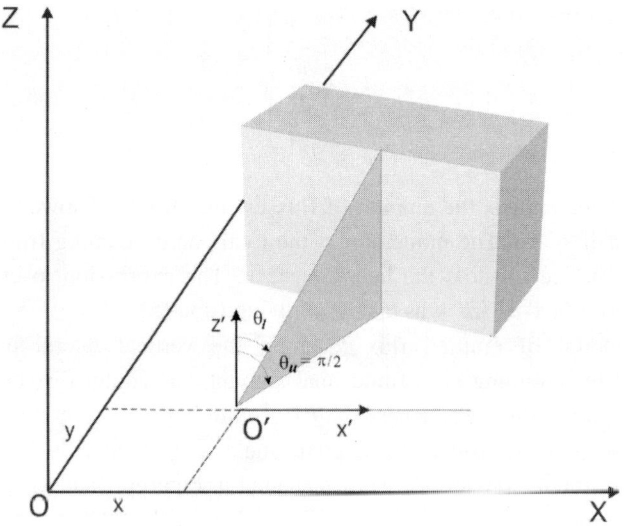

Figure 2. Definition of the boundaries for the integration over the local zenithal angle.

the lower and upper boundaries, respectively. Following Liou [5], the part of total energy coming from the lower horizontal surface towards the lateral surface is given in the form

$$\left(\frac{dE}{dt}\right)_l = I \iint_S dxdy \int_{\phi_1(\vec{r})}^{\phi_u(\vec{r})} d\phi \int_{\theta_l(\vec{r},\phi)}^{\theta_u(\vec{r},\phi)} \cos\theta \sin\theta d\theta \tag{4}$$

In this expression I is the total intensity of radiation (obtained from the monochromatic intensity by integrating it in the range of the whole spectrum); $\cos\theta$ describes the direction of the radiation stream, while $\sin\theta d\theta d\phi$ is the element of solid angle within which our differential amount of energy is confined. Using the same expression, with boundaries for the complete upper half-space, we calculate the amount of flux emitted from the lower horizontal surface into the upper space as $\left(dE/dt\right)_h = IS_l\pi$. The loss coefficient k needed for calculating the average albedo can be evaluated by combining Eqs. (3) and (4).

2.3. *A numerical approach*

As the next step, we present a version of the Monte Carlo (MC) approach. The number of simple geometries which allow analytic solutions is rather restricted, so it was necessary to develop an efficient numerical procedure for calculating the loss coefficient. We have chosen the well known ray-tracing MC method.

Monte Carlo method is a technique for estimating the solution of a numerical mathematical problem by performing statistical sampling experiments. It is a method for iteratively evaluating a deterministic model using sets of random numbers as inputs, turning it into a stochastic one. The result is obtained within the limits of a large number of numerical experiments.

The method is useful for obtaining numerical solutions to problems which are too complicated to be solved analytically. The main advantage of Monte Carlo is that this technique can provide an approximate answer quickly and to a higher level of accuracy with increasing dimension, as opposed to deterministic methods. Also, other numerical methods can be more time consuming, less efficient and expensive.

Monte Carlo techniques can be applied to many different forms of problems. One of the most important uses of Monte Carlo methods is in evaluating difficult integrals, especially multi-dimensional integrals. In these situations Monte Carlo approximations become a valuable tool to use, as they may be able to give a reasonable approximation in a much shorter time in comparison to other formal techniques.

A version of the MC approach used in this study is the so-called ray-tracing Monte Carlo method where we retrace the ray destiny. The light transport equation is solved using Monte Carlo methods to simulate the propagation of light in a scenery.

The main idea is to follow the appropriately chosen ray of light after it had undergone diffuse, homogeneous single scattering from the lower surface of the grid cell. We determine the value of "loss-coefficient" k within a given grid cell geometry by averaging the observed behaviour over a large number of followed light paths.

We actually perform our numerical experiment as follows. First, two random numbers, r_1 and r_2, uniformly distributed in the interval $(0,1)$ are generated in order to randomly sample the point $A(x, y)$ which belongs to the lower surface and represents a point of intercept of this surface and the incoming beam. Then, a random direction is chosen in the upper half-space (θ, ϕ) to simulate the trace of scattered beam. Additionally, two more random numbers, r_3 and r_4, are generated in order to determine the azimuthal $\phi \in (0, 2\pi)$ and zenithal $\theta \in (0, \pi/2)$ range within which the scattered beam is found [$\phi = r_3 \times 2\pi$ and $\theta = r_4 \times \pi/2$]. Further approach was based on the idea of line-plane intersection. The reflected beam was treated as a line, while the vertical area had the role of a plane. The point of intersection was determined by combination of line and plane equations. The case was positive for absorption if the diffusively scattered beam reached the vertical boundary. This procedure was repeated $N = 10^6$ times and the "loss-coefficient" was estimated as

$$k = N_a / N \tag{5}$$

where N_a is the number of cases which were positive for absorption and N is the number of conducted numerical experiments.

As the first test of the method, we reproduced the analytical results for the simplest geometry ("step-like" patch) Kapor *et al.* [3], Mihailovic *et al.* [8]. A grid cell was divided in two subregions having a rectangular form (Figure 3). The area of the lower surface was $S_1 = L \times l$, while the area of the surface lying on a higher level was $S_2 = L \times (L - l)$. Each one of them had a coresspponding albedo, α_1 and α_2, respectively. The boundaries of the integration are

$$\phi_l = arctg[(l - y)/(L - x)] \qquad \phi_u = \pi/2 + arctg[x/(l - y)] \tag{6}$$

$$\theta_l = arctg[(l - y)/(h \sin \phi)] \qquad \theta_u = \pi/2 \tag{7}$$

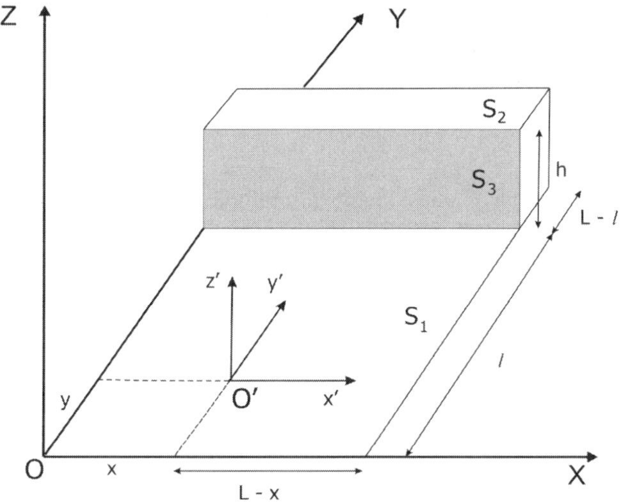

Figure 3. Schematic representation of the square grid-cell consisting of two surfaces of the relative height h .

Flux emitted from the lower surface into the upper half space was calculated as $(dE/dt)_h = ILl\pi$. The expresion for the flux that reached the lateral surface $(dE/dt)_l$ had the following form

$$(dE / dt)_l = I \int_0^l dy \int_0^L dx \int_{\phi_l}^{\phi_u} \int_{\theta_l}^{\theta_u} \cos\theta \sin\theta \, d\theta \, d\phi \tag{8}$$

For simplicity's sake, the reduced dimensionless quantities are introduced

$$\hat{x} = x/L, \quad \hat{y} = y/L, \quad \hat{l} = l/L, \quad \hat{h} = h/L \tag{9}$$

The final expression for the loss coefficient as a function of reduced length $\hat{l}$ and reduced height $\hat{h}$ is

$$k_l\left(\hat{l},\hat{h}\right) = \frac{1}{\hat{l}\pi}\left\{ \hat{l}\,arctg\,\frac{1}{\hat{l}} - \sqrt{\hat{h}^2 + \hat{l}^2}\,arctg\,\frac{1}{\sqrt{\hat{h}^2 + \hat{l}^2}} + \hat{h}\,arctg\,\frac{1}{\hat{h}} + \right.$$

$$\frac{1}{4}\left(1-\hat{l}^2\right)\left[\ln\left(1+\hat{l}^2\right) - \ln\left(1+\hat{h}^2+\hat{l}^2\right)\right] + \frac{1}{4}\hat{h}^2\ln\left(1+\hat{h}^2+\hat{l}^2\right) + \tag{10}$$

$$\left. \frac{1}{4}\left(1-\hat{h}^2\right)\ln\left(1+\hat{h}^2\right) + \frac{1}{4}\hat{l}^2\left[\ln\hat{l}^2 - \ln\left(\hat{h}^2+\hat{l}^2\right)\right] + \frac{1}{4}\hat{h}^2\left[\ln\hat{h}^2 - \ln\left(\hat{h}^2+\hat{l}^2\right)\right]\right\}.$$

    D. Kapor, A. Cirisan and D.T. Mihailovic

Since the numerical results are in an agreement with the analytical ones for various parameters of the patches, we concluded that the application of the ray-tracing MC method is justified in this case.

3. Numerical Tests

In this chapter we have considered an example of central geometry, the so-called "propagating building". Let us note that an analogous example of central geometry, called the "propagating hole", was previously discussed by Mihailovic *et al.* [8]. It is assumed that the region consists of a two-patch grid cell, a grid cell of size L and a "propagating building" with the edge l and height h, both taking various values, simulating the propagation of that area over the grid cell. The basis of building is a square centred at the centre of the cell. Preliminary results were reported previously in Kapor *et al.* [4].

In our study the dimension of the grid cell was $100\times100\,m$. The fractional cover of the central area was $\sigma = (l/L)^2$, while for the rest of the cell it was $1-(l/L)^2$. Calculations of albedo were done using reduced dimensionless quantities $\hat{l} = l/L$ ratio and $\hat{h} = h/L$ ratio. We have calculated analytically values of average albedo according to both the conventional approach and ours. The analytical expression for the loss coefficient is given as

$$k_l(\hat{l},\hat{h}) = 4\left\{ \hat{l}\left((1-\hat{l})/2\right) arctg\ \hat{l}/\left((1-\hat{l})/2\right) - \hat{l}\sqrt{\hat{h}^2 + \left((1-\hat{l})/2\right)^2}\right.$$

$$arctg\ \hat{l}/\sqrt{\hat{h}^2 + \left((1-\hat{l})/2\right)^2} + \hat{l}\hat{h}\,arctg\ \hat{l}/\hat{h} + \left(\hat{h}^2 \ln\hat{h}^2\right)/4 +$$

$$\left(\hat{l}^2 - \left((1-\hat{l})/2\right)^2\right)\ln\left(\hat{l}^2 + \left((1-\hat{l})/2\right)^2\right)/4 - \left(\hat{l}^2 - \hat{h}^2 - \left((1-\hat{l})/2\right)^2\right)$$

$$\ln\left(\hat{l}^2 + \hat{h}^2 + \left((1-\hat{l})/2\right)^2\right)/4 + \left(\hat{l}^2 - \hat{h}^2\right)\ln\left(\hat{l}^2 + \hat{h}^2\right)/4 - \left(\hat{l}^2 \ln\hat{l}^2\right)/4 +$$

$$\left((1-\hat{l})/2\right)^2 arctg\ 1 - \left((1-\hat{l})/2\right)^2 \ln\left(\left((1-\hat{l})/2\right)^2 + \left((1-\hat{l})/2\right)^2\right)/2 + \left((1-\hat{l})/2\right)^2$$

$$\left((1-\hat{l})/2\right)^2 \ln\left((1-\hat{l})/2\right)^2/4 - \left(\hat{h}^2 + \left((1-\hat{l})/2\right)^2\right)\ln\left(\hat{h}^2 + \left((1-\hat{l})/2\right)^2\right)/4$$

$$\ln\left(\left((1-\hat{l})/2\right)^2\right)/2 + \left((1+\hat{l})/2\right)\left((1-\hat{l})/2\right) arctg\left((1+\hat{l})/2\right)/\left((1-\hat{l})/2\right) +$$

$$\left((1+\hat{l})/2\right)^2 \ln\left(\left((1+\hat{l})/2\right)^2 + \left((1-\hat{l})/2\right)^2\right)/2 - \left((1+\hat{l})/2\right)^2 \ln\left(\left((1+\hat{l})/2\right)^2\right)/2$$

$$/2 - \hat{i}\left(\left(1-\hat{i}\right)/2\right)\operatorname{arctg}\hat{i}/\left(\left(1-\hat{i}\right)/2\right) - \hat{i}^2\ln\left(\hat{i}^2 + \left(\left(1-\hat{i}\right)/2\right)^2\right)/2 + \left(\hat{i}^2\ln\hat{i}^2\right)/2$$

$$+\left(\left(1-\hat{i}\right)/2\right)\sqrt{\hat{h}^2 + \left(\left(1-\hat{i}\right)/2\right)^2}\,\operatorname{arctg}\left(\left(1-\hat{i}\right)/2\right)\Big/\sqrt{\hat{h}^2 + \left(\left(1-\hat{i}\right)/2\right)^2} -$$

$$\left(\left(1-\hat{i}\right)/2\right)\hat{h}\operatorname{arctg}\left(\left(1-\hat{i}\right)/2\right)\Big/\hat{h} + \left(\left(1-\hat{i}\right)/2\right)^2\ln\left(\left(\left(1-\hat{i}\right)/2\right)^2 + \hat{h}^2 +\right.$$

$$\left.\left(\left(1-\hat{i}\right)/2\right)^2\right)\Big/2 - \left(\left(1-\hat{i}\right)/2\right)^2\ln\left(\left(\left(1-\hat{i}\right)/2\right)^2 + \hat{h}^2\right)\Big/2 - \left(\left(1+\hat{i}\right)/2\right) \qquad (11)$$

$$\sqrt{\hat{h}^2 + \left(\left(1-\hat{i}\right)/2\right)^2}\,\operatorname{arctg}\left(\left(1+\hat{i}\right)/2\right)\Big/\sqrt{\hat{h}^2 + \left(\left(1-\hat{i}\right)/2\right)^2} + \left(\left(1+\hat{i}\right)/2\right)$$

$$\hat{h}\operatorname{arctg}\left(\left(1+\hat{i}\right)/2\right)\Big/\hat{h} - \left(\left(1+\hat{i}\right)/2\right)^2\ln\left(\left(\left(1+\hat{i}\right)/2\right)^2 + \hat{h}^2 + \left(\left(1-\hat{i}\right)/2\right)^2\right)\Big/2 +$$

$$\left(\left(1+\hat{i}\right)/2\right)^2\ln\left(\left(\left(1+\hat{i}\right)/2\right)^2 + \hat{h}^2\right)\Big/2 + \hat{i}\sqrt{\hat{h}^2 + \left(\left(1-\hat{i}\right)/2\right)^2}\,\operatorname{arctg}\hat{i}\Big/$$

$$\sqrt{\hat{h}^2 + \left(\left(1-\hat{i}\right)/2\right)^2} - \hat{i}\hat{h}\operatorname{arctg}\hat{i}/\hat{h} + \hat{i}^2\ln\left(\hat{i}^2 + \hat{h}^2 + \left(\left(1-\hat{i}\right)/2\right)^2\right)\Big/2 -$$

$$\hat{i}^2\ln\left(\hat{i}^2 + \hat{h}^2\right)\Big/2 + \left(\left(1-\hat{i}\right)/2\right)^2\ln\left(2\left(\left(1-\hat{i}\right)/2\right)^2\right)\Big/2 - \left(\left(1-\hat{i}\right)/2\right)^2$$

$$\ln\left(\left(\left(1-\hat{i}\right)/2\right)^2\right)\Big/2 - \left(\left(\left(1-\hat{i}\right)/2\right)^2 + \left(\left(1+\hat{i}\right)/2\right)^2\right)\ln\left(\left(\left(1-\hat{i}\right)/2\right)^2 +\right.$$

$$\left.\left(\left(1+\hat{i}\right)/2\right)^2\right)\Big/4 + \left(\left(1+\hat{i}\right)/2\right)^2\ln\left(\left(\left(1+\hat{i}\right)/2\right)^2\right)\Big/4 + \left(\left(\left(1-\hat{i}\right)/2\right)^2 + \hat{i}^2\right)$$

$$\ln\left(\left(\left(1-\hat{i}\right)/2\right)^2 + \hat{i}^2\right)\Big/4 - \left(\hat{i}^2\ln\hat{i}^2\right)\Big/4 - \left(\hat{h}^2 + 2\left(\left(1-\hat{i}\right)/2\right)^2\right)$$

$$\ln\left(\hat{h}^2 + 2\left(\left(1-\hat{i}\right)/2\right)^2\right)\Big/4 + \left(\hat{h}^2 + \left(\left(1-\hat{i}\right)/2\right)^2\right)\ln\left(\hat{h}^2 + \left(\left(1-\hat{i}\right)/2\right)^2\right)\Big/2 -$$

$$\hat{h}^2\ln\hat{h}^2/4 + \left(\hat{h}^2 + \left(\left(1-\hat{i}\right)/2\right)^2 + \left(\left(1+\hat{i}\right)/2\right)^2\right)\ln\left(\hat{h}^2 + \left(\left(1-\hat{i}\right)/2\right)^2\right)$$

$$+\left(\left(1+\hat{i}\right)/2\right)^2\Big)\Big/4 - \left(\hat{h}^2 + \left(\left(1+\hat{i}\right)/2\right)^2\right)\ln\left(\hat{h}^2 + \left(\left(1+\hat{i}\right)/2\right)^2\right)\Big/4 -$$

$$\left(\hat{h}^2 + \left(\left(1-\hat{i}\right)/2\right)^2 + \hat{i}^2\right)\ln\left(\hat{h}^2 + \left(\left(1-\hat{i}\right)/2\right)^2 + \hat{i}^2\right)\Big/4 +$$

$$\left(\hat{h}^2 + \hat{i}^2\right)\ln\left(\hat{h}^2 + \hat{i}^2\right)/4\Big\}\Big/\left(1 - \hat{i}^2\right)\pi .$$

The values calculated in this manner were also confirmed by the ray-tracing MC method, as will be demonstrated in the Figures 4-6. They were utilized then as average values of albedo in the prognostic equation for the evolution of the effective surface temperature over a heterogeneous grid cell.

Changes in albedo lead to a significant change in partitioning of the energy at environmental interface. The most remarkable changes are in the values of sensible and latent heat fluxes and surface temperature. These values are calculated using the LAPS parameterisation scheme and then compared to the values obtained with conventional parameterisation of the albedo.

The Land Air Parameterization Scheme (LAPS) is designed as a software package and can be run as part of an atmospheric model or as a stand-alone model. It describes mass, energy and momentum transfer between the land surface and the atmosphere. This scheme has been developed at the Faculty of Agriculture of the University of Novi Sad and has been intensively tested on different kinds of plants and crops. A detailed description and explanation of the scheme's structure, governing equations, the representation of energy fluxes and radiation, model hydrology *etc.* can be found in Mihailovic [6] and Mihailovic *et al.* [7] [9] [11].

The governing prognostic equation for the evolution of the effective surface temperature over a heterogeneous grid cell is

$$C_g \frac{\partial T}{\partial t} = R_{net} - \overline{H}_{agr} - \overline{\lambda E}_{agr} - \overline{G}_{agr} \qquad (12)$$

where T is the effective surface temperature, t is the time, C_g is the effective heat capacity (heat capacity per unit area), R_{net} is the net radiation, $\overline{H}_{agr}$ is the sensible heat flux, $\overline{\lambda E}_{agr}$ is the latent heat flux, and $\overline{G}_{agr}$ is the soil heat flux. Let us note that we used the standard meteorological terminology. Actually, using the correct terminology, all these are not "fluxes" but flux densities. Net radiation, as a sum of incoming and outgoing radiation, could be determined through the following terms

$$R_{net} = (1-\alpha)R_{short} + R_{long} - \varepsilon\sigma T^4 \qquad (13)$$

where R_{short} is the shortwave radiation and R_{long} is the total longwave radiation, α the aggregated albedo, ε the emissivity of the observed surface, and σ the Stephan-Boltzmann constant. The surface fluxes aggregation approach was applied to obtain the mean surface fluxes over a heterogeneous grid cell, after fluxes over a particular area were obtained from the LAPS. Actually, LAPS was run for every area like a homogeneous one, giving the values of fluxes, effective heat capacity and net radiation loss due to the soil heat flux for a particular surface. Using the combination of parameter and flux aggregation methods, the mean values of these parameters are obtained over the heterogeneous cell. Since we analysed a two-patch grid cell, the expression for an aggregated sensible heat flux was

$$\bar{H}_{agr} = H_1\sigma_1 + H_2\sigma_2 . \tag{14}$$

where $\sigma_i\,(i=1,2)$ is the fractional cover. The same expression was used for calculating the latent heat flux

$$\overline{\lambda E}_{agr} = \lambda E_1\sigma_1 + \lambda E_2\sigma_2 . \tag{15}$$

The soil heat flux was calculated, as it has already been mentioned, as a loss of net radiation, expressed as percentage, depending on the surface type (5% in a grass case, 10% for forest and 25% in a case of concrete)

$$\bar{G}_{agr} = \left(G_1\sigma_1 + G_2\sigma_2\right) R_{net} . \tag{16}$$

We have performed the sensitivity tests using the meteorological data for July 17, 1999 in Philadelphia, PA. The grid cell used in these simulations consisted of the urban area (0.85), agricultural field (0.10) and deciduous forest (0.05). Details about the simulations (initial, boundary conditions, soil parameters, etc.) can be found in Mihailovic *et al.* [10].

Simulations were done for several patch areas, with different surface types and a time step of 10 min. Accordingly, each of the subregions had different cover and corresponding albedos. Initial conditions were slightly changed with respect to preliminary tests, leading to better results.

Three situations were analyzed. In all of them, the central area was a concrete building, having the albedo value of 0.30. The other patch of the grid cell in these three simulations was covered with grass, orchard and concrete, having albedo values of 0.20, 0.15 and 0.30, respectively.

The albedo was calculated taking the following values of the reduced lengths 0.50, 0.70 and 0.870. The reason for such choice is that it produced each type of surface with 75%, 50% or 25%. On each plot we presented the effective surface temperature calculated using: (i) aggregated albedo approach (AA) evaluated by ray-tracing MC method, (ii) simple averaging over the grid cell (CA) and (iii) Eq. (11) that will be referred as AE. The altitude of the central patch was case sensitive.

In the grass-concrete case, the concrete building was 2 m high, while the grass was 0.5 m tall. The results are shown in Figure 4. In the case of the orchard-concrete grid cell, the orchard was taken to be 2.5 m high, while the concrete was 10 m tall. The same height of the central area was applied when both patches were concrete covered. Figures 5 and 6 show the respective results obtained.

The presented plots show that the aggregated approach in every situation gives a higher daily maximum of the effective surface temperature. The significant difference between temperature profiles obtained by the aggregated and conventional approaches is noticed in the case of orchard-concrete and especially in the concrete-concrete case. Effective surface temperature deviation depends on albedo value but also on the fractional cover of both areas. The largest deviation occurs when both patches occupy the same areas. It can be explained by the competition of two basic factors: size of the area from which the reflected light comes from and area on which it is absorbed. Figure 7 shows how "loss-coefficient" varies with different values of reduced length and height. As we can see, it has the largest value when the central area occupies 75% of the grid. Although this geometry was not that simple, we made an analytical calculation, and as we can see in Figures 4-6, the ray-tracing Monte Carlo simulation reproduced the obtained results quite well, which gave a reliable base for further research in the case of more complicated geometry.

In order to determine more precisely the difference between temperature profiles obtained by aggregated and conventional approaches, the root mean square error (RMSE) is found. Results in Figure 8 indicate that the maximum increase in the temperature and decrease in albedo is for 50%-50% area coverage when both patch areas are covered with concrete. In the orchard-concrete case we also obtained a significant RMSE, especially for the case when the central area occupies 25% of the cell.

4. Conclusions

The aim of this study was to show further achievements related to the new approach for the aggregation of albedo over the heterogeneous grid cell. We have elaborated the effect for a central geometry for the most general case using only two patches distinguished by relative height of the central one. Simulations were done for different surface types, with different horizontal dimensions and heights.

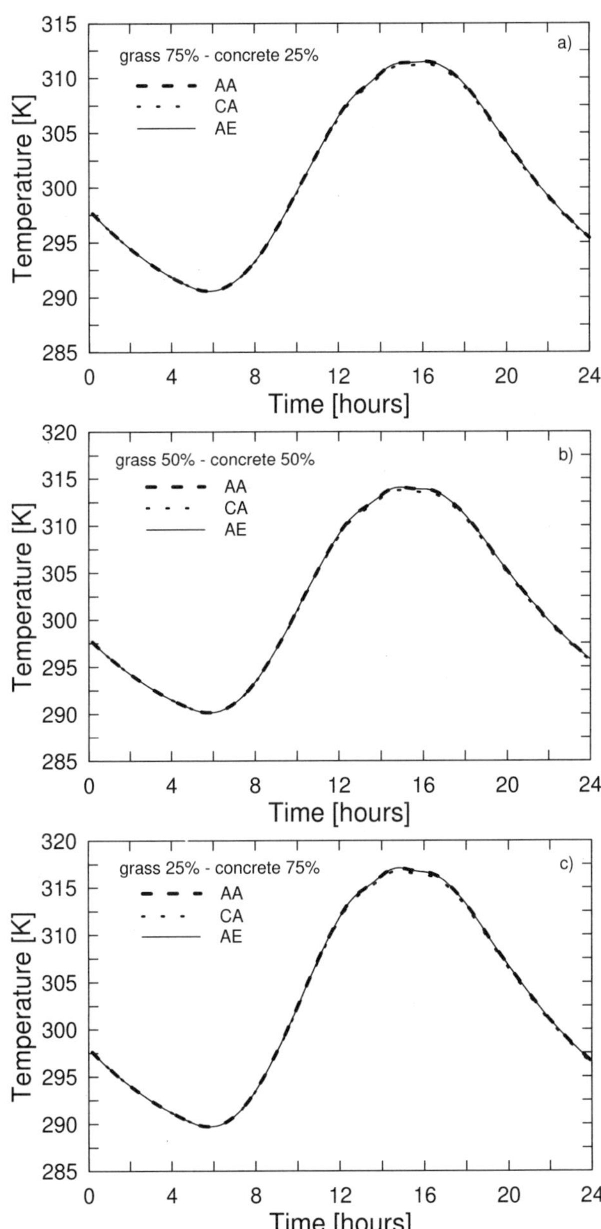

Figure 4. Daily variation of surface temperature in a case of grass-concrete covered grid-cell, with (a) grass 75%, (b) concrete 50%, (c) concrete 75% using both approaches.

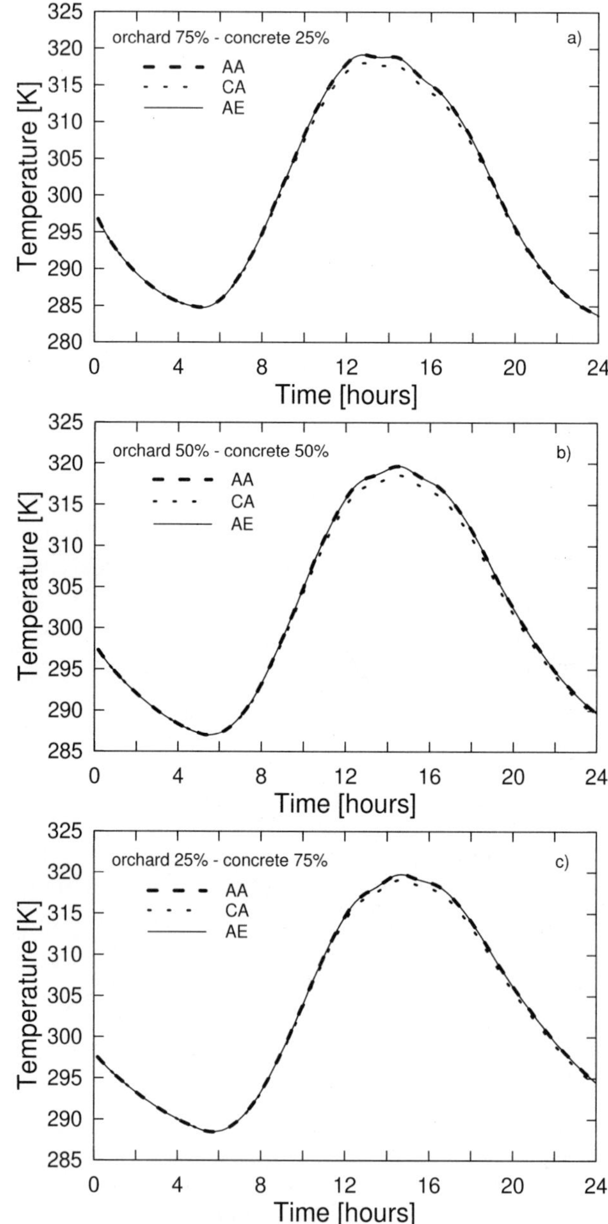

Figure 5. Daily variation of surface temperature in a case of orchard-concrete covered grid-cell, with (a) orchard 75%, (b) concrete 50%, (c) concrete 75% using both approaches.

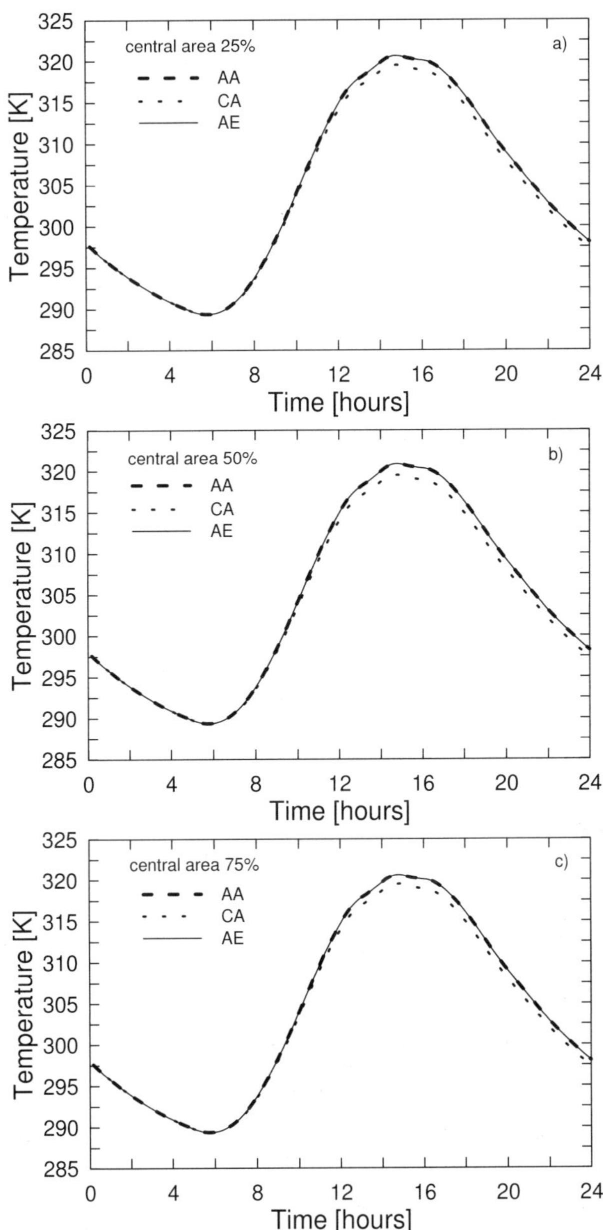

Figure 6. Daily variation of surface temperature in a case of concrete-concrete covered grid-cell, with (a) central area 25%, (b) central area 50%, (c) central area 75% using both approaches.

    D. Kapor, A. Cirisan and D.T. Mihailovic

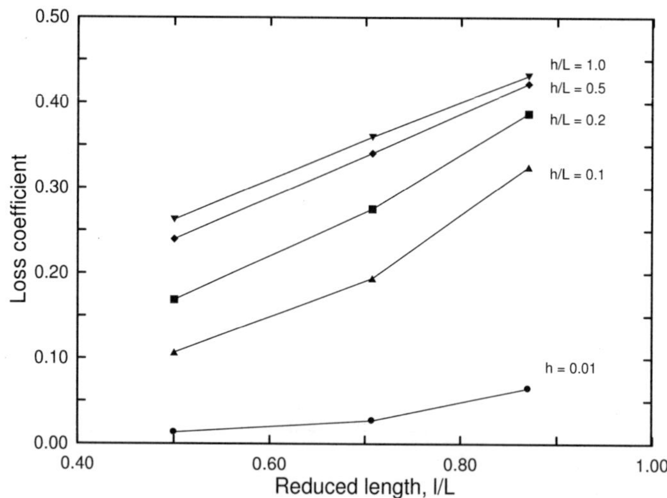

Figure 7. Dependence of "loss-coefficient" on reduced length l/L and height h/L.

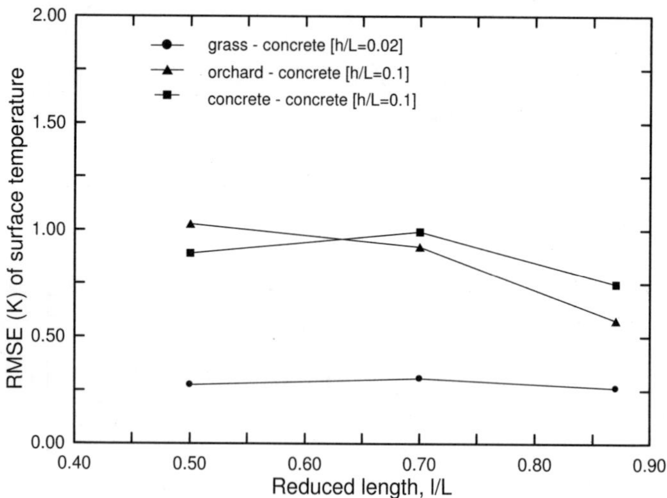

Figure 8. Dependence of RMSE of effective surface temperature on the reduced length l/L.

After a short review of the general approach to the calculation of the flux that is lost due to the absorption on the vertical lateral boundaries, a theory was applied to a simple model of square patches. Because of a rather cumbersome analytical solution, we exposed an efficient numerical procedure using the ray-tracing Monte Carlo method for albedo calculations. The results were compared with the results of the conventional method as well as with those of the analytical ones and were applied in the calculation of effective surface temperature over the heterogeneous area. The land surface scheme LAPS was used to compute the effective surface temperature.

Some of the key findings we obtained through this study is that depending on the relative size of the patches and their heights, and especially depending on the surface type, the decrease of albedo can be significant. Accordingly, it also leads to differences (increase) in the effective surface temperature calculations. This indicates that geometrical effect plays an important role in albedo estimation. Our further work will be devoted to more complex and more demanding geometries.

Acknowledgements

The research presented in this paper has been funded by the Serbian Ministry of Science and Environmental Protection under the project "Modelling and numerical simulations of complex physical systems", No. ON141035 for 2006-2010.

List of Symbols

List of Symbols		
Symbol	**Definition**	**Dimensions or Units**
C_g	effective heat capacity (heat capacity per unit area)	$[M \cdot K^{-1} \cdot T^{-2}]$
E	radiant energy	$[M \cdot L^2 \cdot T^{-2}]$
G_{agr}	aggregated soil heat flux	$[M \cdot T^{-3}]$
H_{agr}	aggregated sensible heat flux	$[M \cdot T^{-3}]$
I	total intensity of radiation	$[M \cdot L^2 \cdot T^{-3}]$
L	size of a grid cell	$[L]$
N	number of conducted numerical experiments	
N_a	number of cases positive for absorption	
R_{long}	total longwave radiation	$[M \cdot T^{-3}]$
R_{net}	net radiation	$[M \cdot T^{-3}]$
R_{short}	shortwave radiation	$[M \cdot T^{-3}]$
S	the total grid cell area	$[L^2]$

S_1, S_2	the areas of the subregions of the grid-cell with corresponding albedos α_1 and α_2	$[L^2]$
T	effective surface temperature	[K]
h	the relative height of the higher surface	[L]
k	loss coefficient	
l	the relative length of the higher surface	[L]
r_1, r_2, r_3, r_4	random numbers	
t	time	[T]
α	aggregated albedo	
α_c	conventional albedo	
α_i	albedo of particular patch	
α_n	new approach (aggregated) albedo	
ε	emissivity	
θ	zenithal angle	
λ	latent heat of vaporisation	$[L^2 \cdot T^{-2}]$
λE_{agr}	aggregated latent heat flux	$[M \cdot T^{-3}]$
σ	Stephan-Boltzmann constant	$[M \cdot T^{-3} \cdot K^{-4}]$
σ_i	a partial fractional cover	
φ	azimuthal angle	
$d\Omega$	the element of solid angle within which our differential amount of energy is confined to	$[L^2 \cdot L^{-2}]$

References

1. Y. Delage, L. Wen and J.M. Belanger, *Atmos. Ocean.* **37**, 157-178 (1999).
2. R.E. Dickinson, *Adv. Geophys.* **25**, 305-353 (1983).
3. D. Kapor, D.T. Mihailovic, T. Tosic, S.T. Rao and C. Hogrefe, An approach for the aggregation of albedo in calculating the radiative fluxes over heterogeneous surfaces in atmospheric models. *The International Environmental Modeling and Software Society (iEMSs)*, Lugano, Switzerland, 2, 389-394 (2002).
4. D.V. Kapor, A.M. Cirisan and D.T. Mihailovic, The impact of geometrical effects on calculated value of albedo on the heterogeneous environmental interfaces. *Fourth Biennial Meeting: International Congress on Environmental Modelling and Software (iEMSs 2008)*, Barcelona, Spain, 1, 114–121 (2008).

5. K.N. Liou, An introduction to atmospheric radiation. Second Edition (Academic Press Inc. 2002).

6. D.T. Mihailovic, *Global Planet. Change.* **13**, 207-215 (1996).

7. D.T. Mihailovic, S.T. Rao, K. Alapaty and J.Y. Ku, A study of the effects of land-use types on the boundary layer evolution. Abstracts of Scientific Conference on Environmental Monitoring, Evaluation, and Protection in New York: Linking Science and Policy, Albany, New York, USA, 39 (2001).

8. D.T. Mihailovic, D. Kapor, C. Hogrefe, J. Lazic and T. Tosic, *Environ. Fluid. Mech.* **4**(1), 57-77 (2004a).

9. D.T. Mihailovic, K. Alapaty, B. Lalic, I. Arsenic, B. Rajkovic and S. Malinovic, *J. Appl. Meteorol.* **43**(10), 1498-1512 (2004b).

10. D.T. Mihailovic, S.T. Rao, K. Alapaty, J.Y. Ku, I. Arsenic and B. Lalic, *Environ. Modell. Softw.* **20**(6), 705-714 (2005).

11. D.T. Mihailovic and D. Kapor, Fluid Mechanics of Environmental Interfaces. Gualtieri C., Mihailovic D.T. Eds., Taylor & Francis, 71– 95 (2008).

12. T.R. Oke, *Boundary Layer Climates.* 2nd edn. (Methuen 1987).

PART TWO

APPLICATIVE, SOFTWARE AND EXPERIMENTAL ISSUES

Chapter 8

LOCATING A POSSIBLE SOURCE OF AIR POLLUTION USING A COMBINATION OF MEASUREMENTS AND INVERSE MODELING

BORIVOJ RAJKOVIĆ[1], MIRJAM VUJADINOVIĆ[1] and ZORAN GRŠIĆ[2]

[1]*Institute of Meteorology, Faculty of Physics, University of Belgrade, Serbia*
[2]*Institute of Nuclear Sciences Vinča, Belgrade, Serbia*

Two methods for locating a possible source of air pollution that combine measurements and inverse modeling based on Bayesian statistics are proposed in this work. In both methods a puff model was used to generate the pollutant concentration field and the synthetic observations in predefined measuring points using real meteorological data. In the first approach the position of the possible source was found with an iterative process, defined as the maximum of probability density function from an ensemble of possible sources. The simplest form of the method was used, with a single source of known strength and known starting moment of the release. The simplicity of this approach and its numerical efficiency make it especially applicable for use in operational mode. Similar to the first method, the second method uses a library of records and scenarios, with combinations of values for the meteorological and emission parameters in the problem. This is done once, well in advance of the actual search for the source and can be computationally expensive. The second step is fully operational and is accomplished by calculating a marginal probability function from the observed concentrations of a pollutant at sampling points and those from appropriate scenarios.

1. Introduction

It initially seems possible to infer parameters of the pollution field only from measurements obtained from a network of measuring stations that have pollutant detection capabilities. If a quantitative assessment is desired, measurements of the amount of pollutant are necessary, i.e. its concentration at several (the more the better) stations. Further consideration shows that in the case of localized sources and for pollutants released in the atmosphere, this is not really possible. The most important reasons for this are the turbulent nature of the motions in the Planetary Boundary Layer and the very large gradients characteristic of the concentration fields coming from a localized source or sources. In that case, it is difficult to synthesize (analyze in meteorological terminology) concentration fields and therefore it is impossible to search for the source(s). An example is given in the Appendix, which shows a highly artificial situation and yet, even then, it was not possible to get a clear picture of the positions of the sources.

167

This problem can be overcome by introducing modeling as an equally important component of the solution. This combination of modeling and measurements is called the inverse method or modeling, since it usually results in the determination of one or several parameters of the process under consideration.

Inverse modeling in general, as well as for the problem of finding the position of the source or its strength, must minimally have the ability to detect the presence of the pollutant. Measurements of the amount of pollutant are even better. Inverse modeling can be considered a better method of designing an optimal network for measuring points and stations relative to the envisioned or possible sources of pollution. The most basic prerequisite for this is the local wind climatology. Considering our problem, the most difficult situation is when the standard meteorological network for meteorological parameters must be relayed, like two meter temperature and ten meter winds, where even the closest station can be quite far away [18]. In addition to the dispersion modeling approach, various statistical and optimization models were applied to the question of air quality measurement position [5] [6] [11] [14] [17] [19] [26]. Apart from these relatively straightforward approaches to obtaining information from monitoring points and checking their quality, there are several other more sophisticated methods. One method is to use the spatial covariance structure of the quality patterns from the so called "sphere of influence" around each site [as in 4, 13, 28]. Another approach is to generate data in a relatively large set of points (dense grid) and then analyze the average concentration, its clustering, frequency, clustering of the average concentration, *etc.* [as in 27]. This study focuses only on the detection aspect of the inverse problem assuming the existence of measurements of the concentrations of a passive substance.

Inverse modeling was chosen to describe the dispersion or transport of a passive substance. The general form of the transport equation relates the rate of change of the concentration of a passive mater to its advection, diffusion, sources, and sinks. In the case of physically or chemically unstable substances, possible transformations are included in the sink term. One way of solving this problem is by setting a 3D grid, finding the appropriate form for each term and treating it as an initial/boundary value problem. Such models belong to the class of Eulerian models or Computational Fluid Dynamics (CFD) models. The other class is Lagrangian models where particles are followed as they are carried by the wind. The statistics of the trajectories are computed for a large number of particles. Both approaches are very computationally expensive and in practice quite difficult to solve. For instance, in the case of the Eulerian approach, boundary conditions are complicated to define since they involve knowledge of the wind and the thermodynamic state of the atmosphere in a localized region usually far below the resolution of routine meteorological networks. On the other end of modeling spectrum are Gaussian plum models as the "bulkiest" ones with the largest amount of parameterizations (simplifications). Wind is treated as a known constant, while vertical and horizontal dispersion is heavily parameterized through stability classes [16] [17] [21]. Calculation of the dispersion can be improved by using the Monin–Obuhkov theory or even second-order closure theories [25]. The most problematic part in the Gaussian approach is the constant wind assumption. The next generation of "bulk" models avoids this simplification by introducing a series of puffs, whose superposition gives the concentration field.

The most efficient, complete and accurate approach is a combination of direct modeling and inverse techniques with the Bayesian probabilistic approach. This is done by constructing a probability density function (*pdf*) for the problem, and from its marginal distribution, calculate the value of considered parameter. In examples of the Bayesian method [3] [9] [30] there is a combination of the observed data with possible *additional* prior knowledge used to calculate the value of the unknown model parameters. Such knowledge is inherently probabilistic and therefore more complete, thus closer to the real world. Another aspect of this approach is that, in principal, it can take into account all of the uncertainties in the measurements and modeling. An almost universal assumption is that distributions of uncertainties are Gaussian. This assumption greatly simplifies the mathematics of constructing joint probabilities from the probabilities over the subspaces in the problem.

Calculating *pdf* without these assumptions can be done in several ways: the Markov Chain Monte Carlo method or the better suited sequential Monte Carlo approach [1] [2] [7] [10] [14], but these methods are computationally expensive and therefore not suitable in practical application. Therefore, the Bayesian approach and the assumption of Gaussianity of relevant *pfds* are used in this work.

In many realistic problems the number of parameters and their ranges make the problem of parameter(s) estimation computationally expensive. This is particularly true if there is a time limit involved when decisions must be made based on the results. Hence, one can do part of the work well in advance and use the results in making the final computations. This approach has been used by [8] [23] [24]. Although their applications are in a different context, the main idea can be applied for our problem. Their main concern is the speed with which parameter estimations are obtained, while in our case the efficiency means a substantial reduction in the computing time.

2. Theoretical Background

In general, the scientific procedure for the study of a physical system (process) may be divided into three steps: the formation of a set of parameters (m) characterizing the system, forward modeling (represented by the operator F) that enable predictions for given values of the model parameters to be made, F(m), often referred to as simulated measurements (d_m), and the last step, inverse modeling, where one can infer the value of a parameter from the measurements (d), [26 among others]. If we denote the space of parameters as M, the space of simulated measurements as D_m (spanned by forward predictions for all possible values of parameters), and the space of actual measurements as D then three probability density functions (*pdf*) can be defined, which enables us to completely describe the system. The first function is $f(m,d_m)$ defined by the Descartes product of M and D_m, (M x D_m), the second is $\rho(m,d)$ defined by the Descartes product of M and D, (M x D) and the third is $p(m,d,d_m)$ defined by M x (D x D_m) as a conjunction of probabilities f and ρ

$$p(m,d,d_m) = const \cdot \frac{\rho(m,d)f(m,d_m)}{v(m,d)}. \tag{1}$$

Since the spaces of the parameters and actual measurements are independent, the probability ρ can be written as the product of $\rho_m(m)$ and $\rho_d(d)$,

$$\rho(m,d) = \rho_m(m) \cdot \rho_d(d) \tag{2}$$

$\rho_m(m)$ is referred to as a prior probability, while $\rho_d(d)$ is the probability for a certain outcome of a measurement, containing errors in the measurement process. Similarly, the probability $v(m,d)$ can be written as

$$v(m,d) = v_m(m) \cdot v_d(d) \tag{3}$$

which are homogenous distributions and we can assume they are constants. Direct determination of $f(m,d_m)$ is not easy, so we will rewrite it, with the help of conditional probabilities, as

$$f(m,d_m) = f(d_m/m) \cdot v(m). \tag{4}$$

The probability $f(d_m/m)$ can be estimated numerically: for given values of parameters (m), the model computes values for the simulated measurements (d_m). This gives us the final form of p, the posteriori probability

$$p(m,d,d_m) = const \cdot \frac{p_m(m)\rho_d(d)f(d_m/m)}{v_d(d)} .$$ (5)

3. Short Description of the Puff Model

In our case the forward modeling was done using a puff model [11] [22]. Its main characteristic is that emissions from a continuous source are treated as a series of puffs that are released at every time step. The center of each puff is advected by the wind that can be spatially variable and different from one time step to another. This possibility of varying wind in time and space is the main advantage over the standard assumption of constant wind used in Gaussian plume models. Concentration in a cell with coordinates (x_{ic}, y_{ic}, z_{ic}) is the *sum* of the contributions of all puffs, released up to that moment. If the index of the receiving cell is denoted by ic and the index of the puffs is ipf, then the contribution of all puffs is:

$$q_{ic}(x_{ic}, y_{ic}, z_{ic}, n \cdot \Delta t) = \frac{Q \cdot \Delta t}{\sigma_{ipf}} \sum_{ipf=1}^{Npf} \exp\left\{ -\left[\frac{(x_{ic} - x_{ipf})^2}{2\sigma_{x,ipf}^2} + \frac{(y_{ic} - y_{ipf})^2}{2\sigma_{y,ipf}^2} \right] \right\}$$

$$\times \left\{ \exp\left[-\frac{(z_{ic} - z_{ipf})^2}{2\sigma_{z,ipf}^2} \right] + \exp\left[\frac{(2z_{inv} - z_{ipf})^2}{2\sigma_{z,ipf}^2} \right] \right\}$$ (6)

where σ_{ipf} is defined as:

$$\sigma_{ipf} \equiv (2\pi)^{2/3} \sigma_{x,ipf} \sigma_{y,ipf} \sigma_{z,ipf} .$$ (7)

Parameterization of dispersion, i.e. the specification of the values of σ's, comes from the Pasquill–Turner stability categories and wind intensity. A small point can be made here that even if a puff (its center) leaves the model domain and if the wind direction changes in a sufficiently short time, the puff may return to the domain and continue to contribute to the concentration field. The model is capable of differentiating between warm (lighter) and cold (denser) gases by raising the lighter and sinking the denser one. There are standard procedures for wind extrapolation with height. The topography can be included and the concentration filed adjusted to it. This adjustment takes into account the local stability of the gas advection when approaching the obstacle and going over it. The model domain had 301x301 points, with a spatial distance of 60 meters thus spanning the area of about 324 square kilometers. The time interval between two consecutive puffs was one minute.

4. The Source Localization Procedures

4.1. *Direct approach*

As explained in the theoretical background section, our goal is to construct a *pdf* for the possible source location. The maximum of that function provides the desired location. In practice, inverse modeling starts with measurements at one or several locations that are then used to obtain the *pdf* and its maximum. However, in the development stage of the method, measurements are usually simulated and used as surrogates for the actual measurements. Therefore the first step was for a point to be assigned a source of a given strength. In addition, we specify several points as "measurement" points.

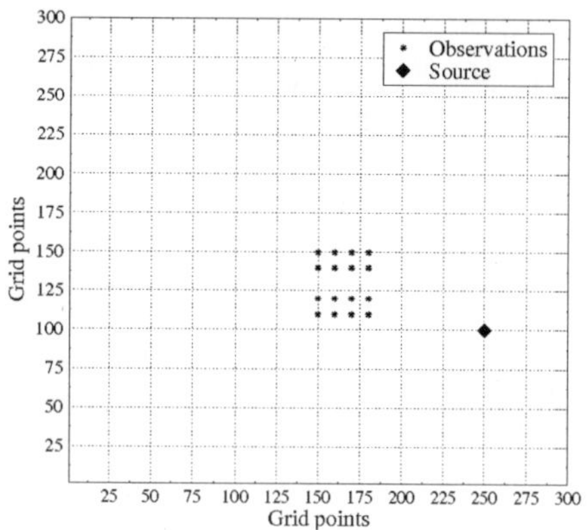

Figure 1. Relative position of the measurement points (stars) and the actual air pollution source (diamond).

A possible case of relative positions of the assumed source and measurement points are presented in Figure 1. The source is designated with a filled diamond while crosses represent the measuring points. After that, releasing it from the source for 120 minutes generates a field of passive substances. The values of simulated concentration in the "measurement" points are designated as measurements after modification of the concentration values by 5% or more. This perturbation mimics the measurement errors. This means we have the set of measurements we would have for the real scenario and can begin the search for the source location.

The first step is to find the point with the highest concentration among all of the measured points. An iterative procedure consisting of several steps follows. In the first step we create the cluster of points covering a relatively large area and treat them as possible sources. At this stage we had a relatively low resolution of 30 grid points between neighboring cluster members. The position of the first cluster is given in Figure 2 in the left panel, whose members are denoted by black crosses. For each member of the cluster we calculated probability of deviation of simulated value from the observed ones using:

$$p(ic, jc) = \frac{1}{\sigma\sqrt{2\pi}} \cdot \exp\left[-\sum_{n=1}^{Nobs} \frac{(q(iobs, jobs) - q_{obs}(iobs, jobs))^2}{\sigma^2} \right], \quad (8)$$

where (*iobs,jobs*) are indices of measurement points, and (*ic,jc*) are indices of a cluster member that is in general a posteriori probability function. The value of σ should take into account the expected (or assumed) measurement errors and uncertainties in a priori knowledge and modeling. The value of $\sigma = 0.01$ used was an order of magnitude above the value that corresponds to the assumed measurement error in the range 5%–30%. The point that had a maximum value of p and became the center of the second cluster with identical geometric characteristics, meaning that the first cluster was translated to its new location (Figure 2, right panel, squares). The calculation is repeated, and the new maxima located. If the new maximum was *inside* the area of the previous cluster, then in addition to the translation we halve the distance between the cluster's members, thus effectively doubling the resolution. If the maximum is at the edge of the cluster we simply move the center to that position without changing the resolution. Positions of the members of the third cluster are given also in the left panel of Figure 2 (circles). Again the fourth cluster was constructed and again the resolution was increased by a factor of two. The same is done for the fifth cluster and so on. If the maximum value in the next iteration is smaller than in the previous, the iteration process is halted. In this case it was iteration number five in Figure 3, the triangle denoted with Ms. This position was only two grid points away from the real source (black diamond).

                    B. Rajković, M. Vujadinović and Z. Gršić

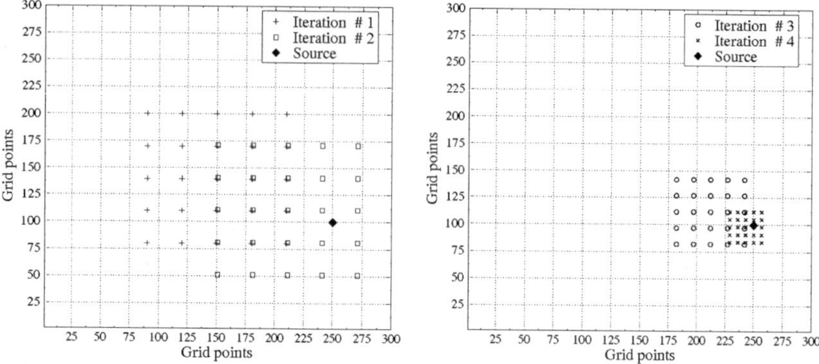

Figure 2. The left panel shows the position of the cluster's members in the first and second iteration. The right panel shows their position in the third and fourth iteration. On both panels, the black diamond represents the actual source location.

The whole procedure takes about 12 minutes on a relatively new personal computer thus clearly fulfilling the efficiency requirement. The accuracy of two grid points equal or less than 120 meters is also acceptable.

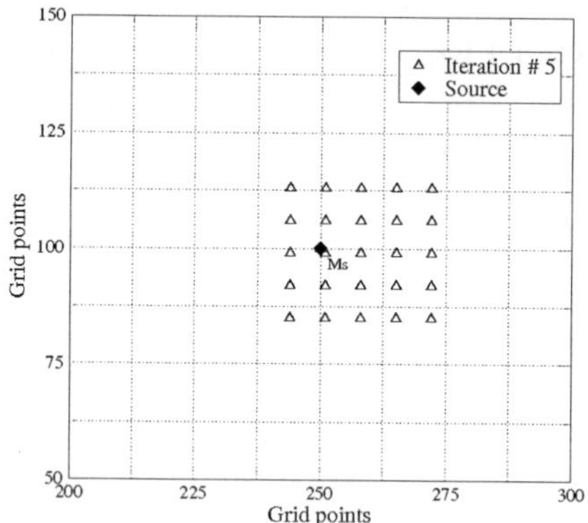

Figure 3. Position of the fifth (final) cluster and the real source location (black diamond). The triangle denoted by Ms represents the position of the source obtained by the iterative procedure.

4.2. *The library approach*

The second method also uses the Bayesian approach and the assumption of Gaussianity, but divides the process of the value estimation for a parameter into two phases. The first is the formation of many combinations of parameters relevant to the problem of dispersion in the atmosphere. They can be generally divided into two groups: meteorological parameters concerning wind and stability and the release of the passive substance, position of the source, duration of the episode, state of the substance, etc. Once the library is generated we can get an estimate for a particular parameter by finding the maximum, which is its marginal *pdf*.

The formation starts by defining classes of values for the relevant meteorological parameters. In principal they should cover all possible values of each parameter. For the wind intensity, we proposed a logarithmic scale, with winds of 0.1, 0.2, 0.5, 1, 2, 5, 10, 20 and 50 m/s, which covers most wind intensities. The stability classes were quantized having values from 1 to 6. During the method testing, we found a very large sensitivity to the wind direction. In response we substantially narrowed the wind direction interval to 1 degree instead of either 1/8 or even 1/16 of the full circle, as is standard for characterizing wind direction. For the source locations we take positions of all possible sources in the particular configuration. For the source strength the exponential increment from 1 to 10^6 for the appropriate units was used. Finally, we assumed that the accident duration is between 1 minute and 3 hours, which should be enough to cover a realistic accidental release that can be treated using either a Gaussian plum or puff model. So let N_{wi} designate the number of different wind intensities, N_{dir} the wind directions, N_{sc} the number of stability classes (in the simplest case 6), N_Q the possible values for the source strength (logarithmic scale), N_{dur} the possible values for the length of accident duration, and finally N_{pos} the number of possible positions. A local parameter that depends on a specific site, rather than the number of records in the pre-computed library is:

$$N_{tot} = N_{wi} \cdot N_{dir} \cdot N_{sc} \cdot N_Q \cdot N_{pos} . \tag{9}$$

In a conservative estimate the library should have several million records or more. At the end, we note that all of the above values should be reexamined in each concrete case. The above choices are viewed as typical rather than all possible. This initial phase can be quite computationally expensive but it is done only once for a particular site. Please note that if we change the model or any of its components or parameterizations, the library must be re-computed.

The second phase can be conditionally called "operational". To carry it out, we must have meteorological observations of wind, its intensity and direction, a decision on which the stability category is relevant for that moment. We do not wish to discuss how this should be done in this manuscript, only that it can be done in many ways that are simple manner or involve some sub-model of small or quite high sophistication. Again using the analogue of Eq. (8) we can calculate *pdf* as

$$p = \frac{1}{\sigma\sqrt{2\pi}} \cdot \exp\left[-\sum_{n=1}^{Nobs} \frac{(q_{obs} - q_{lib})^2}{\sigma^2} \right] \tag{10}$$

but now for *all* members of the library whose meteorological parameters have been observed. Since wind intensity does not vary significantly, once we are above the spectral gap, we assumed that it was constant and equal to its average value during the accident. However, due to the very high sensitivity of the model to wind direction, we should account for the entire range of observed directions. The standard deviation (σ) measures the spread between the observations (q_{obs}) and values from the library (q_{lib}). In general, it takes into account the uncertainty in our prior knowledge, uncertainties in measurements (σ_D) and in modeling (σ_T). In this case, there is no prior knowledge, so the total σ is the sum of σ_D and σ_T. Again as in the first method we have assumed that measurement errors were in the range of 5%–30%, which, together with the values of simulated measurements, gives an estimate for σ_D of up to 10^{-4}, while we took the total σ to be equal to $5\cdot10^{-2}$, which is two orders of magnitude larger than σ_D. The *pdfs* are computed through the summation

$$p_{mar} = \frac{1}{n}\sum_{1}^{n} p \tag{11}$$

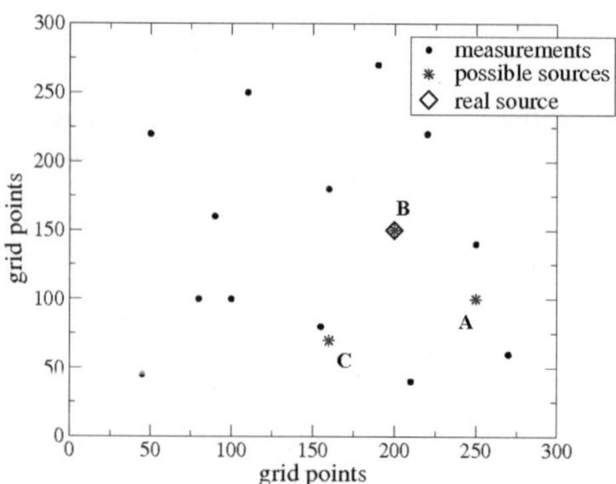

Figure 4. The position of measuring points (dots), possible sources A, B, C (stars) and real source (diamond).

over all library records, but the desired parameter gives the marginal probability for that parameter and its maximum gives its best estimate. The number n is the ratio of all the records corresponding to the values of the observed meteorological parameters N_{tot} and N of all records in the library.

As an illustration of the procedure we present a limited case where we have created a small library with close to 20x103 members by reducing the possible values in all categories, such as having 3 possible positions, a changing wind direction interval of 30 degrees, etc. In Figure 4 we present positions of the assumed sources designated by the letters A, B and C. Black dots show the positions of the assumed measuring network that is responsible for detection of a possible source of pollution. The observed wind data is used to simulate emissions from one of the sources (black diamond). The calculated concentrations were than perturbed up to 30% to mimic the errors in measurements that are always present in such measurements in even larger amounts. Assuming that a sampling average over a 30-minute interval, we calculated marginal probabilities for all three possible sources and the one with the maximum *pdf* was the source. Probability changes with time for each of the sources (Figure 5, left panel).

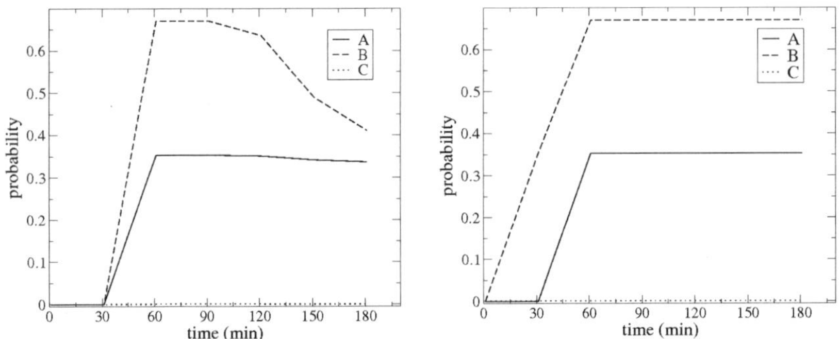

Figure 5. Marginal probability function for each of the three sources (A, B, C) in 30-minute intervals. On the left panel is its change in the case of measurements simulated using observed values, while the right panel is for the case of measurements simulated using a constant wind direction.

Each source has a different history line and the correct source (in this case source B) has the largest value, during the entire integration. At the beginning, probability was very small and reached its maximum, thus indicating the position, and then it decreased. The reason for this is the wind in the simulated measurements was realistic, changing its direction during the simulation, while the library records have a constant wind direction. To verify this, we redid our simulation of concentration with a constant wind direction. Indeed, as shown in Figure 5 (the right panel), once the probability reaches its maximum, it stays constant.

In the end we propose that a combination of both methods is optimal. If we have the problem of locating the position of the source(s) we can first use the second approach to make an estimate of all other unknown parameters with some accuracy, such as the released amount of material, duration of the accident, *etc.*, and then through several iterations of the first approach obtain the position of the source inside the area of interest.

5. Conclusions

A combination of simulated measurements and several iterations of the Bayesian statistical approach results in a quite accurate and yet efficient method of detecting a possible point-like source of air pollution whose strength was assumed to be known. If we relax that condition the method in principal stays but the actual computational time will increase depending on the range of the possible values for the source strength that we will consider. The assumption of the perturbation is only 5% and was not crucial in obtaining a good estimate of the source position since we used perturbations as large as 30% and obtained the same result. Five percent is simply the value often used in literature in the context of simulated measurements.

The second approach can be viewed as a generalization of the first and gives the possibility of relaxing the need for values of many (theoretically all) parameters in the problem. In our mind its bigger advantage that makes it superior to many other methods in solving this problem and finding other parameters is that once the library is formed, the necessary calculations become extremely short, thus making the method very suitable for operational use.

6. Appendix

Here we show that even a high number of measurements can not solve the problem of locating the source position. Let us generate a concentration field with a release from 17 sources. The geometry resembles the situation in the petrochemical zone of the city Pančevo near Belgrade. In Figure A1 in the right panels is a field of 24 hour-average concentration obtained using a Gaussian plum diffusion model [22] for a continuous source, with 10 minute averages of the wind field and stability parameters from Pasquill–Gifford stability categories. The only deviation from the actual situation is the assumption that all sources are at the same height. The grid used for calculations was a very dense one, with 90601 points (301x301). The idea is that such a high density grid will reduce problems with the accuracy of the model and thus the calculated field can be regarded as an "observed" one, i.e. we treat values in grid points as measured ones. Formation of the isolines (graphical representation of the field) was done using the Kriging method, a common method for this purpose. In the case of a very dense grid we can clearly recognize the positions of all sources as the positions of the local maxima.

If we now reduce the number of grid points, and therefore the number of "sampling" points, by two orders of magnitude (31x31), we obtain a situation depicted in the first row of the left panel in Figure A1. We use the Kriging method again to form the "continuous" field. Comparison with the starting situation (right panel of the same figure row) shows that we still have a very high resemblance with the original field. But this situation is still far from a realistic one regarding the possible density of the actual sampling points and tells us more about the possibility of reducing the computation effort.

A grid with 36 points (6x6) is closer to the regulatory recommendations and economical feasibility. Again, with the help of the Kriging method, we arrive at the result in Figure A1 in the second row. Similar to the first row, on the left we have the original concentration field and on the right is the result from the 36 points. Dots on the left panel are positions of the "sampling" locations. We now see that the distribution field is very different from the starting one, showing only the gross characteristics of the pollution concentration fields, but with complete loss of its local details. This still may be acceptable as a very rough assessment of the long term influences of pollution sources for that area, but totally insufficient in an accident situation.

Finally, we created an irregular set of "sampling" stations that qualitatively represent the actual situation in that area. As is often the case in the real-life situations, several points are very close to each other, while others are relatively far away (Figure A1, the third row, left panel). Using the same analysis method, we get the distribution shown in the third row of the right panel in Figure A1. The general structure is similar to the previous case with a few exceptions. This distribution is somewhat closer than the one with 36 points, due to the fact that at several points "samplings" are very close to the position of the sources and thus accidentally giving "better" results, meaning that at least one source is relatively visible. It is also possible that the particular wind direction has "helped" in this case.

     B. Rajković, M. Vujadinović and Z. Gršić

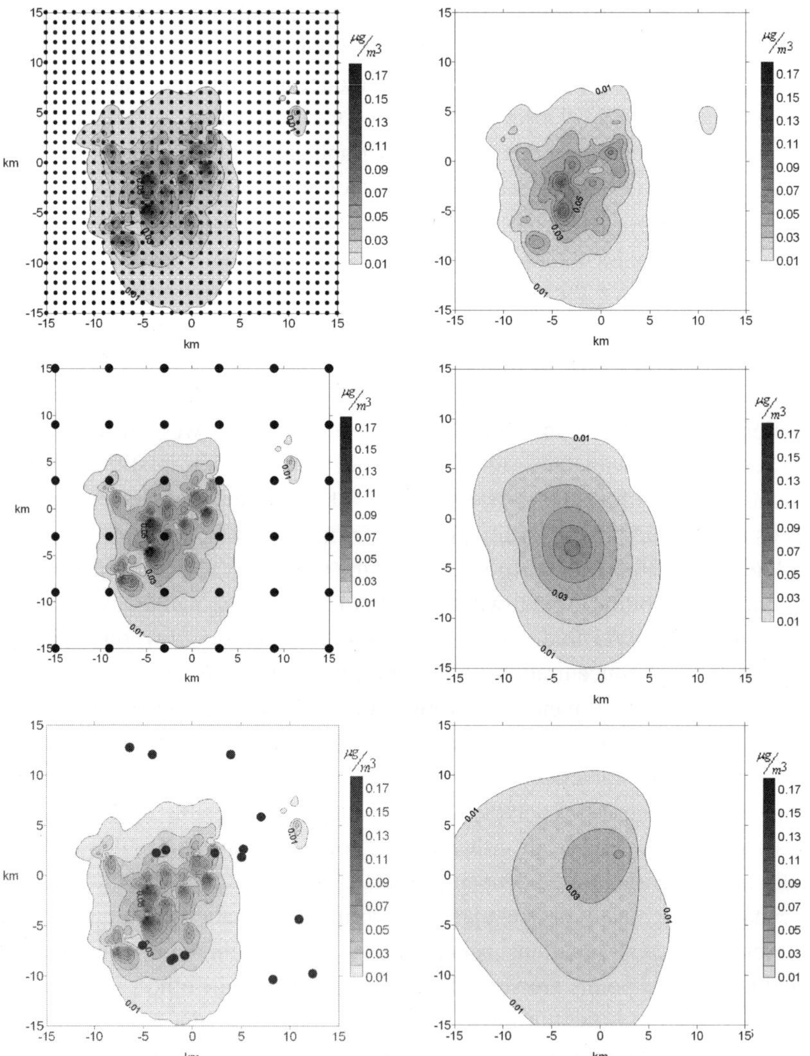

Figure A1. The right panel in every row shows the same concentration field, obtained from a release of 17 sources using a Gaussian plum model with a very dense grid (301x301 grid point). Black dots on each right panel represent "sampling" points. In the first row there is a 31x31 "sampling" point, in the second row 6x6, and in the third row, there are 16 irregularly spaced "sampling points". The left panels show concentration fields interpolated, using the Kriging method from the "sampling" points on the right.

APPENDIX - LIST OF SYMBOLS

	List of Symbols	
Symbol	**Definition**	**Dimensions or Units**
D	space of measurements	
D_m	space of simulated measurements	
F	model operator	
M	space of parameters	
N	number of records in the library	
N_{dir}	number of wind direction categories in the library	
N_{dur}	number of accident duration categories in the library	
N_{obs}	number of observation points	
N_{pf}	number of puffs in the domain	
N_{pos}	number of possible source positions in the library	
N_Q	number of source strength categories in the library	
N_{sc}	number of stability classes in the library	
N_{tot}	number of records in the library corresponding observed situation	
N_{wi}	number of wind intensity categories in the library	
Q	source strength	$[M\ T^{-1}]$
d	measurements	
d_m	simulated measurements	
f	probability density function on the spaces D_m and M	
m	parameters that characterize the system	
p	a posteriori density function	
p_{mar}	marginal probability density function	
q	calculated concentration of a pollutant	$[M\ L^{-3}]$
q_{ic}	concentration of a pollutant in the *ic* grid point	$[M\ L^{-3}]$
q_{lib}	concentration of a pollutant from the library	$[M\ L^{-3}]$
q_{obs}	measured concentration of a pollutant	$[M\ L^{-3}]$
x_{ic}	x-coordinate of the *ic* grid point	$[L]$
x_{ipf}	x-coordinate of the center of *ipf* puff	$[L]$
y_{ic}	y-coordinate of the *ic* grid point	$[L]$

y_{ipf}	y-coordinate of the center of *ipf* puff	[L]
z_{ic}	z-coordinate of the *ic* grid point	[L]
z_{ipf}	z-coordinate of the center of *ipf* puff	[L]
Δt	integration time step	[T]
ν	probability density function on the spaces M and D	
ν_d	homogenous probability density function on the space D	
ν_m	homogenous probability density function on the space M	
ρ	probability density function on the spaces M and D	
ρ_d	probability density function on the space D	
ρ_m	probability density function on the space M	
σ	standard deviation	$[M\,L^{-3}]$
σ_D	uncertainties in measurements	$[M\,L^{-3}]$
σ_T	uncertainties in modeling	$[M\,L^{-3}]$
σ_{ipf}	standard deviation of concentration for the *ipf* puff	$[L^3]$
$\sigma_{x,ipf}$	standard deviation in the x-direction of concentration for the *ipf* puff	[L]
$\sigma_{y,ipf}$	standard deviation in the y-direction of concentration for the *ipf* puff	[L]
$\sigma_{z,ipf}$	standard deviation in the z-direction of concentration for the *ipf* puff	[L]

Acknowledgement

This work has been partially funded by the Republic of Serbia, Ministry of Science, grants no. 1197 and 141035, and the Italian ministry of Environment and territories through its two projects, SINTA and ADRICOSM-STAR. An earlier version of this work has been presented at 29/NATO International Technical Meeting on Air Pollution. The first two authors acknowledge financial aid for participation at that conference. The second author holds the fellowship from the Ministry of Science, Republic of Serbia. The third author is partially funded through the project of decommission of the nuclear reactor in Vinča.

The idea of using pre-computed libraries of possible scenarios was presented by Dr. Ashok Gadgil at the 29/NATO Int. Thech. Meeting on Air

Pollution in Aveiro, Portugal. It was presented for a different case of the possible chemical accident/attack in housing buildings where scenarios were about exits/entrances, elevators windows, etc. However, the basic idea was to evaluate all possible scenarios of chemical substance movement inside the building, and this is what gave us the idea to compile the possible scenarios for the meteorological and source parameters.

References

1. C. Andrieue, N. da Freitas, A. Doucet and M.I. Jordan, *Mach. Lear.*, **50**, 5-43 (2003).
2. M.S. Arulampalam, S. Maskell, N. Gordon and T. Clapp, IEEE *Trans. Signal Process.*, **50 (2)**, 174-188 (2002).
3. J.M. Bernardo and A.F.M. Smith, *Bayesian theory* (Wiley, 1994).
4. C.E. Buell, The topography of the empirical orthogonal functions, 4th *Conference on Probability and statistics*, AMS, Boston, USA, 188-193 (1975).
5. J.C. Chow, J.P. Engelbrecht, N.C.G. Freeman, J.H. Hashim, M. Jantunen, J.P. Michaud, S. de Tajeda, J.G. Watson, F. Wei, W.E. Wilson, M. Yasuno and T. Zhu, *Chemosphere*, **49(9)**, 873-901 (2002).
6. B. Croxford and A. Penn, *Atmos. Environ.*, **32(6)**, 1049-1057 (1998).
7. A. Doucet, N. de Freitas, N.J. Gordon, *Sequential Monte Carlo methods in practice* (Springer-Verlag, 2001).
8. A.J. Gdagil, Rapid data assimilation in the indoor environment theory and examples from real-time interpretation of indoor plumes of airborne chemical, Borrego C. and Miranda A.I. Eds., **263**, Springer (2008).
9. A. Gelman, J.B. Carlin, H.S. Stern and D.B. Rubin, *Bayesian data analysis* (Chapman and Hall, 2004)
10. W.R. Gliks, S. Richards and D.J.E. Spieglhalter, *Markov Chain Monte Carlo in practice* (Chapman and Hall, 1996).
11. Z. Grsic and P. Milutinovic, Automated meteorological station and the appropriate software for air pollution distribution assessment, Gryning S.E. and Batchvarova E. Eds, **781**, Springer (2001).
12. T.C. Haas, *Atmos. Environ.*, **26(18)**, 3323-3333 (1992).
13. J. Langstaff, C. Seigneur, M.K. Liu, J.V. Behar and J.L. McElroy, *Atmos. Environ.*, **21**, 1393-1410 (1987).
14. J.S. Liu, *Monte Carlo strategies in scientific computing* (Springer-Verlag, 2001).
15. N.A. Mazzeo and L.E. Venegas, *Int. J. Environ. Pollut.*, 14, 246-259 (2000).
16. P. Melli and E. Runca, *J. Appl. Meteorol.*, **18(9)**, 1216-1221 (1979).
17. G. Mocioaca and S. Stefan, *Int. J. Environ. Pollut.*, **19(1)**, 32-45 (2003).

18. K.E. Noll and S. Mitsutomi, *Atmos. Environ.*, **17**, 2583-2590 (1983).
19. K.E. Noll, T.L. Miller, J.E. Norco and R.K. Raufer, *Atmos. Environ.*, 11, 1051-1059 (1997).
20. G.W. Oehlert, *Atmos. Environ.*, **30(8)**, 1347-1357 (1996).
21. F. Pasquill and F.B. Smith, *Atmospheric Diffusion* (Wiley, 1983).
22. B. Rajkovic, I. Arsenic and Z. Grsic, Point source atmospheric diffusion, Gualtieri C. and Mihailovic D.T. Eds., **17**, Taylor and Francis (2008).
23. M.D. Sohn, P. Reynolds, N. Singh and A.J. Gadgil, *J. Air Waste Manage. Assoc.*, **52**, 1422-1432 (2006).
24. P. Sreedharan, M.D. Sohn, A.J. Gadgil and W.W. Nazaroff, *Atmos. Environ.*, **40**, 3490-3502 (2006).
25. R.I. Sykes and R.S. Gabruk, *J. Appl. Meteorol.*, **36(8)**, 1038-1045 (1997).
26. A. Tarantola, *Inverse problem theory and methods for model parameter estimation* (SIAM, 2005).
27. C.C. Tseng, N.B. Chang, *Environ. Int.*, **26(7)**, 523-541 (2001).
28. N.D. Van Egmond and D. Onderdelinden, *Atmos. Environ.*, **15**, 1035-1046 (1981).
29. H.W.Y. Wu and L.Y. Chan, *Atmos. Environ.*, **31(7)**, 935-945 (1997).
30. M. West and J. Harrison, *Bayesian forecasting and dynamic models* (Springer, 1997).

Chapter 9

LONG-TERM MEASUREMENTS OF ENERGY BUDGET AND TRACE GAS FLUXES BETWEEN THE ATMOSPHERE AND DIFFERENT TYPES OF ECOSYSTEMS IN HUNGARY

TAMÁS WEIDINGER[1], LÁSZLÓ HORVÁTH[2], ZOLTÁN NAGY[2] and ANDRÁS ZÉNÓ GYÖNGYÖSI[1]

[1]*Department of Meteorology Eötvös Loránd University, Budapest, Hungary*
[2]*Hungarian Meteorological Service, Budapest*

Following a presentation of the brief history and present status of micrometeorological research activity in Hungary, the field measurement programs, instrumentation and flux calculation methodology are presented. The improvement in micrometeorological measurements and data acquisition system is also illustrated. Field experiments using the gradient and profile methods for trace gas fluxes were initiated in 1990. First, daily and annual variation in deposition velocities of ozone, nitrogen oxides and sulfur dioxide above grassland and spruce forest were investigated. The main goal of the investigations was the determination of the dry deposition velocity over different types of vegetation in different seasons. The measurements of dry deposition processes of nitrogen compounds, first of all of ammonia, over semi-natural grasslands, have been performed since year 2000. The ammonia fluxes are bidirectional, with net deposition on the annual time scale. Continuous measurements of the budget of energy, water, carbon and nitrogen, respectively, were carried out in the framework of EU5 (Greengrass) and EU6 (NitroEurope) projects in the central part of the country ("Bugac-puszta") over grassland since 2002. One of the main objectives of this investigation is to reduce the large uncertainty in the estimation of CO_2, N_2O and CH_4 fluxes into and from plots of grassland under different climatic conditions. The final part of the paper illustrates the development of the new Hungarian basic climatological network for the detection of the possible effects of future climate change, which includes standard climate station measurements, soil profile (of temperature and moisture), radiation, energy budget components and CO_2 flux measurements. Based on continuous measurements of micrometeorological elements, trace gas concentrations and turbulent fluxes, more detailed information have been obtained of the structure and possible future changes of the surface layer and the intensity of the surface-biosphere-atmosphere interactions.

Keywords: micrometeorology, surface energy budget, trace gas flux, heat flux, Bowen ratio, eddy covariance, gradient measurement method, eddy accumulation technique, band pass covariance method, turbulent diffusion coefficient, dry deposition velocity, basic climatological network

1. Introduction

The subject of micrometeorology is the analysis of small-scale features taking place in a shallow layer above the Earth's surface, the so called Planetary Boundary Layer (PBL), where the momentum, energy, water vapor and the

185

atmospheric part of the biogeochemical cycles (carbon, nitrogen, sulfur, etc.) initiate and end up. There are only a few parts of meteorology which preserved such a close relationship between measurement and modeling, theory and application as micrometeorology. This close relation is present even in the work of each individual researcher or international integrated project setup (such as the EU6 CarboEurope and NitroEurope programs [59] or the FLUXNET observation network [6]).

Besides the observation of standard meteorological parameters of the near surface layer (using operational observation network) the detection of micrometeorological features such as the vertical profiles of meteorological elements (wind speed, temperature humidity, etc.) atmospheric stability, surface energy budget components, trace gas and aerosol fluxes are also essential. For these purposes special long-term monitoring projects have been initiated. Such kinds of measurements are also present in the measurement program of very few central meteorological observatories [3]. Results have been applied in both meteorological model development and ecological or environmental issues [64].

2. Micrometeorological Measurements in Hungary

2.1. *Short history of the early years of micrometeorological studies*

Following studies of Schmidt, Geiger and Prandtl, micrometeorological investigations in Hungary have been initiated in the twenties and thirties of the past century. Discovering the link between micrometeorology and ecology was crucial from the beginning [5].

Near-surface profile measurements of temperature, humidity and wind speed began at the meteorological observatory of the Department of Meteorology, Eötvös University, at Erdőhát at the end of the 1940s [62]. This station was operational till the end of the seventies [7].

In the beginning of the seventies, investigations were initiated in the region of the Bükk Mountains in order to explore the meteorological background of deforestation (due to acid rain or frequent droughts). Studying the forest ecosystem at Síkfőkút Experimental Forest International LTER site (SIK) lasted till the mid nineties. Continuous near-surface profile measurement of wind, temperature and humidity over woodland was performed [34] [35].

The agrometeorological station at Szarvas was built in the sixties. Till the end of the nineties it was operated by the Hungarian Meteorological Service (1963-1999). Wind speed, temperature and humidity measurements were performed at six levels on a 30 m tall tower [33] [66]. The agrometeorological station of Keszthely was founded in 1972 [1] [38]. Radiation and hydrological measurements, besides crop land experiments have been performed at both sites. These stations are now operated by local universities: Tessedik Samuel College, Szarvas and University of Pannonia Georgikon Faculty of Agriculture, Keszthely.

Continuous measurements of radiation budget, soil temperature and soil moisture have been made at the Agrometeorological Observatory of the University of Debrecen from the sixties. Wind speed and direction, temperature and humidity profile sensors are installed on a 20 m high mast. More than two decades long carbon dioxide dataset is also associated to this station. Quality controlled hourly data set from this observatory is also available for the scientific community [2] [61].

The 120 m high tower of the nuclear power plant at Paks has been operating since the end of the seventies. Measurements are performed at four levels (2 m, 20 m, 50 m and 120 m above ground level, respectively) with a fundamental purpose to be used for the assessment of the wind profile, atmospheric stability and turbulent diffusion [33] [63].

Measurements of radiation and hydrological features near and over the great Hungarian lakes (Lake Balaton, Fertő and Velencei) with micrometeorological methods were also established in the seventies and eighties [9] [39] [40].

The background air quality monitoring station at K-puszta, which is now a part of the EMEP monitoring network [72], has been working since the beginning of the seventies. Dry deposition rate of trace gases were obtained on the basis of the dataset of the well established station and with deposition velocities from the literature [24] [47] [48] [50]. Locations of the listed sites and of the most relevant measurement projects are shown in Figure 1.

2.2. *Measurement projects since 1990*

The Hungarian Meteorological Service (HMS) joined the EUREKA BIATEX framework program at the end of the eighties [18]. The aim of this program was the description of exchange processes between surface-biosphere-atmosphere over short and tall vegetation surfaces. Aerosol exchange studies were added to this topic in 1996 (BIATEX-2). This research initiated Hungarian inter-institutional cooperation that gave an opportunity to join integrated national and international research projects (Figure 1 and Table 1) and established investigations focused on the nitrogen budget [27]. The assessment of surface energy balance components, besides standard meteorological measurements, is included in the measurement program of all projects introduced in this chapter.

Two special sites among present research projects are the measurement systems at Bugac and Hegyhátsál, where fluxes of greenhouse gases have also been measured. Several projects were devoted to turbulent exchange processes of ammonia. Besides the long term daily ammonia series from K-puszta (denuder technique, [30]), the ammonia gradient was also measured with the high precision AMANDA system [29], and with the photoacoustic equipment for long-term, automatic concentration and flux monitoring developed by the University of Szeged [32], which is used at the NitroEurope site "Bugac" since the summer of 2008.

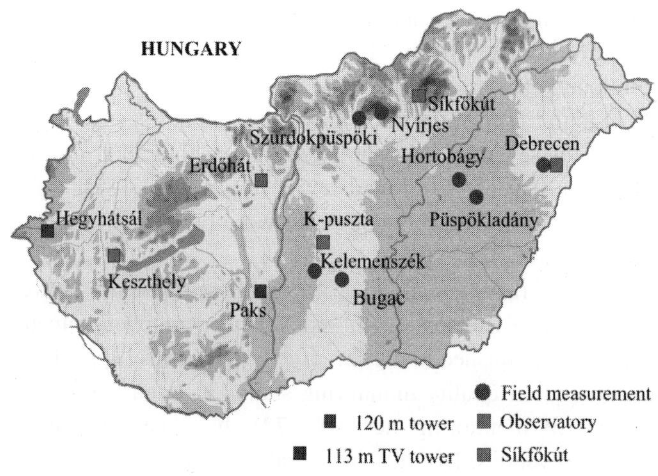

Figure 1. Micrometeorological measurement sites in Hungary.

Various measurement programs required the development of flux assessment schemes. Gradient, profile, Bowen-ratio, direct flux (eddy covariance) methods, static and dynamic chamber measurements [31] have been applied and, in addition, experimental application of the eddy accumulation technique is also performed [56]. Development of a state-of-the-art measurement and data acquisition system is highly pronounced. All measured raw data (both from slow sensors and from the eddy covariance system) are being logged.

In addition, micrometeorological measurements provide valuable information for climate studies and also input data for the operation and development of numerical forecast models and of parameterization schemes. These targets were taken into account in the formation of the climate monitoring network of four Hungarian stations with the aim of providing a high-precision and long-term detection of the local effects of a possible climate change in the near surface layer. Besides the scope of a synoptic station's duties, soil temperature and moisture, vertical profile of meteorological parameters, radiation and energy budget components are also measured for this purpose. Turbulent fluxes have been determined using the eddy covariance technique at the central station in Debrecen [52].

Since our research is focused on the calculation of turbulent fluxes, its methodology is summarized below. Our experimental work, in this chapter, will be illustrated though three examples. First we present daily variation of ozone and sulfur dioxide in different seasons and over short and tall vegetation (BIATEX). Then we will analyze the ammonia measurements (GRAMINAE). Finally, we will describe the measurement program of the basic climatological network, because the authors of this chapter have mainly been involved in the above projects.

3. Turbulent Exchange Processes of the Surface Layer: The Turbulent Fluxes

The transport of different scalar quantities (momentum, energy, water vapor and the concentration of trace gases) in the PBL is performed by turbulent eddies of a size proportional to their heights above ground level. Turbulence is assumed to be stationary, horizontally homogeneous and isotropic. Disturbances due to advection, instationarity, convection, etc., are not considered here. Issues rising from such disturbances are discussed in details in [19] and [20].

3.1. *Raw fluxes*

The turbulent flux (F_c) is the amount of a quantity, for example concentration (c), flowing through a unit surface in a unit time interval. In mathematical form, this is the covariance of the vertical velocity (w) and the c quantity (Fig. 2):

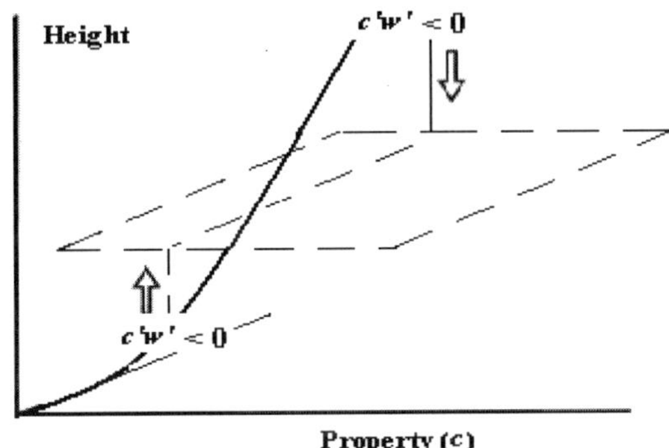

Figure 2. Schematic diagram of turbulent exchange processes. The surface is a sink of scalar c.

                                T. *Weidinger et al.*

$$F_c = \overline{w'c'} , \tag{1}$$

where variables indicated by primes are fluctuations $(c = \bar{c} + c')$, i.e., instantaneous deviations from the mean values, and those indicated by overbars are averages. The average vertical velocity is assumed to be zero ($\bar{w} = 0$). Note that this requirement is strictly satisfied only in case of dry air, however, for trace gases or aerosol this is not fully true. The bias is described by the Webb correction [41] [65].

Using standard notation, the fluxes of momentum (τ), sensible (H) and latent heat (LE), respectively, can be written as follows:

$$\tau = -\rho_m \overline{w'u'} , \quad H = c_{pm}\rho_m \overline{w'T'} , \quad LE = L\overline{w'\rho'_v} , \tag{2}$$

where ρ_m is the average moist air density, c_{pm} is the isobaric heat capacity of moist air, L is the latent heat of vaporization, while T', ρ'_v, u' and w' are the fluctuations of temperature, vapor density, and the horizontal and vertical velocities, respectively. In the calculation of raw fluxes, detrending has not been applied, and sonic temperature (a quantity that is close to virtual temperature and that can be calculated from the sound speed data of the anemometer) is considered instead of temperature fluctuations. Fluxes can be obtained with the application of different corrections and quality control procedures on the raw data [4] [17] [20] [45] [46] [68] resulting in sometimes up to 20-30% modification on the raw data. Note that even in the best flux calculation schemes "there are a lot of corrections and choices between correction methods". Even in the flux calculation method itself, there is a high level of uncertainty.

3.2. *The generalized Bowen ratio*

In the following a short review of other flux calculation algorithms will be given. Introducing the generalized form of the Bowen-ratio (B) after [57] as:

$$B = \frac{F_c}{F_T} , \tag{3}$$

which is the fraction of the flux of scalar c and the reference heat flux (F_T). This ratio can be determined on many ways. It can be measured with the most precise *direct flux (eddy covariance) method* as

$$B = \frac{\overline{w'c'}}{\overline{w'T'}} , \tag{4}$$

or by *gradient measurement* [42] [69] (from data available at level z_1 and z_2)

$$B = \frac{K_c \left[c(z_2) - c(z_1) \right]}{K_H \left[T(z_2) - T(z_1) \right]} \; ,$$ (5)

and in case of multilevel data by *profile method* (profile fitting). Here, in the first approximation, the turbulent diffusion coefficients for heat and for the scalar c are assumed to be equivalent ($K_c = K_H$).

3.3. *The eddy accumulation and the band pass covariance technique*

Fluxes can be obtained even from a single level measurement data using for example the *eddy accumulation method*. This measurement technique is based on a simultaneous application of a sensor with slow respond characteristic (which is usually used for trace gas concentration measurement) and of a fast respond sensor (sonic anemometer). Concentrations (c^+, c^-) at upward (w^+) and downward (w^-) drifts, respectively, have to be separated by parallel measurements of vertical velocity and concentration. The turbulent flux F_c will be proportional to the differences of them and to the standard deviation (σ_w) of vertical velocity, which is an indicator of the intensity of turbulence [12] [56]:

$$F_c = b \sigma_w \left(c^+ - c^- \right) .$$ (6)

The typical value of b is approximately constant having a value of about 0.6. The algorithm for the choice of c^+ and c^- episodes is demonstrated in Figure 3, where horizontal bars over and below the *x*-axis are representing positive and negative vertical velocity intervals, respectively. Note that this case has been picked up from a windy ($u = 10$ m/s) and thermally unstable situation resulting in unusually high vertical motion ($w' > \pm 2$ m/s) at such a low height (4 m AGL).

Naturally, a critical vertical velocity (w_k) can be chosen for better separation of upward and downward drifts, and only actual concentrations in case of above and below respective c^+ and c^- cases are considered (for example one have to analyze air sample collected in different tanks, or we try to log the data of the slow sensor with higher resolution but lower accuracy, interlocked with sonic anemometer data). The generalized Bowen ratio can be written as:

$$B = \frac{b_c \sigma_w \left[c(w > w_k) - c(w < -w_k) \right]}{b_H \sigma_w \left[T(w > w_k) - T(w < -w_k) \right]} \; .$$ (7)

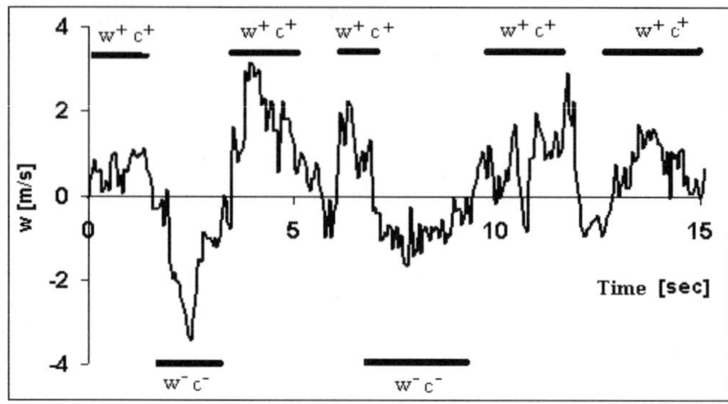

Figure 3. Schematic diagram of the eddy accumulation measurement technique. The presented vertical velocity values are from the test data series of the EU5 Greengrass program.

In order to separate ascending and descending eddies, a critical value of the temperature flux $(wT)_k$ can also be used. This is the so called *hyperbolic eddy accumulation method* [11]:

$$B = \frac{b_c \sigma_w \left[c\left(wT > (wT)_k\right) - c\left(wT < -(wT)_k\right) \right]}{b_H \sigma_w \left[T\left(wT > (wT)_k\right) - T\left(wT < -(wT)_k\right) \right]} . \tag{8}$$

As a first assumption, the constants for two different scalar quantities are assumed to be equal ($b_c = b_H$).

Some experiments have been made by us with the application of the eddy accumulation technique to the calculation of raw fluxes for sensible and latent heat and also the CO_2 fluxes. Let us consider the following example: 21 Hz data of the Gill R4 sonic anemometer and the LI-7500 H_2O/CO_2 sensors from the site "Bugac" of the EU5 Greengrass program [51] have been analyzed. The height of measurements was 4 m AGL. Raw data from the eddy covariance and the eddy accumulation methods have been compared. Simulation of a slow signal has been generated from the LI-7500 data applying 5 second averaging. Temperature flow has been obtained from the eddy accumulation method with the same averaging interval. The correlation of the results from the above mentioned two methods is presented in Figure 4. Raw fluxes have been underestimated by the eddy accumulation method, but the degree of correlation is high ($R^2 = 0.94$).

The raw flux of CO_2 has been calculated with three different methods: eddy covariance, eddy accumulation and a *modified eddy accumulation technique* that is considering the ratio of two different temperature flux calculations (Figure 5):

$$\left(\overline{w'c'_{CO_2}}\right)_{accum,\text{mod}} = \frac{\left(\overline{w'T'}\right)}{\left(\overline{w'T'}\right)_{accum}}\left(\overline{w'c'_{CO_2}}\right)_{accum}.\tag{9}$$

All plots show similar behavior, but the one calculated with the modified eddy accumulation technique fits better to direct flux measurements.

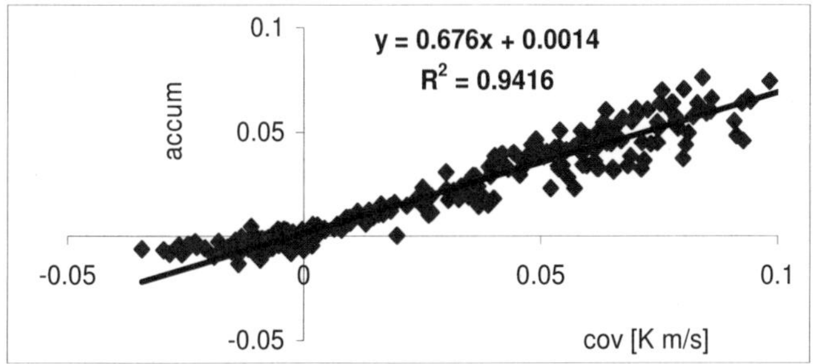

Figure 4. Raw temperature fluxes ($\overline{w'T'}$) from eddy covariance (cov) versus eddy accumulation (accum) techniques (with averaging interval is 5 s) every half hour calculated from the data of Gill R4 sonic anemometer (Bugac, August 2003).

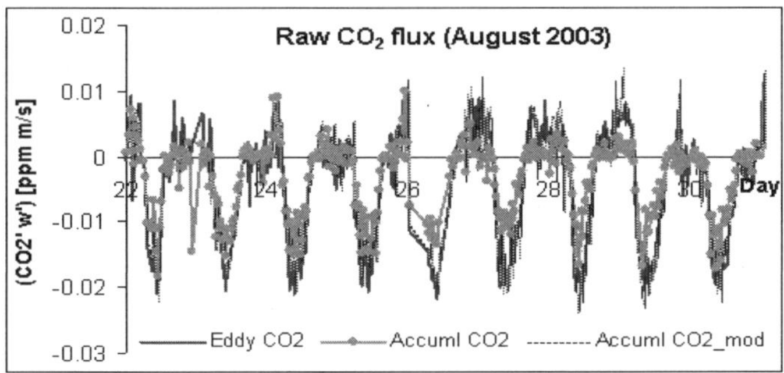

Figure 5. Raw CO_2 fluxes from eddy covariance, eddy accumulation and the modified eddy accumulation techniques, respectively (Bugac, August 2003).

The generalized Bowen ratio can be obtained from the turbulent spectra of trace gas and temperature fluxes, respectively. The assumption is that the low frequency ($f < f_{krit}$) part of each spectrum – measurable with slow sensors and long averaging interval – have similar behavior. This is the so-called *band-pass covariance technique* [10]. The ratio of the two covariances in the low frequency part of the spectra is assumed to be equal to the generalized form of Bowen ratio, i.e., the ratio of the covariances in the whole spectrum (Eq. 4)

$$B = \frac{\overline{w'c'}}{\overline{w'T'}} \cong \frac{Cov(wc)(f < f_{krit})}{Cov(wT)(f < f_{krit})} \ . \tag{10}$$

The turbulent flux of scalar quantity c can be obtained with different formulations of the generalized Bowen ratio using slow respond sensors too. Calculation is always based on the heat flux and on the generalized Bowen ratio (i.e., the ratio and not the value of the two fluxes).

3.4. *Chamber techniques*

Finally, we will briefly introduce the chamber measurement technique, which can be used for trace gas fluxes. The static and the dynamic chambers can be described in the following way. In the first one there is no flow, while in the dynamic chamber there is a continuous air stream. The basic equation of the *static chamber measurement* is:

$$V\frac{dc}{dt} = S_1 - S_2, \tag{11}$$

where V is the volume, S_1 is the surface source intensity, which is the product of the flux (F_c) and the area of the chamber (A), and it is negative in case of deposition. S_2 is representing the effect of chemical processes on concentration. Important issues are (i) the size of the chamber, i.e., the volume of the layer including the upper porous soil layer and (ii) the dynamics of concentration change, which determines the optimal sampling strategy. A common assumption is that concentration change is linear in the first stage. The duration of this for N_2O or CH_4 is about 10 to 30 minutes – depending on the size of the chamber [31].

For a *dynamic chamber*, the basic equation of concentration change is [58]:

$$V\frac{dc}{dt} = S_1 - S_2 + v_{air} \cdot A_{tube} \cdot (c_{out} - c), \tag{12}$$

where v_{air} is the velocity of in- and outflowing air, A_{tube} is the cross section of the inlet tube and c_{out} is the concentration outside the chamber. Once the equilibrium concentration (c_{int}) has been established, the flux can be calculated as follows:

$$F_c = \frac{v_{air} \cdot A_{tube} \cdot (c_{out} - c_{int}) - S_2}{A} .$$ (13)

A hidden assumption is that the air stream through the chamber sweeps over the entire covered soil surface with a uniform velocity, and its direction is parallel to the enclosed soil surface.

4. Case studies

Following the presentation of micrometeorological measurements in Hungary and the methodology of turbulent flux calculation, we will show results of some research activity. A few results of gradient measurements of trace gases in the 1990's are presented first, followed by contributions in the research of ammonia exchange and the description of the concept and measurements of the new basic climatological network in Hungary.

4.2. *Turbulent exchange of ozone, sulfur-dioxide and nitrogen oxides over tall vegetation*

One of the main goals of the studies in the EUREKA BIATEX-1 and BIATEX-2 programs in Hungary was the determination of the dry deposition velocities (v_d) of tracers (O_3, NO_x, SO_2) over various surfaces, in seasonal separation during the 1990's years. Dry deposition can be obtained from the deposition velocity by the following formula [26] [49]:

$$F_c = -\frac{c_{ref} - c_0}{R_a + R_b + R_c} = -(c_{ref} - c_0) \cdot v_d ,$$ (14)

where c_{ref} is the concentration of the given trace gas over vegetation, c_0 is the concentration at the surface, which, for example in case of vapor, is equal to the saturation vapor pressure over soil or leaf surface, and in case of ozone (which is absorbed by vegetation) is equal to zero. R_a, R_b and R_c are the aerodynamic, quasi-laminar and canopy-surface resistances, respectively. While the determination of R_a and R_b are relatively simple, the procedure to determine R_c is a major challenge, since it significantly depends on season, vegetation characteristics, etc. The canopy-surface resistance can be separated into three further terms:

196 *T. Weidinger et al.*

$$\frac{1}{R_c} = \frac{1}{R_{st}} + \frac{1}{R_s} + \frac{1}{R_w} , \tag{15}$$

where the respective resistances are of the stoma (R_{st}), the soil (R_s) and the cuticula (leaf surface) (R_w). These resistances are acting in parallel. Analyses for Hungary included two different types of surface vegetation: (1) spruce forest in the mountain "Mátra" and (2) grassland in the "Hortobágy" National Park. An example of the measured data is depicted in Figure 6.

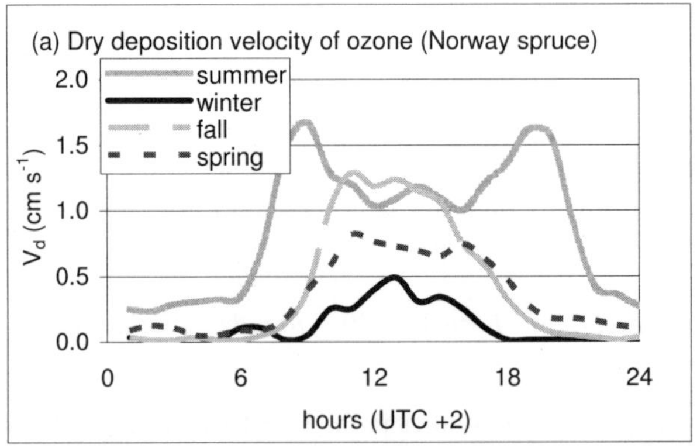

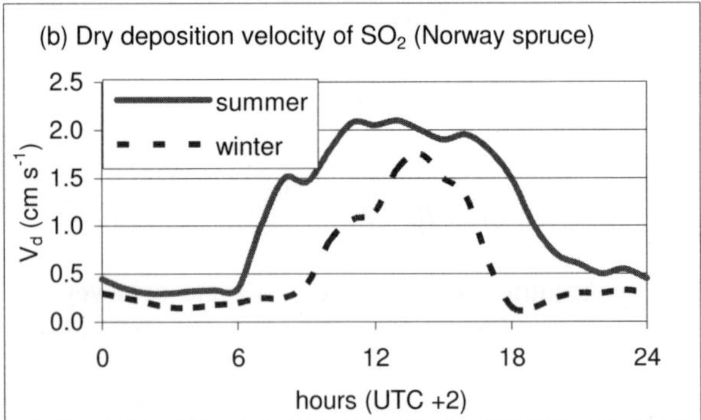

Figure 6. Daily variation of the dry deposition velocity of ozone (a) and sulfur-dioxide (b) over spruce canopy.

The uptake by stoma plays an important role in the deposition of ozone. During a summer day, by intensive insolation, the stoma is closed to protect the plant from dehydration. Therefore, R_{st} increases, which results in lower dry deposition velocity. In case of SO_2 such an effect cannot be observed, which indicates the fact that SO_2 which is solvable in water, is preferably absorbed on the wet cuticula instead of the stoma. The interaction between sulfur-dioxide and ammonia is also important [30]. This has been investigated under the NitroEurope program. Continuous ammonia flux measurements started in the summer of 2008. In the next section, our studies of ammonia exchange processes will be demonstrated on a case study.

4.2. *Ammonia exchange over short vegetation*

NH_3 exchange processes between the atmosphere and biosphere over grassland have been investigated during the EU4 Graminae project (1999-2002) on a European cross section [60]. The Hungarian station was located near Püspökladány (for the location of the site see Figure 1 and Table 1), operated by the Forest Research Institute. Ammonia emission occurred only in the daytime during the vegetative period, when leaf stoma is open, in all other cases deposition was observed (Figure 7a). The pH factor of the soil was around 6, therefore soil emission was impossible. The effect of fertilizers was also investigated. After fertilization (100 kgN/ha) high release was detected, but only during the daytime, which supports the theory of stoma emission (Figure 7b). Ammonia is emitted through the stoma after fertilization of the acid soil [29]. Modeled fluxes were in a good agreement with measurements.

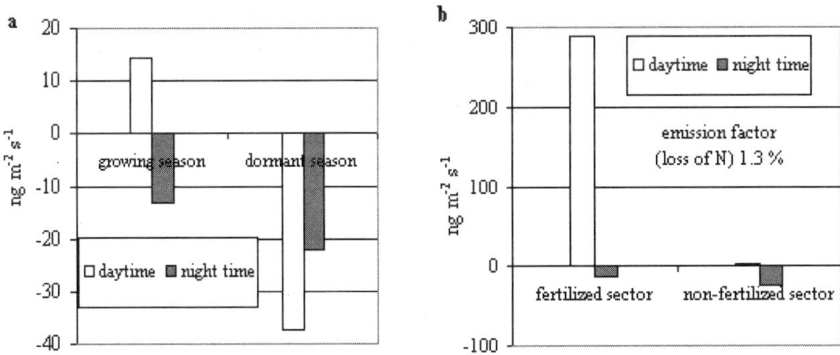

Figure 7. Ammonia flux over grassland. Averaging done for: two years (a) and two weeks in May 2001 after fertilization (b) (Püspökladány, GRAMINAE program).

4.3. *Design of the basic climatological network for the detection of long-term effects of climate change*

In addition to the high precision monitoring of greenhouse gases [23], there was a growing demand for a target-oriented climate observation system in Hungary.

The objective of the Hungarian Meteorological Service (HMS) is the establishment of the basic climatological network integrated into the automatic surface observatory system. These are such climate stations where measurement conditions, spatial representativity, the applied measurement methods and devices as well as quality control and maintenance procedures permit high accuracy, reliability and temporal stability of measurements, which could not be reached earlier. This makes the system suitable for the long-term and operational survey of the local effects of climate change in Hungary, similarly to such systems of other countries [36] [37]. Four stations are being installed at locations with different surface and climatic characteristics taking into account the existing observatory network. Stations Budapest and Kékestető have an extended measurement program for the radiation budget components (Figure 8).

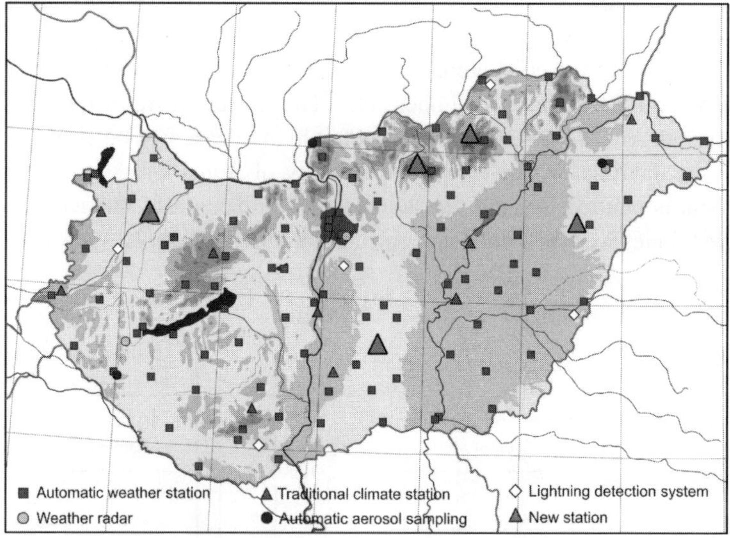

Figure 8. The location of existing weather stations of the HMS, planned climate stations and the extended radiation measurements at Kékestetö.

The center of the system is the Agrometeorological Observatory of the University of Debrecen as depicted in Figure 9. At this station measurements have been performed since April, 2008 according to the following points:

- operation as a standard climate station with high precision measurement of the basic climate parameters (air temperature, moisture, wind speed and direction, pressure, global radiation, soil temperature, precipitation, etc.) based on methodological research [54],

- measurement of temperature, wind and humidity at 1, 2, 4 and 10 m height AGL, respectively,
- assessment of energy budget components with eddy covariance and/or Bowen ratio methods [45] [43],
- CO_2 concentration and flux measurement,
- measurements of short and long wave radiation budget components,
- assessment of ground heat flux in the upper layer of soil using ground heat flux, soil moisture, soil temperature measurements.

Figure 9. Central station of the targeted climatological network at the Agrometeorological Station at Debrecen-Kismacs; 4 m mast for eddy covariance and 10 m mast for gradient measurement, automatic weight measurement ombrometer in the background (upper panels) and the equipment for the measurement of radiation budget components and soil heat flow (temperature, humidity, heat flux measuring plate) (lower panels).

The complex measurement program of the central station is based on the sonic anemometer usable for fast sampling rate for wind speed and temperature (for momentum and sensible heat flux), and on the LI-7500 device usable for the assessment of air humidity and CO_2 concentration. At the three other stations that have to be installed, similar measurement program excluding the above two instruments will be implemented. The extended measurement gives opportunity to the estimation of energy budget components with eddy covariance, gradient, profile and Bowen-ratio methods and also provides measurement background for different types (e.g., soil science, air quality) of studies and research. The assessment of the receipt term in the radiation budget is emphasized. This determines the total energy balance of the surface. This is ensured by separate measurements of each components of the budget: global, reflex, long wave radiation of the atmosphere and the long wave emission of the surface. For the assessment of the radiation budget components the best currently available devices are applied. Recalibration to the International Radiation Scale is performed regularly.

Instrumentation of the stations for gradient measurements is based on sensors used in the network of the HMS. The accuracy, calibration and regular inspection of the instruments suits into the quality control system of the HMS. Pre-installation calibration of the equipment is already done. For air temperature and air humidity equipments, full on-site calibration is possible. Based on the analyses of recent measurements (see Figure 10 and 11) the findings are the following:

1. Wind, temperature and humidity profiles are sufficiently precise, and reflect the characteristics of the surface layer, so they can be used for the estimation of atmospheric stability and with the Monin-Obukhov similarity theory for the estimation of surface fluxes.

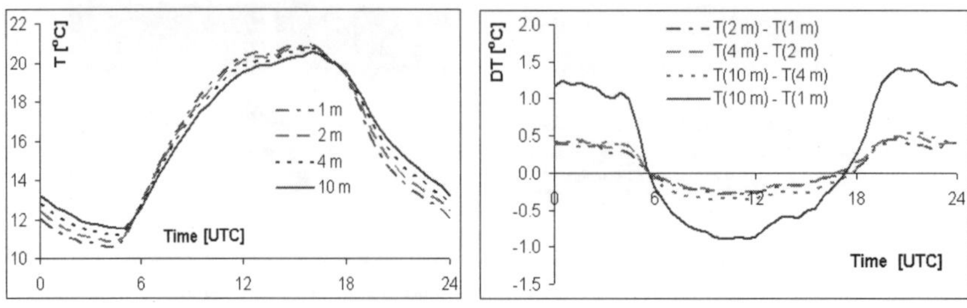

Figure 10. Daily variation of air temperature (left) and temperature differences in each layer (right) (May, 2008)

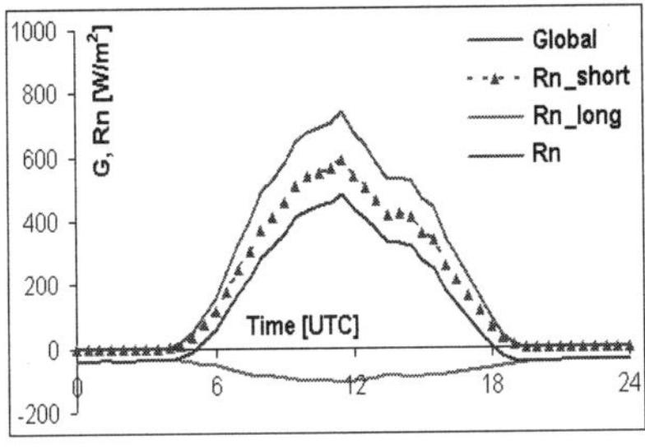
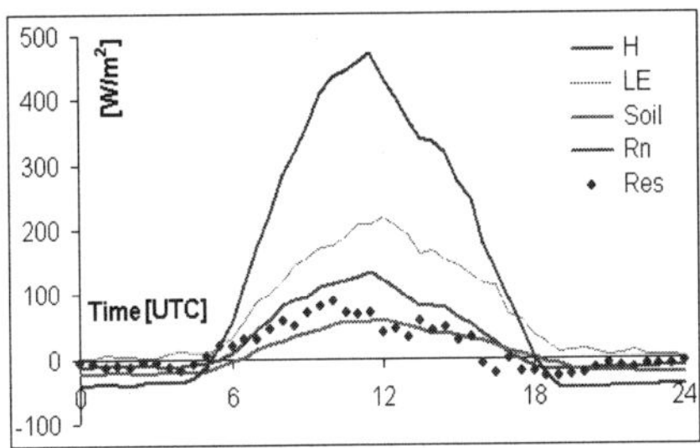

Figure 11. Daily variation of the energy budget components (May, 2008). *Rn*: net radiation; *Soil*: ground heat flux; *H*: sensible and *LE*: latent heat flux; *Res*: residual term.

2. Analyses of the energy budget components, and the closure of the energy balance indicates that the created database is suitable for the investigation of the surface layer and for climatological analyses and, in addition, it is also suitable for the development of flux assessment methodology.

3. The database provided by the installed monitoring system is applicable for further research activity (e.g., micrometeorological and air quality protection applications) in addition to the detection of possible local effects of the global climate change [70].

5. Conclusion

A brief description of micrometeorological studies in Hungary has been given. Measurement projects, devoted to surface-biosphere-atmosphere exchange processes of trace gases, surface energy budget components, carbon and nitrogen budget assessment, have been reviewed. Due to their high operational and equipment costs and the continuous need for data analysis and algorithm development, such research can only be efficient under national and international cooperation.

The methodology of the turbulent flux assessment has been analyzed in detail. Our results have been illustrated through three examples.

The installation and continuous operation of the basic climatological network for the detection of long-range effects of climate change in Hungary is emphasized. The system provides profile and energy budget data besides standard meteorological parameters.

6. Acknowledgement

This study has been prepared under the framework of the NKFP 6-028-2005 project and the EU6 IP NitroEurope program. The authors thank the staff of the

                              T. Weidinger et al.

Scientific Institute of Forestry, Szent István University, Debrecen University, Department of Meteorology Eötvös Loránd University and Hungarian Meteorological Service for the participation in the research projects. Special thanks to Dr. Agnes Havasi and Attila Machon from the Department of Meteorology for they help in the preparation of the manuscript.

List of symbols

List of Symbols		
Symbol	**Definition**	**Dimensions or Units**
A	area of measurement chamber	$[\text{m}^2]$
A_{tube}	cross section of the inlet tube of dynamic chamber	$[\text{m}^2]$
B	generalized form of the Bowen ratio	$[\text{kg m}^{-3}\,\text{K}^{-1}]$
F_c	turbulent flux of a scalar quantity	$[\text{kg m}^{-2}\,\text{s}^{-1}]$
F_T	reference temperature flux	$[\text{K m s}^{-1}]$
H	sensible heat flux	$[\text{W m}^{-2}]$
K_c	turbulent diffusion coefficients for scalar quantity	$[\text{m}^2\,\text{s}^{-1}]$
K_H	turbulent diffusion coefficients for sensible heat	$[\text{m}^2\,\text{s}^{-1}]$
L	latent heat of vaporization	$[\text{J kg}^{-1}]$
LE	latent heat flux	$[\text{W m}^{-2}]$
R_a	aerodynamic resistance	$[\text{s m}^{-1}]$
R_b	quasi-laminar resistance	$[\text{s m}^{-1}]$
R_c	canopy-surface resistance	$[\text{s m}^{-1}]$
R_s	soil resistance	$[\text{s m}^{-1}]$
R_{st}	stoma resistance	$[\text{s m}^{-1}]$
R_w	cuticula resistance	$[\text{s m}^{-1}]$
S_1	surface source intensity of the area in the measurement chamber	$[\text{kg s}^{-1}]$

S_2	source of scalar quantity due to chemical reactions in the measurement chamber	$[\text{kg s}^{-1}]$
T	temperature of air	$[\text{K}]$
V	volume of flux measurement chamber	$[\text{m}^3]$
b	fitting parameter used in the eddy accumulation technique	
c	concentration of a scalar quantity	$[\text{kg m}^{-3}]$
c_0	concentration of trace gas at the surface	$[\text{kg m}^{-3}]$
c_{out}	concentration of scalar quantity outside the chamber	$[\text{kg m}^{-3}]$
c_{pm}	isobaric heat capacity of moist air	$[\text{J K}^{-1}\,\text{kg}^{-1}]$
c_{ref}	concentration of trace gas over vegetation	$[\text{kg m}^{-3}]$
f	frequency of turbulent spectra	$[\text{s}^{-1}]$
u	horizontal wind speed	$[\text{m s}^{-1}]$
v_{air}	velocity of air in the inlet tube of dynamic measurement chamber	$[\text{m s}^{-1}]$
v_d	dry deposition velocity	$[\text{m s}^{-1}]$
w	vertical velocity	$[\text{m s}^{-1}]$
ρ_m	average moist air density	$[\text{kg m}^{-3}]$
ρ_v	vapor density	$[\text{kg m}^{-3}]$
σ_w	standard deviation of vertical velocity of air in turbulent flow	$[\text{m s}^{-1}]$
τ	momentum flux	$[\text{kg m}^{-1}\,\text{s}^{-2}]$

References

1. A. Anda, *Időjárás* 106, 137 (2002).
2. F. Ács, G. Szász and M. Drucza, *Időjárás* **109**, 71 (2005).
3. J. Asanuma, H. Ishikawa, I. Tamagawa, Y. Ma, T. Hayashi, Y. Qi and J. Wang, *Water Resour, Res.* **41**, W09407 (2005).
4. M. Aubinet, A. Grelle, A. Ibrom, Ü. Rannik, J. Moncrieff, T. Foken, A.S. Kowalski, P.H. Martin, P. Berbigier, Ch. Bernhofer, R. Clement, J. Elbers, A. Granier, T. Grünwald, K. Morgenstern, K. Pilegaard,

C. Rebmann, W. Snijders, R. Valentini and T. Vesala, *Adv. Ecol. Res.* **30**, 113 (2000).

5. N. Bacsó and B. Zólyomi, *Időjárás* **38** (in Hung) (1934).

6. D.D. Baldocchi, E. Falge, L. Gu, R. Olson, D. Hollinger, S. Running, P. Anthoni, Ch. Bernhofer, K. Davis, J. Fuentes, A. Goldstein, G. Katul, B. Law, X. Lee, Y. Malhi, T. Meyers, W. Munger,W. Oechel, U.K.T. Paw, K. Pilegaard, H.P. Schmid, R. Valentini, S. Verma, T. Vesala, K. Wilson, K. ans S. Wofsyn, *B. Am. Meteorol. Soc.* **82**, 2415 (2001).

7. Z. Barcza, *Long term atmosphere/biosphere exchange of CO_2 in Hungary.* Ph.D. Dissertation, Budapest, Hungary, (2001) http://nimbus.elte.hu/~bzoli/thesis/.

8. Z. Barcza, L. Haszpra., H. Kondo, N. Saigusa, S. Yamamoto and J. Bartholy, *Tellus B* **55**, 187 (2003).

9. B. Béll and L. Takács, (Eds.), *Climate of Lake Balaton* (Hungarian Meteorological Service (in Hung.), (1974).

10. F. Beyrich F., Lindenberg Status Report: A CEOP reference site contribution from BALTEX, *1st Pan-GEWEX Meeting*, Frascati, oct 9/12. 2006.

11. D.R. Bowling, A.C., Delany, A.A. Turnipseed, D.D. Baldocchi and R.K. Monson, *J. Geophys. Res.* **104**, D89121 (1999).

12. J.A. Businger and S.P. Oncley, *J. Atmos. Oceanic Technol.* **7**, 349 (1990).

13. CarboEurope (Integrated Project CarboEurope-IP Assessment of the European Terrestrial Carbon Balance), http://www.carboeurope.org/.

14. Carbomont (Effects of land-use changes on sources, sinks and fluxes of carbon in European mountain areas), http://www.uibk.ac.at/carbomont/.

15. CHIOTTO (Continuous high-precision tall tower observations of greenhouse gases), http://www.chiotto.org/index.html/.

16. Z. Dunkel, Monitoring of Vegetation Period Course of Soil Moisture Based on Measurements and Evaluation in Hungary, *Conference on Climate and Water*, Helsinki, Finland, 262 (1989).

17. L. Erdős, *Időjárás* (in Hung.) **79**, 274 (1975).

18. EUREKA BIATEX-2 (BIosphere ATmosphere EXchange of Trace Gases and Aerosols), http://www.atmosci.ceh.ac.uk/projects/biatex2.htm/.

19. C. Feigenwinter, C. Bernhofer and R. Vogt, *Bound.-Layer Meteorol.* **113**, 201 (2004).

20. Th. Foken and B. Wichura, *Agric. For. Meteorol.* **78**, 83 (1996).

21. GREENGRASS (Sources and Sinks of Greenhouse Gases from managed European Grasslands and Mitigation Strategies), http://www.clermont.inra.fr/greengrass/.

22. L. Haszpra, Z. Barcza, P.S. Bakwin, B.W. Berger, K.J. Davies and T. Weidinger, *J. Gephys. Res.* **106D**, 3057 (2001).

23. L. Haszpra, Z. Barcza, D. Hidy, I. Szilágyi, E. Dlugokencky and P. Tans, Atmos. Environ. **42**, 8707 (2008).

24. L. Horváth and M.A. Sutton, *Atmos. Environ.* **32**, 339 (1998).

25. L. Horváth, L. Bozó, L. Haszpra, J. Kopacz, Á. Molnár, T. Práger and T. Weidinger, *Gradient measurement of Air-Soil Exchange of gases.* In: Precipitation Scavenging and Atmosphere-Surface Exchange **3**, 637 (*Hemisphere Publishing Corporation, Washington*, 1993).

26. L. Horváth, T. Weidinger, Z. Nagy and E. Führer, E., *Measurement of dry deposition velocity of ozone, sulfur dioxide and nitrogen oxides above pine forest and low vegetation in different seasons by the gradient method.* In Biosphere-Atmosphere Exchange of Pollutants and Trace Substances **4** (*Springer, Heidelberg.*, 1996).

27. L. Horváth, S. Christensen, E. Führer, S. Kelecsényi, R. Mészáros, Z. Nagy and T. Weidinger, *Időjárás* **103**, 1 (1999).

28. L. Horváth, J. Pinto and T. Weidinger, *Időjárás* **107**, 249 (2003).

29. L. Horváth, M. Asztalos, E. Führer, R. Mészáros and T. Weidinger, *Agric. Forest Meteorol.* **130**, 282 (2005).

30. L. Horvath, H. Fagerli and M.A. Sutton, *Long-Term Record (1981-2005) of Ammonia and Ammonium Concentrations at K-Puszta Hungary and the Effect of Sulphur Dioxide Emission Change on Measured and Modeled Concentrations*, Sutton M.A., Reis S. and Baker S. Eds., Springer, 181 (2008).

31. L. Horváth, E. Führer and K. Lajtha, *Atmos. Environ.* **40**, 7786 (2006).

32. H. Huszár, A. Pogány, Z. Bozóki, Á. Mohácsi, L. Horváth and G. Szabó, *Sensors and Actuators B: Chemical* **134**, 1027 (2008).

33. J.S. Jánosy, S., Deme and E. Láng, Modeling of Potential Air Dispersants from Normal and Incidental Nuclear Power Plant Operations, *2003 Business and Industry Symposium*, Florida, USA, 51 (2003).

34. J. Justyák and K. Tar, *Időjárás* **106**, 248 (2002).

35. J. Justyák, *Időjárás* **107**, 205 (2003).

36. N. Kalthoff, F. Fiedler, M. Kohler, O. Kolle, H. Mayer and A. Wenzel, *Theor. Appl. Climatol.* **62**, 65 (1999).

37. E. Kevin, T.E. Trenberth, R. Karl. and W. Spence, *B. Am. Meteorol. Soc.* **83**, 1593 (2002).

38. T. Kocsis and J. Bem, History of the meteorological measurements at Keszthely, one of the eldest stations in Hungary, *7th Annual Meeting of the European Meteorological Society (EMS)*, El Escorial, Spain, (2005).

39. Z. Kováts, Z. and E. Kozmáné Tóth (eds.), *Natural capabilities of Lake Fertő* (Hungarian Meteorological Service (in Hung.), 1982).

40. E. Kozmáné Tóth, A. Nagyné Dávid A. and Gy. Szabó, *Vízügyi Közlemények* (in Hung.) **64**, 438 (1982).

41. G. Kramm, R. Dlugi and D.H. Lenschow, J. Hydrol. **166**, 283 (1995).

42. G. Kramm, *Bound.-Layer Meteorol.* **48**, 315 (1989).

43. H. Liu and T. Foken, *Z. Meteorol.* **10**, 71 (2001).

206   *T. Weidinger et al.*

44. A. Machon, B. Grosz, L. Horváth, K. Pintér and Z. Tuba, *Cereal Res. Commun.* **36**, 203 (2008).
45. M. Maunder, and Th. Foken, *Documentation and Instruction Manual of the Eddy Covariance Software Package TK2* (University Bayreuth, Arbeitsergebnisse, Nr. 26. 2004).
46. M. Maunder, S.P. Oncley, R. Vogt, T. Weidinger, L. Ribeiro, C. Bernhofer, Th. Foken, W. Koshiek and H. Liu, *Bound.-Layer Meteorol.* **123**, 29 (2007).
47. E. Mészáros and L. Horváth, *Atmos. Environ.* 18, 1725 (1984).
48. E. Mészáros, I. Mersich and T. Szentimrey, *Atmos. Environ.* 21, 2505 (1987).
49. D.T. Mihailovic, *Global Planet Change* **13**, 207 (1996).
50. Á. Molnár, E. Mészáros, K. Polyák, I. Borbély-Kiss, E. Koltay, Gy. Szabó and Zs. Horváth, *Atmos. Environ.* 29 1821 (1995).
51. Z. Nagy, K. Pintér, Sz. Czóbel, J. Balogh, L. Horváth, Sz. Fóti, Z. Barcza, T. Weidinger, Zs. Csintalan, N.Q. Dinh, B. Grosz and Z. Tuba, *Agric. Ecosyst. Environ.* **121**, 21 (2007).
52. Z. Nagy, G. Szász, T. Weidinger, S. Szalai, Z. Tóth, E. Nagyné Kovács, B. Debreceni, I. Matyasovszky and A.Z. Gyöngyösi, *Geophys. Res. Abstr.* **10**, EGU2008-A-09929 (2008).
53. Z. Nagy, T. Weidinger, Gy. Baranka, Z. Tóth, E. Nagyné Kovács, R. Mészáros, A.Z. Gyöngyösi and O. Török, Baseline climate network in Hungary and application for air quality dispersion models, *EMS 2008*, Amsterdam, The Netherlands, EMS2008-A-00391, (2008).
54. Z. Nagy, Effect of thermometer screens on accuracy of temperature measurements, *WMO Technical Conference on Meteorological and Environmental Instruments and Methods of Observation*, Switzerland (2006). http://www.wmo.int/web/www/IMOP/publications/IOM-94-O2006/P3(14)_Nagy_Hungary.pdf.
55. NitroEurope (The nitrogen cycle and its influence on the European greenhouse gas balance), http://www.nitroeurope.eu/.
56. S.P. Oncley, A.C. Delany, T.W. Horst and P.P. Tans, *Atmos. Environ.* **27**, 1686 (1993).
57. S.P. Oncley, New tools to Study to Motion of Trace Gases in the Atmosphere, *Frontiers in Assessment Methods for the Environment (FAME). A symposium on recent advances in environmental measurement and assessment technologies, modeling, and data processing tools to promote a cleaner environment*, Minneapolis, USA (2003).
58. R. Reichman and D.E. Rolston, *J. Environ. Qual.* **31**, 1774 (2002).
59. L.E. Rustad, *Sci. Tot. Environ.* **404**, 222 (2008).
60. M.A. Sutton, C. Milford, E. Nemitz, M.R. Theobald, P.W. Hill, D. Fowler, J.K. Schjoerring, M.E. Mattsson, K.H. Nielsen, S. Husted, J.W. Erisman,

R. Otjes, A. Hensen, J. Mosquera, P. Cellier, B. Loubert, M. David, S. Genetmont, A. Neftel, A. Blatter, B. Herrmann, S.K. Jones, L. Horvath, E.C. Führer, K. Mantzanas, Z. Koukoura, M. Gallagher, P. Williams, M. Flynn and M. Riedo, *Plant and soil* **228**, 131 (2001).

61. G. Szász, *Időjárás* **106**, 161 (2002).

62. J. Száva-Kováts and D. Berényi, *Micro- and Agroclimatology* (Hungarian Meteoological Service (In Hung.) 1948).

63. K. Tar, Renew. Suit. Energ. Rev. **12**, 1712 (2008).

64. K.E. Trenberth T.R. Karl and T.W. Spence, T.W., *B. Am. Meteorol. Soc.* **83**, 1593 (2002).

65. E. Webb, G. Pearman and R. Leuning, *Q. J. R. Meteorol. Soc.* **106**, 85 (1980).

66. T. Weidinger, *Időjárás* (in Hung.) **90**, 266 (1986).

67. T. Weidinger, Sz. Simon, J. Mádlné Szőnyi, J. and Á. Bordás, *Uncertainties in the estimation of a shallow lake water budget.* Environmental, Health and Humanity Issues in Down Danubian Region: Multidisciplinary Approach. Edited By D.T Mihailovic and M. Miloradov (World Scientific, New York, London, Singapore 2008).

68. T. Weidinger, F. Ács, R. Mészáros and Z. Barcza, 1999, *Időjárás* **103**, 145 (1999).

69. T. Weidinger, J. Pinto and L. Horváth, *Z. Meteorol.* **9**, 139 (2000).

70. T. Weidinger, Z. Nagy Gy., Baranka, R. Mészáros and A.Z. Gyöngyösi, Determination of meteorological preprocessor for air quality models in the New Hungarian Standards. *Croat. Meteorol. J.* (Special issue the 12th International Conference on Harmonization within Atmospheric Dispersion Modelling for Regulatory Purposes, HARMO 12.) 460 (2008).

71. G. Wohlfahrt, M. Anderson-Dunn, M. Bahn, M. Balzarolo, F. Berninger, C. Campbell, A. Carrara, A. Cescatti, T. Christensen, S. Dore, W. Eugster, T. Friborg, M. Furger, D. Gianelle, C. Gimeno, K. Hargreaves, P. Hari, A. Haslwanter, T. Johansson, B. Marcolla, C. Milford, Z. Nagy, E. Nemitz, N. Rogiers, M.J. Sanz, R.T.W. Siegwolf, S. Susiluoto, M. Sutton, Z. Tuba, F. Ugolini, R. Valentini, R. Zorer, *Ecosystems* **11**, 1338 (2008).

72. World Data Centre for Greenhouse Gases (WDCGG) web site http://gaw.kishou.go.jp/wdcgg/.

Table 1. Main micrometeorological measurement projects in the umbrella of national and international collaborations for analyzes of surface-biosphere-atmosphere interactions from 1990.

Measurement site, h, φ, λ	Time period	Research program	Vegetation	Profile, gradient (chamber)	Eddy covariance	Contact person(s), references
Hortobágy, 95 m 47°25', 20°58'	1990–1995	BIATEX-1 MAKA[1]	Natural grassland	4 m tower u, f, T, O_3, NO_x, SO_2	-	Horvath, L. (HMS[4]) [25] [26]
Nyírjes, 560 m 47°50', 19°56'	1991–2000	BIATEX-1, 2 [18] MAKA[1]	Spruce forest	28 m tower, u, f, T, O_3, NO_x, SO_2	τ, H, O_3 (temporary)	Horvath, L. (HMS) [26] [27] [28]
Püspökladány, 88 m 47°20', 21°06'	2000–2001	EU4-GRAMINAE[2]	Semi-natural grassland	u, t, NH_3	τ, H	Horvath, L. (HMS) [29]
Bugac, 113 m 46°42', 19°36'	2002–	EU5-GREENGRASS [21] EU6-CarboEurope [13] EU6-NitroEurope [55]	Semi-natural grassland	O_3, NOx, NH_3 (chamber: CO_2, NO, N_2O, CH_4)	τ, H, LE, CO_2	Horvath, L. (HMS) Tuba Z. (SZIU[5]) [44] [51]
Szurdokpüspöki, 300 m 47°51', 19°43'	2003–	EU5-Carbomont [14]	Grassland	- (chamber: CO_2, N_2O, CH_4)	τ, H, LE, CO_2	Tuba Z. (SZIU) Nagy Z. (SZIU) [71]
Hegyhátsál, 249 m 46°57', 16°39'	1995–	MAKA[1] EU5-CHIOTTO [15] EU6-CarboEurope	mixed agricultural landscape	113 m TV tower u, f, T, CO_2	τ, H, LE, CO_2	Haszpra L. (HMS) Barcza Z. (ELTE) [8] [22] [23]
Kelemenszék, 97 m 46°47', 19°27'	2005–2007	HU-NKFP[3]	Lake water surface	u, f, T	-	Weidinger T (ELTE[6]) [67]
Debrecen, 121 m 47°33', 21°36'	2007–	HU-NKFP[3]	Cultivated grassland	u, f, T	τ, H, LE, CO_2	Nagy Z. (HMS) Szász G. (DE[7]) [52] [53]

[1]US Hungarian Scientific found, [2]GRassland AMmonia INteractions Across Europe (8 institutes from UK, DK, NL, FR, CH, HU and GR), [3]National Research and Development Programmes (NKFP) in Hungary, [4]Hungarian Meteorological Service, [5]Szent István University, Gödöllő, Hungary, [6]Eötvös Loránd University, Budapest, Hungary, [7]University of Debrecen, Hungary.

Chapter 10

INTEGRATION OF SPATIO-TEMPORAL DATA FOR FLUID MODELING IN THE GIS ENVIRONMENT

LUBOS MATEJICEK

Faculty of Science, Charles University in Prague, Czech Republic

Fluid modeling covers a wide range of principles describing the motion of matter and energy in dependence on spatial scales, time scales and other attributes. In order to provide efficient numeric calculations, the information systems have to be developed for management, pre-processing, post-processing and visualization. In spite of that many software tools contain sophisticated subsystems for data management and implement advanced numerical algorithms, there is still need to standardize data inputs/outputs, wide used data analyses, and case oriented computational tools under one roof. Thus, the geographic information system (GIS) is used to satisfy all the requirements. As an example, the case study focused on dust dispersion above the surface coal mine documents the GIS ability to solve all the tasks. The input data are represented by terrain measurements of meteorological conditions and by estimates of the emission rates of potential surface dust sources. Remote sensing helps to identify and classify the coal mine surface in order to map erosion sites and other surface objects. GPS is used to improve the accuracy of the erosion site boundaries and to locate other point emission sources such as excavators, storage sites, and line emission sources such as conveyors and roads. The 3D mine surface for modeling of wind flows and dust dispersion is based on GPS measurements and laser scanning. All data inputs are integrated together with simulation outputs in the spatial database in the framework of the GIS project. In case of dispersion modeling, a few ways can be used to provide numeric calculations together with GIS analyses. The traditionally used way represents using of standalone simulation tools and the input/output data linkage through shared data files. The more advanced way is the implementation of dispersion models in the GIS environment. The methods are demonstrated by using U.S. EPA modeling tools and by linking standalone numerical calculations in the GIS environment with using case oriented programming libraries and GIS development tools.

Keywords: spatio-temporal modeling; geodatabase; GIS; lidar

1. Introduction

Environmental models focused on fluid mechanics require spatio-temporal data for calibration, verifications, setting of boundary conditions, and visualization. The lack of suitable environmental data and modeling development tools have been the serious impediments to the development and use of environmental models. Nowadays, various software tools have been developed to support environmental models based on dynamic modeling or spatial modeling, but they rely on many different data formats and incompatible functionality. Dynamic modeling, based on the numerical solution of algebraic or differential equations [1], is represented by mathematical models implemented in a number of simulation packages or standalone applications [2]. Spatial modeling, attempting to emulate geographic processes, is often provided by Geographic Information Systems (GISs) [3]. Modeling, in the context of GIS, is mostly focused on using formal models that describe how things are located and related in space [4]. In order to integrate dynamic modeling and spatial modeling in the framework of a common platform, GIS is becoming one of the most important software tools [5]. In the past, it has been necessary to couple GIS with special software designed for high performance simulation in dynamic modeling. But with the increasing power of computer hardware and software, GIS can support time consuming numerical calculations. Thus, it is now possible to reconsider this relationship. Spatio-temporal modeling in GIS raises a number of important issues, including fluid modeling and the roles of scale and accuracy, which can influence the design of infrastructure to facilitate sharing of spatio-temporal data and models. Historically, there are two main ways for data analysis and simulation in the framework of environmental modeling.

From the spatial modeling point of view, existing GISs can be extended by data structures and numerical methods that can assist in solving dynamic models. The strategies can range from the simple pre- and postprocessor linkage through shared data files to building dynamic models as fully functional extensions by using programming development tools in the GIS environment. In the case of dynamic modeling, there are case oriented extensions for the display of simulation outputs. However, the data structures are dedicated for the calculation of numerical models [6]. In spite of the fact that the graphic user interface allows the models to be created as a work flow by stringing individual processes together, and running each model via its dialog box or the command line [7], the interconnection with the original spatial environment is quite limited.

As noted earlier, GIS was designed to mainly support spatial data management, spatial analysis and visualization in the two-dimension (2D) space or, lately, in the three-dimension (3D) space. But, the power of GIS for these applications has been greatly enhanced by the availability of graphic interfaces that deal with user interaction. For example, ESRI ModelBuilder or user tools implemented in the ERDAS IMAGINE software allow the user to interact with various stages of the modeling process through interactive geoprocessing tools. Thus, users can create a model of their geoprocessing work flow by stringing processes together and run the model with a single click. Then by setting model parameters, the user of the model can supply values for these parameters when the model is run via its dialog box or the command line [8].

2. Materials and Methods

2.1. *Spatio-temporal data*

Data managed by GIS store information about the location, shape and attributes of real-world entities. But, a number of various data structures are used to provide an efficient storage, to enable geostatistical analysis, and to support environmental modeling. In general, raster data structures and vector data structures are used to represent spatio-temporal phenomena.

In the raster structure, the data are arranged in a grid of cells with spectral or attribute data. They can represent imaged or continuous data. Each cell in a raster is a measured quantity. The data source for a raster dataset is mostly the satellite image or the aerial photograph. A raster dataset can also be a result of a geostastistical interpolation such as dispersion of selected pollutants over surface. Raster datasets excel in working with continuous data.

The vector structures are more complex. They are usually formed by a collection of discrete features, such as points, lines, and polygons, which is best applied to discrete objects with defined shapes and boundaries. The features have a precise shape and position based on the coordinate precision, whereas the raster have not, because it depends on the cell size. Vector datasets are preferred for bounding homogenous patches or fields, such as contour lines, street networks or land cover boundaries. Vector datasets can better support geometric operations such as calculating length and area, identification of overlaps and intersections or finding other features that are adjacent or nearby.

In addition to the raster datasets and vector datasets, there are other data structures used in the framework of GIS or CAD (computer-aided design). In order to create the precise terrain surface, the triangulated irregular network (TIN) is created from remote sensing data, GPS measurements or laser scanning. TIN datasets contain sets of triangulated irregularly located points with elevations sampled from a surface. In spite of that TINs are most often used to model the earth's surface, they can also be used to study the distribution of a continuous environmental factor such as pollutant concentrations.

The described datasets are included into geographic data models that are an abstraction of the real world. They employ a set of data objects that support 2D or 3D map display, query, editing, analysis and visualization. The very first computerized spatio-temporal data models were usually created with CAD software. A second generation of geographic data models was based on ESRI's coverage data models. These data models can manage spatial information in datasets together with temporal attribute data and topological relationships.

2.2. *Object-oriented data models*

In order to create more complex data models, a new generation of tools has been developed to support upcoming ESRI's object-oriented data models called the geodatabase data models [9] [10]. The tools are included in ESRI's ArcGIS that contains a suite of integrated applications: ArcMap, ArcCatalog, and ArcToolbox. Using these applications together with their key extensions 3D Analyst, Geostatistical Analyst and Spatial Analyst [11] offers advanced data management and geoprocessing based on spatial interpolations and sharing of simulation data with standalone fluid numerical models. The geodatabase by itself offers object-oriented data modeling that can characterize features more naturally by defining user objects and their interactions. A principal advantage of applications based on geodatabase data model are:
-a uniform repository of geographic data that can be stored and centrally managed in one database,
-more accurate data entry and editing prevented by advanced validation behavior,
-more intuitive data objects, generic points, lines and polygons are transformed, for example, to the point, line or area sources of pollution,
-topological associations, spatial representation, general relationship,
-more efficient scripting and creation outputs by map layers,
-feature properties on a map display are linked to data and can respond to their changes,
-shapes of features are better defined by other graphic entities like curves and Bézier splines,
-management of large data sets by implementing geodatabase in the framework of relational database management systems
-sharing data simultaneously, which enables pre-processing data, solving fluid models and visualization by many users.

A geodatabase can manage raster and vector datasets, TIN and address locators. In case of fluid modeling, the wind flow and pollutant concentrations are arranged by time series measurements taken at meteorological stations and sampling points. Some temporal data are measured at regular intervals, such as minute mean wind speed and wind direction, and other temporal data are captured irregularly in time, such as satellite or aerial images, mapping of potential emission sources at particular locations on a surface. The surface terrain, images, and wind flow fields are represented by the raster datasets. The vector datasets are used for mapping of emission sources and surface measurements that are archived in their attribute tables. The TIN dataset support visualization of the surface terrain.

2.3. *Spatial data from passive remote sensors*

The kind of remote sensing data supporting fluid modeling is devoted to observation of the earth's land and water surfaces by means of reflected or emitted electromagnetic energy. The energy spectrum recorded by the camera lens is much different than the sunlight energy spectrum that entered the atmosphere. Mostly, the blue light is scattered by the atmosphere, and reaches the camera without being reflected from the earth's surface. The blue, red, green, and infrared light is used to be reflected from the canopy, but reach the camera lens in energy spectrum that is different from the energy spectrum that is intercepted by the canopy. The sensor systems transform these kinds radiation in different colors, which usually do not match those of the radiation they represent. Thus, data within the image have to be translated by computer programs into valuable information through the process of image analysis [12] [13].

In case of the fluid modeling application, the satellite scene from the Landsat 7 Enhanced Thematic Mapper (ETM+) is used for giving an overview of the whole mining area, and for the mapping of land degradation features related to soil erosion [14]. A true color composite and pseudocolor composites combining the bands 1-7 at a resolution of 30 meters support display of the coal mine and its surroundings. The aerial images at a resolution of 0.5 meters are draped on the digital terrain model (DTM) to display the surface of the local areas of interest, the temporary storage site and the nearby residential zone.

2.4. *Surface laser scanning and GPS mapping*

For the last twenty years, remote sensing focused on using laser technology has promised to revolutionize land management by delivering spatial information critical for land surface characterization. Its implementation in the framework of lidars brings new possibilities in exploration of the surface structure, and in building the DTMs. Lidar (**l**ight **d**etection and **r**anging) is an active remote sensing technique, analogous to radar, but using laser light. A pulse of laser energy is fired towards the surface. This incident pulse of energy then interacts with the surface, reflecting off of branches, leaves and ground, and back to the instrument where it usually is collected by a sensor. The time of flight of the pulse, from initiation, to return to sensor, is measured and provides a distance estimates [15].

The same principle is implemented in the terrain surface laser scanners. Their new generation significantly reduces field costs and increases phase-based data quality for many types of as-built and site surveys where users want to take advantage of ultra-high speed of phase-based laser scanning.

In order to provide more accurate simulation of wind flows over the potential emission sources, terrain surface laser scanning by Leica HDS systems was chosen for acquisition of the DTM in an area of the temporary coal storage site and its neighboring slopes. It represents the most important parts that influence the dust transport by wind flows drifting from the mining areas to the residential zones. Multiple scans from different locations have to be provided to obtain complete data sources from shielded parts of the terrain. Finally, the clouds of points are georeferenced in the framework of the GIS project together with satellite images and aerial images. In spite of the fact that airborne laser scanning is becoming the standard technique for acquisition of DTMs, surface scanning can also capture spatial points from shielded areas. But, more data processing is needed to filter redundant points or incorrect points caused by vegetation interference and reflection of the mining equipment [16].

Global Positioning System (GPS) is a part of the satellite-based navigation system developed by the U.S. Department of Defense under its NAVSTAR satellite program. In order to improve estimated positions on the earth's surface, differential GPS (DGPS) technique is used for reducing the error in GPS-derived positions by using additional data from a reference GPS receiver at a known position. Thus, there is a network of the local reference GPS receivers (CZEPOS) managed by the national authority in the Czech Republic.

In order to extend spatial data captured by surface laser scanning, a set of spatial points and lines is used to be collected by GPS for mapping slopes and surface objects such as excavators, conveyors, sorting places, man-made barriers, and pylons. The accuracy of the GPS measurements is improved during the post-processing phase by data from the nearest reference stations in the territory of the selected surface mine in the Czech Republic.

2.5. *Pre-processing of meteorological data*

Utility programs are used to transfer archived measurements from local meteorological stations, and to convert meteorological data from national weather services. In the framework of fluid modeling, the wind speed and wind direction over the surface together with other meteorological data are needed for setting model parameters. The utility WRPLOT assists to generate wind rose statistics and plots for user-specified time ranges. Wind roses help to define the dominant transport direction of the winds for a selected location. These data are imported into a geodatabase data model. After preprocessing by PCRAMMET, data can support modeling by U.S. EPA's short term air quality dispersion models such as ISCST3, or by standalone numerical models.

2.6. *Dispersion modeling*

A number of software packages focused on air dispersion modeling can support processing of spatio-temporal data in the GIS. For example, the AERMOD with its U.S. EPA models (ISCST3, ISC-PRIME, AERMOD, and AERMOD-PRIME) can assess via GISs information about a wide variety of emission sources, archived terrain measurements, and terrain characteristics. Then, simulation results can be included backward into GISs in order to provide spatio-temporal analysis, final visualization, and printing. For simulation of the dust dispersion, meteorological data together with a DTM complemented by surface objects such as buildings and mining equipment must be imported and pre-processed by tools included in the modeling package. Though the package includes display tools, export of the data into the GIS is needed to provide spatio-temporal modeling and visualization. In this case, a pre- and postprocessor linkage through shared data files is needed to run the simulation and to import data into the GIS [17].

Similar approach is represented by a standalone software application that shares data by the simple pre- and postprocessor linkage through shared data files. This approach can be also made more comfortable by providing a direct implementation of numerical models in the GIS environment. Thus, the pre- and post processing procedures can be reduced through the sharing of data in the framework of a common database. But, in cases of time consuming numerical operations, actual implementation by scripts or macros is not computationally efficient. The numerical models can be based on similar principles like the U.S. EPA models or specific numerical tools such as numerical modeling by Reynolds averaged Navier-Stokes (RANS) equations for incompressible flows with turbulent closure of the model by an algebraic turbulence model where the numerical simulation is solved by a semi-implicit finite-difference scheme [18].

2.7. *GIS analysis and visualization*

The ESRI's ArcGIS was selected to provide a spatio-temporal framework for fluid modeling. In addition to spatio-temporal data management, advanced spatial analysis is used to create the DTM, to provide pre-processing and post-processing for numerical simulation, to construct interpolations for thematic map layers, and to create virtual scenes for visualization [19] [20].

For these reasons, a new object-oriented data model is developed in the framework of the ESRI's geodatabase data model. Features represented by vectors, rasters, TINs or other user defined data structures are stored and centrally managed in one database together with the defined spatial relationship.

3. Results and Discussion

3.1. *A case study: spatio-temporal modeling of the dust transport over a surface coal mine*

This study is focused on dust emissions from the selected dominant source, a temporary coal storage site and other minor sources. The procedures provided in the framework of the GIS project are illustrated in Figure 1.

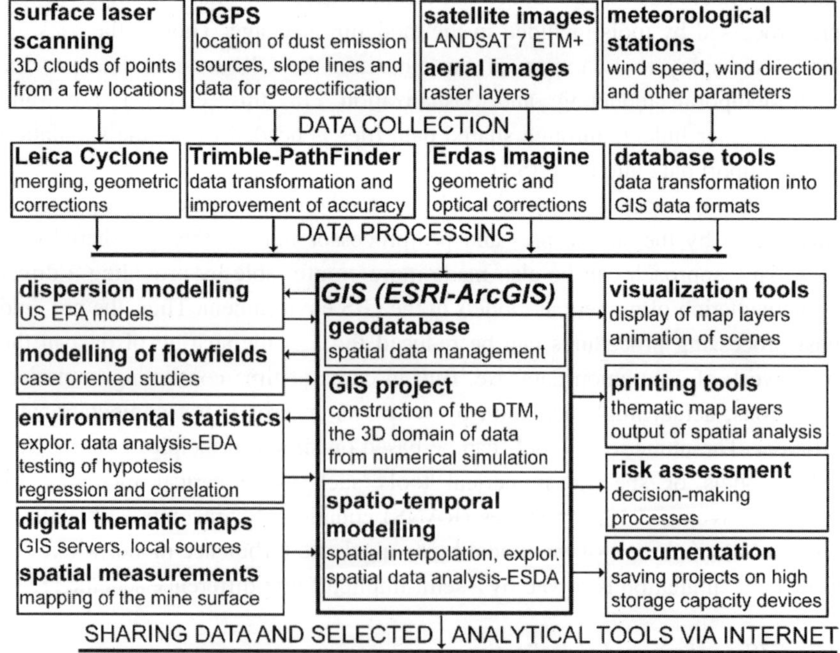

Figure 1. Spatio-temporal data processing in the framework of the GIS project.

The upper part of the scheme in Figure 1 illustrates the spatial data sources and their processing for creation of the DTM with draped images. After the surface laser scanning, the 3D clouds of points from a few locations must be merged, filtered, and put through geometric corrections. Slope lines together with potential emission sources, and other surface features are located by DGPS and processed by Path Finder Office in order to improve accuracy by data from the nearest reference stations of the CZEPOS in the Czech Republic. The satellite images and aerial images are put through geometric and optical corrections in ERDAS Imagine.

The local meteorological stations assist in estimating of the meteorological conditions: wind velocity, wind direction, temperature, and atmospheric pressure. In case of meteorological data, after the correction and filtering, the selected ranges of data series from meteorological stations are imported into the geodatabase data model.

The key unit is represented by the GIS (ESRI-ArcGIS) and its tools for construction of the DTM, management of the 3D domain for data from numerical simulations, exploratory spatial data analysis (ESDA), and spatial interpolations. The links between the GIS and software tools for dispersion modeling, modeling of flowfields in the framework of case oriented studies, environmental statistics, and existing digital map sources are on the left side. The data exchange is provided by sharing files or functionality in the framework of ArcObjects programming library that represent development tools in ArcGIS and other ESRI's software tools. The output modules form visualization tools for display of map layers and animation of 3D scenes, printing tools for creating of thematic map layers and spatio-temporal analysis outputs, case oriented tools for support of the decision-making processes in risk assessment, and tools for saving of the whole project. In addition to the local processing, the sharing data and functionality enables implementation of the geodatabase data model in the framework of a relational database management system (RDBMS) extended by ESRI's spatial database engine.

As an example, the satellite image from Landsat 7 ETM+ in Figure 2 illustrates a true color composite combining the bands 1-2-3. The area of interest is marked and visible such as the light-colored site. A better overview is given by a pseudocolor composite with the bands 2-3-4 in Figure 3. Using of the near-infrared band 4 enables mapping of land degradation features related to soil erosion such as surface mines in a more appropriate way. It can support methods of supervised and unsupervised classification.

The geodatabase data model includes spatial data for the DTM creation. In order to display the surface in a more realistic way, the DTM can by draped by the aerial images and other thematic map layers. While the DTM represents the spatial data, the time series of meteorological measurements are included into temporal data. The link between meteorological measurements and locations of meteorological stations is established by relationship classes. The DTM and meteorological data are complemented by data about emission sources. All the data are used for setting of dispersion models that are solved by standalone software applications included in the AERMOD View package. The simulation results are backward imported into the geodatabase data model. As examples, the data processing in ArcCatalog and in ArcScene is in Figure 4 and in Figure 5.

The final DTM complemented by the thematic layers focused on the dust dispersion is shown in Figure 6. The aerial images in the background are draped on the DTM in order to provide a more realistic view of the dust transport over the surface mine and the neighboring residential zone. The bottom layer illustrates the dust transport with the estimated predominant wind direction. In addition to visualization in the ArcGIS environment, Figure 7 shows, as an example, the PM_{10} dispersion from the surface emission sources in the GoogleEarth over the aerial images. The simulation results were also exported directly into this mapping application from the AERMOD View. A view on the dust dispersion from a temporary coal storage place in Figure 6 is also exported in the Google Earth in Figure 8.

3.2. *Discussion*

There are a number of ways the spatio-temporal data focused on fluid modeling can be managed and several different reasons for doing so. Historically, the fluid modeling tools were often represented by standalone software packages that required predefined input data structures and generated specific data outputs. Thus, the simple manual file exchange and data correction between the environmental modeling system and the GIS were widely used for sharing both the temporal and spatial data. In order to simplify these steps, the widely used U.S. EPA models based on FORTRAN programs and their input/output data structures were integrated into the software systems that are user friendly and can also particularly manage spatial data. For example, it is a case of the next generation air dispersion modeling system that consists of a few components - AERMOD (atmospheric dispersion model), AERMET (meteorological data preprocessor) and AERMAP (terrain preprocessor) [21].

Figure 2. A true color composite combining the bands 1-2-3 of the satellite Landsat 7 ETM+ (captured in May, 2000). The surface mining areas are visible such as the light-colored site.

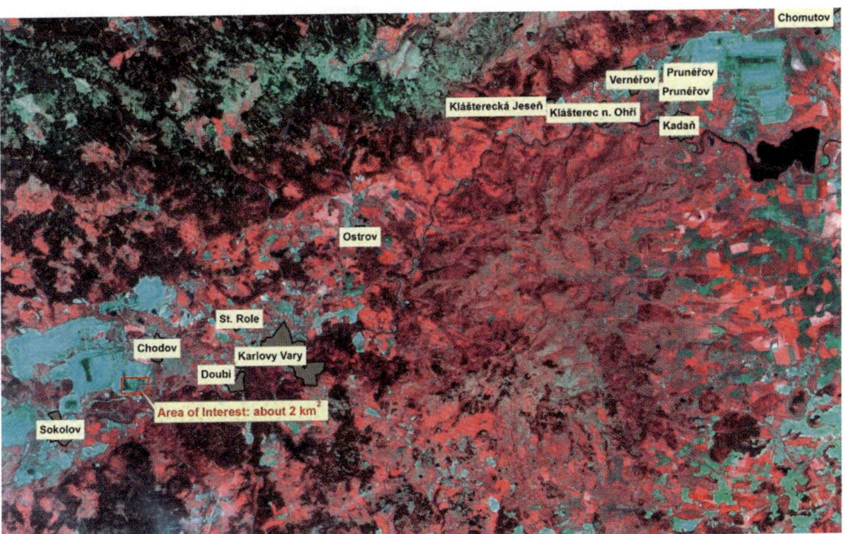

Figure 3. A pseudocolor composite combining the bands 2-3-4 of the satellite Landsat 7 ETM+ (captured in May, 2000). The near-infrared band 4 (760–900 nm) enables a better mapping of land degradation features. The area of interest, the surface coal mine, is marked by rectangle.

Figure 4. Display of the map layers extended by wind roses in the ArcCatalog environment. The main structure of the geodatabase data model is on the left.

Figure 5. Visualization of the simulation outputs in the ArcScene environment.

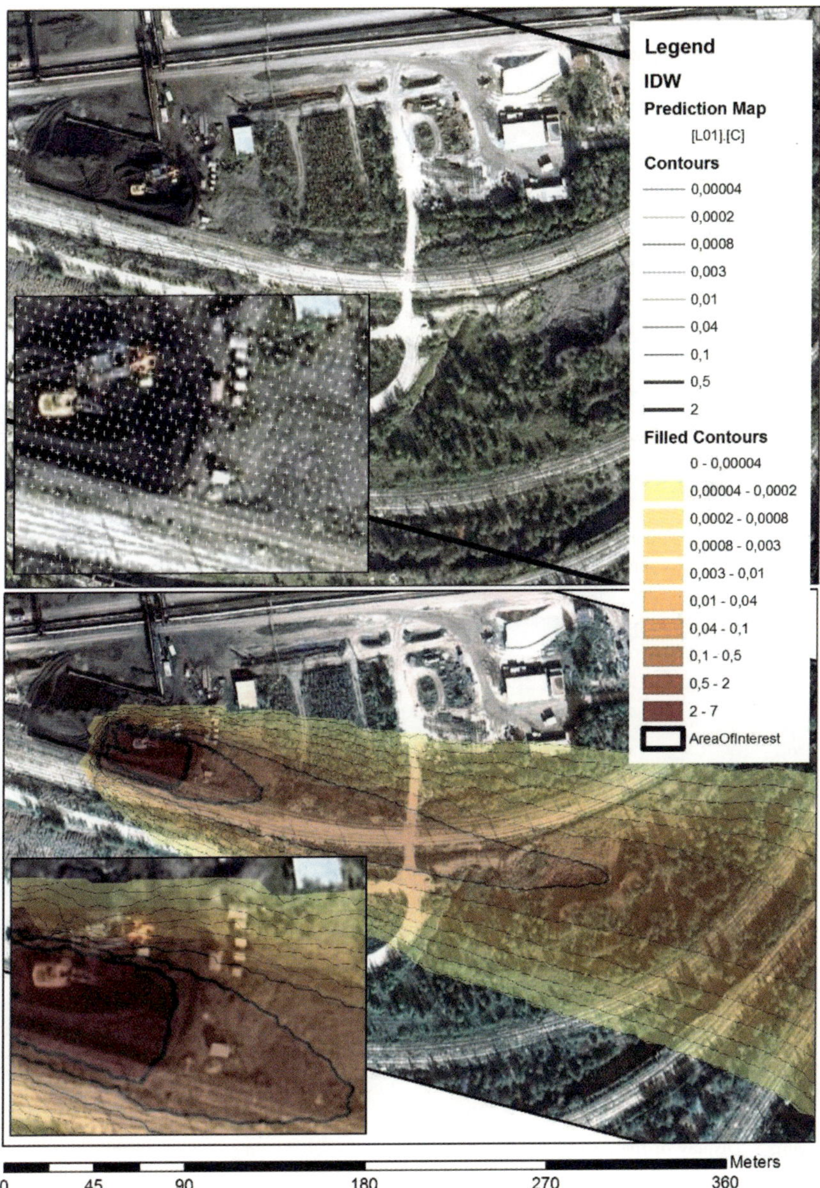

Figure 6. A view on the dust dispersion from a temporary coal storage place. The upper part shows the site and the detailed view on the calculation grid for dust concentrations. The bottom part includes spatial interpolations based on relative data attached to the grid from a numerical modeling standalone system.

L. Matejicek

Figure 7. Display of the dust dispersion from emission sources (solved with the AERMOD View and exported as a map layer into the GoogleEarth).

Figure 8. Display of the dust dispersion from a temporary coal storage place (pre-processed in the ArcGIS and exported as a map layer into the GoogleEarth).

Similarly, many case oriented simulation tools for emission estimates and dispersion modeling were developed to help manage and optimize air pollution by PM_{10} [22] [23] [24] [25] and other air pollutants [26] [27]. It also documents modeling outputs from AERMOD View exported into GIS in Figure 5 and into Google Earth in Figure 7.

The next level integration of spatio-temporal data for fluid modeling can be represented by building of spatio-temporal models in GIS with support of spatial database tools such as ESRI's geodatabase [28] [29] [30]. It involves linking geospatial data describing the physical environment with fluid process models describing how air pollutants moves through the environment. Measured time series of meteorological variables such as the wind velocity and wind direction are needed to set numerical models. Besides sharing the GIS functionality, geodatabase data models provide a robust environment for integrating geospatial and time series data for air resources with simulation models for fluid processes. By building interface tools for accessing geodatabase data, an efficient way was developed to link process models with GIS. It demonstrates modeling outputs in Figure 6 and their display in Google Earth in Figure 8.

In terms of the GIS, the highest level of integration is full implementation of the model solvers into GISs besides object data models. Thus, the model solvers can also use full functionality of data management tools, spatial analytical functions and visualization methods. In spite of that it is the most efficient way of processing spatio-temporal data, transfer of the source codes into the GIS programming environment will approximately take a few years. In case of presented examples focused on dispersion modeling from a dominant emission source, the U.S. EPA models such as ISCST3, ISC-PRIME or AERMOD need to be implemented in the ESRI's GIS modules such as ArcMap, ArcScene and ArcGlobe or developed as the standalone applications based on ESRI's ArcObjects programming library. Similarly, in case of solving the fluid models based on Navier-Stokes equations, there is need to re-program and debag all the time consuming numerical calculations in the GIS programming environment.

In this paper, two scenarios were tested to demonstrate capabilities of ESRI's ArcGIS in fluid modeling. In case of using the U.S. EPA models implemented in the AERMOD View, the object data model in geodatabase assisted in management of PM_{10} dispersion modeling inputs/outputs. The emission sources included the temporary storage site as the main emission source and emission sources caused by mining operations and transport. The higher concentrations above 50 $\mu g/m^3$ were observed only in the short distance from the emission sources. The simulated concentrations and observed data were highly variable in dependence on meteorological conditions and mining activities. In case of using the standalone numerical models based on Navier-Stokes equations, only one dominant emission source, the temporary storage site, was selected to explore wind flows over the surface and transport of the dust particles.

In the both cases, the wind direction and velocity were estimated by processing data from meteorological stations. While the tested U.S. EPA models simulated short term concentration values from multiple emission sources on specified grid locations, the model based on Navier-Stokes equations [31] included the influence of one dominant emission source, the temporary storage site, explored in the 3D grid concentration layers. In order to display the concentration fields, spatial interpolations by ordinary kriging assisted in creating final inputs for the 3D visualization.

4. Conclusion

The presented case studies focused on the integration of spatial data and the outputs/inputs of dispersion modeling demonstrate more complex tools for the risk assessment of the dust transport over a surface coal mining area. Compared with some previously published dust transport studies [23] [24], sharing of the spatial data based on surface laser scanning together with the wind flow fields can result in a more detailed view of the processes of wind erosion. The processes depend strongly on local wind flow characteristics that are affected by terrain geometry. In case of this studied area and the temporary coal storage site in particular, the prediction of the wind flows over stockpiles represents a key stage in the assessment of potential erosion that causes significant environmental impacts. Integration of the dispersion modeling in the GIS environment allowed more complex insights into the studied processes. Thus, other similar studies will be carried out for the testing of both man-made barriers as well as different flow conditions, in order to minimize the dust emission and deposition in the neighboring residential zones.

5. Recommendation and Perspectives

Additional research is needed to acquire further understanding of fluid processes in living environment. But, developing of the acceptable and efficient decision–making approaches will require using of more complex modeling tools, which are able to manage a wide range of spatio-temporal data captured by new generation of sensors and presented to a wider research community [32]. Thus, the GIS has been tested and presented as a candidate to support fluid modeling together with spatio-temporal data management, spatial analysis and visualization.

Spatio-temporal modeling has been carried out by conducting field studies at the Nove Sedlo mine in the Czech Republic. The next research is going to be focused on implementation of the simulation results into the decision-making processes in order to minimize deposition in the neighbor residential zones.

Acknowledgements

This paper was carried out in the framework of a project supported by the Academy of Sciences in the Czech Republic (AVCR 1ET400760405) in the GIS Laboratory at the Faculty of Natural Science, Charles University in Prague. The author is grateful to the mining management for their support during field mapping and monitoring. Modeling support from the research group at the Czech Technical University in Prague Tomas Bodnar and Ludek Benes is also gratefully acknowledged.

APPENDIX - LIST OF SYMBOLS

List of Symbols		
Symbol	**Definition**	**Dimensions or Units**
2D	two-dimension space	
3D	three-dimension space	
AERMAP	atmospheric dispersion modeling terrain preprocessor	
AERMET	atmospheric dispersion modeling meteorological data preprocessor	
AERMOD	atmospheric dispersion modeling	
CZEPOS	multipurpose positioning system for the Czech Republic	
DGPS	differential global positioning system	
DTM	digital terrain model	
ESDA	exploratory spatial data analysis	
ETM+	enhanced thematic mapper	
CAD	computer-aided design	
ESRI	Environmental Systems Research Institute	

GIS	geographic information system	
GPS	global positioning system	
HDS	high-definition surveying	
ISCST3	industrial source complex - short term model	
Lidar	light detection and ranging	
NAVSTAR	navigation signal timing and ranging	
RAMMET	meteorological preprocessor program	
PRIME	plume rise model enhancements	
RANS	Reynolds averaged Navier-Stokes	
RDBMS	relational database management system	
TIN	triangulated irregular network	
U.S. EPA	United States Environmental Protection Agency	
WRPLOT	wind rose plots for meteorological data	

References

1. S.E. Jørgensen and G. Bendoricchio, *Fundamentals of ecological modelling* (Elsevier 2001 3rd edition).
2. E. Holzbecher, *Environmental Modeling: Using MATLAB* (Springer-Verlag 2007).
3. D.J. Maguire, M. Batty and M.F. Goodchild, *GIS, Spatial Analysis and Modeling* (ESRI Press 2005).
4. M. Zeiler, *Modeling our world: The ESRI guide to geodatabase design* (ESRI Press 1999).
5. M.F. Goodchild, L.T. Steaert, B.O. Parks, C. Johnston, D. Maidment, M. Crane and S. Glendinning, *GIS and Environmental Modeling: Progress and Research Issues* (GIS World Books 1996).
6. B.L. Littlefield and D.C. Hanselman, *Mastering Matlab 7* (Prentice-Hall 2004).
7. J.B. Dabney and T.L. Harman, *Mastering Simulink* (Prentice-Hall 2003).
8. J. McCoy, *ArcGIS 9 Geoprocessing in ArcGIS* (ESRI Press 2004 reprint).
9. A. MacDonald, *Building a Geodatabase* (ESRI Press 2001).
10. M. Zeiler and D. Arctur, *Desiginig Geodatabases* (ESRI Press 2004).
11. S. Crosier, B. Booth, K. Dalton, A. Mitchell and K. Clark, *Getting Started With ArcGIS* (ESRI Press 2005).
12. J.B. Campell, *Introduction to Remote Sensing* (Taylor and Francis 2002 3rd edition).

13. D.A. Landgrebe, *Signal Theory Methods in Multispectral Remote Sensing* (Wiley 2003).

14. G.I. Metternicht and A. Fermont, *Remote Sens. Environ.* **64**, 254 (1998).

15. W.G. Rees, *Physical Principles of Remote Sensing* (Cambridge University Press 2001 2nd edition).

16. G. Vosselman, P. Kessels and B. Gorte, *Int. J. Appl. Earth* **6**, 177 (2005).

17. L. Matejicek, Spatio-temporal modeling of the dust emissions from an opencast coal mining area. Dubois AN. Ed., NOVA Publishers (2008).

18. L. Matejicek, Z. Janour and L. Benes, Integration of environmental models based on fluid mechanics in GIS, *International Congress on Environmental Modelling and Software Integrating Sciences and Information Technology for Environmental Assessment and Decision Making (iEMSs)*, Barcelona, Spain, 122 (2008).

19. A. Mitchell, *GIS Analysis, Volume 1: Geographic Patterns & Relationships* (ESRI Press 1999).

20. A. Mitchell, *GIS Analysis, Volume 2: Spatial Measurements & Statistics* (ESRI Press 2005).

21. J.L. Thé, C.L. Thé and M.A. Johnson, *ISC-AERMOD View* (Lakes Environmental Software 2007).

22. T. Badr and J.T. Harion, *Atmos. Environ.* **39**, 5576 (2005).

23. M.K. Chakraborty, M. Ahmad, R.S. Singh, D. Pal, C. Bandopadhyay and S.K. Chaulya, *Environ. Modell. Softw.* **17**, 467 (2002).

24. M.K. Ghose and S.R. Majee, *Atmos. Environ.* **34**, 2791 (2000).

25. H. Lu and Y. Shao, *Environ. Modell. Softw.* **16**, 233 (2001).

26. A.D. Bhanarkar, S.K. Goyal, R. Sivacoumar and C.V.C Rao, *Atmospheric Environment* **39**, 7745 (2005).

27. Q. Zhang, Y. Wei, W. Tian and K. Yang, *Atmos. Environ.* **42**, 5150 (2008).

28. D.R. Maidment, *ArcHydro: GIS for Water Resources* (ESRI Press 2002).

29. L. Matejicek, L. Benesova and J. Tonika, *Ecol. Model.* **170**, 245 (2003).

30. L. Matejicek, P. Engst and Z. Janour, *Ecol. Model.* **199**, 261 (2006).

31. L. Matejicek, Z. Janour, L. Benes, T.Bodnar and E.Gulikova, *Sensors* **8**, 3830 (2008).

32. J.C. Ascough II, H.R. Maier, J.K. Ravalico and M.W. Strudley, *Ecol. Model.* **219**, 383 (2008).

Chapter 11

MODELING PATHOGEN INTRUSION ON SAFE DRINKING WATER: CFD VERSUS PHYSICAL MODELS

P. AMPARO LÓPEZ-JIMÉNEZ, J. JESÚS MORA-RODRÍGUEZ,
VICENTE S. FUERTES-MIQUEL and F. JAVIER MARTÍNEZ-SOLANO

Centro Multidisciplinar de Modelación de Fluidos
Universidad Politécnica de Valencia.
Camino de Vera s/n, 46022 Valencia, Spain

Water distribution systems usually conduct good quality water, which is considered safe drinking water, to supply the population and thus satisfy its basic consumption needs. Water quality in the system depends on the quality level, which is normally controlled by the water treatment plant. Moreover, the possibility of elements surrounding the system entering the main itself (pathogen intrusion) can become an additional problem. This chapter describes the representation of pathogen intrusion in water distribution systems through experimental and numerical modeling in order to study one of the phenomena that cause drinking water contamination. Pathogen intrusion occurs when negative pressure conditions are achieved in the systems, allowing the entrance of water around a leak, causing a problem of water quality. The modeling process is based on experimental and computational procedures: an analysis of the behavior of intrusion considering hydrodynamic principles and transportation of pollutant is presented. In order to compare the results of the measurements and to visualize many other aspects, this case has been implemented in the CFD (Computational Fluid Dynamics) software FLUENT Inc. The computational model numerically solves the governing laws of Fluid Dynamics. These equations, taking into account turbulent phenomena, are solved in a geometric domain when a number of suitable boundary conditions are given. In CFD, the relevant magnitudes (velocity, pressure and concentration) are calculated in a discrete manner at the nodes of a certain mesh or grid and they are represented along the mesh. These computational models are especially useful when they have been validated, and we discuss the process for the calibration of both modeling techniques in this chapter. Through experimental and numerical models, we want to study steady state conditions of the leak and the subsequent mixture, entry and diffusion of the pollutant within the pipe to examine the phenomenon in detail and complement the experiences that will be developed in the laboratory. Computational Fluid Mechanics techniques become a powerful tool in this special problem to be used to obtain a deeper knowledge of the problem of pathogen intrusion into drinking water.

1. Introduction

Water quality is a technical term that is based upon the characteristics of water in relation to guideline values of what is suitable for human consumption and for all usual domestic purposes, including personal hygiene. Components of water quality include biological, chemical, and physical aspects that are habitually controlled by the water treatment plant. Nevertheless, the main parameters of water quality can be altered. In this sense, contamination of drinking water due to exposure to biological and chemical pollutants is a major cause of illness and mortality. The possible entry of elements surrounding the main into the system (pathogen intrusion) can become an additional problem [8]. Contaminant intrusion into a water distribution network may be more frequent and of a greater importance than previously suspected [7].

Water quality and hydraulic performance in networks are linked by the intrusion, as leaks represent a potential influx of undesirable substances into the system [4]. The risk of water pollution is related to several factors. Kirmeyer et al. classified pathogen entry routes in 2001 that focused on the level of risk considering the causes resulting from the intrusion. The routes that have high risk were water treatment breakthrough, transitory contamination, cross connection and water main repair/break.

The aim of this contribution is to approach the problem of pollution occurring in water supply systems based on the pathogen intrusion, as previously stated. The modeling of the leakage from a hydraulic point of view can be done as an analysis of flow through an orifice [9] and remain valid in cases of overpressure and depression, which can favor the entrance of pollutants. The leakage flow depends on the differences in pressure, the resistant feature of the structural defect, the pipe material, and type of defect [12]. All these aspects have been considered in the making of the models, both physical and computational, referred to in this contribution.

1.1. *The Pathogen Intrusion phenomenon*

The physical integrity of pipes is the ability of the distribution system to act as a physical barrier that prevents external contamination from affecting the quality of the internal drinking water supply.

Studies have demonstrated that the soil surrounding buried pipe can be contaminated with fecal indicator micro-organisms and pathogens [5]. Besides contaminated soil, runoff from streets and agricultural land can contain high concentrations of microbiological and chemical contaminants [11], and this runoff can contaminate pipes during a main break, during the unprotected storage of replacement pipe materials, and even during pipe installation in the trench. Breaches in physical and hydraulic integrity can lead to the influx of contaminants across pipe walls, through breaks, and via cross connections. These external contamination events can act as a source of entrants: introducing nutrients and sediments, or decreasing disinfectant concentrations within the distribution system, thus resulting in a degradation of water quality.

The problem of external contaminant intrusion is more aggravated in developing countries where the pollution sources come into contact with water distribution systems and intermittent water supplies are prevalent [2], [14], [13].

Pathogen intrusion causes serious problems in the quality of the supplied water. Such problems lead to increased health risks as water becomes contaminated with pathogens due to intrusion from surrounding foul water sources (e.g. sewers, ditches) through joints and cracks in the deteriorated water distribution pipes. Thus, in developing countries the distribution network has become a point at which contamination frequently reaches unacceptably high levels, posing a threat to public health [16], [1] [17].

The scheme shown in Figure 1 is going to be represented by physical and computational models in order to study this phenomenon. Those representations are focused on developing the factors that are involved in this kind of pathogen intrusion. Then, the design of the prototype and the different models that have been simulated in the process of intrusion are presented.

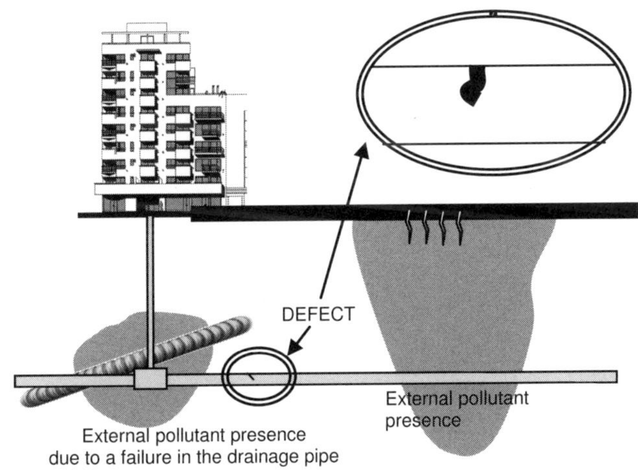

Figure 1. Conditions for pathogen intrusion

The development of accurate models considering aids in understanding the processes in any real system. This also applies to hydraulic models occurring in water distribution networks [10]. The model has to be consistent; it requires a compromise between complexity and simplification.

2. Models for Pathogen Intrusion in Water Mains

According to the case being studied, the developed device is shown here, in order to create the prototype (Figure 2). First, as mentioned in the introduction, the pathogen intrusion requires three elemental components in order to affect drinking water quality in water distribution networks.

The first component that has to be represented is a defect in the main, which, in an ordinary operational situation, will generate a leak. In the prototype, this component will be represented by a circular orifice.

The second component that is represented by the prototype is contaminant flow. The exterior flux that could enter the main is conducted by a secondary tube that was kept at a constant level and a particular concentration in order to be distinguished from the potable flow.

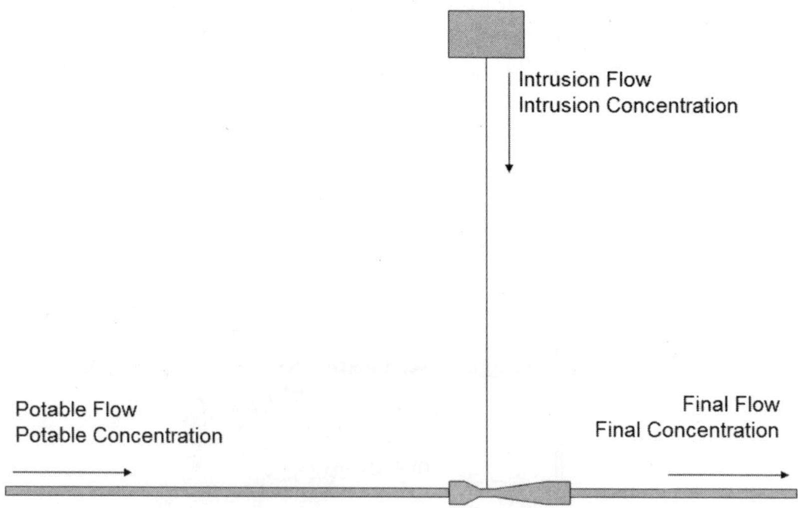

Figure 2. Prototype of the phenomenon of pathogen intrusion

And finally, the third component is a mechanism that is required in order to generate the inflow to the principal tube represented by a Venturi tube. The contaminated flow around the main tube will be injected with this element. In this case, the flux of the secondary tube is connected to the section representing the defect of the principal tube.

For the experimental model, a physical representation of the prototype was built in the laboratory simulating the leakage of the pipe with an orifice placed in the throat of the Venturi tube. Numerical modeling simulates the prototype in three dimensions, using a program based on Computational Fluid Dynamics (CFD), which displays the fields of hydrodynamic components.

Further, an analysis of the behavior of intrusion is presented considering the transportation of pollutant modeled through a conservative parameter, in this case water salinity.

Using both mathematical and physical models, it is intended to have better knowledge of quantities that cannot be measured, such as velocity fields, aspects of turbulence, pressure fields, and concentrations. Also, mixing processes related to external intrusion can be discovered. Through computational modeling, we seek to study conditions under steady state of the leak and the subsequent mixture, entry and diffusion of the pollutant within the pipe to see in detail the phenomenon that occurs and complement the experiments that will be developed in the laboratory.

2.1. *The physical model*

In order to represent the studied flows, and the intrusion phenomenon in stationary state, the first step to be considered here is a suitable assembly developed in the laboratory (Figure 3).

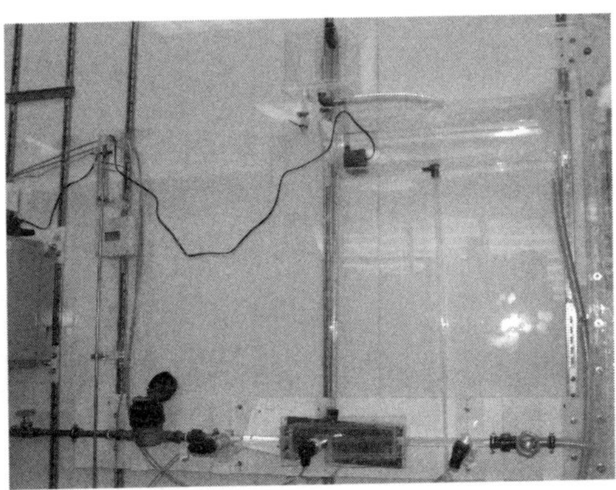

Figure 3. Experimental model for the analysis of the flow into a Venturi tube

In Figure 3, the description of the different elements contained within the model can be observed in order to represent the prototype. It is designed to calibrate the computational model with a completely controlled environment in which the measurements can be precise.

The depression has been represented by a Venturi tube in order to force the intrusion inside the main conduct. Thus, we can observe and model the entrance of external flow into the principal flow [15]. Figure 4 shows the Venturi tube used to generate the depression. Due to this correlation between the fall in pressure on the zones with different diameters and the volume circulating through the conduit, we can introduce and control the external flow inside the main water flow.

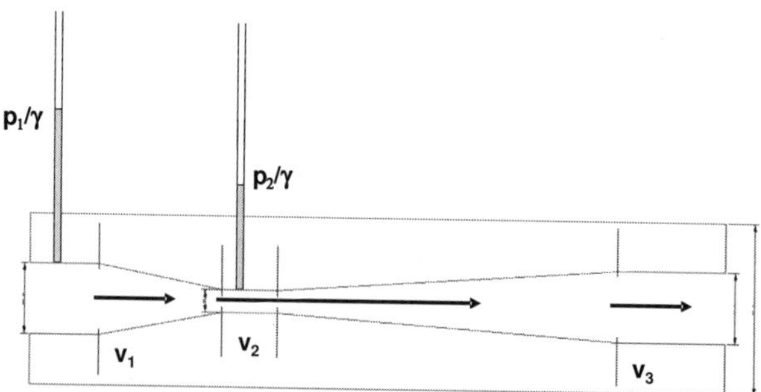

Figure 4. Bernoulli's principle employed in Venturi tube

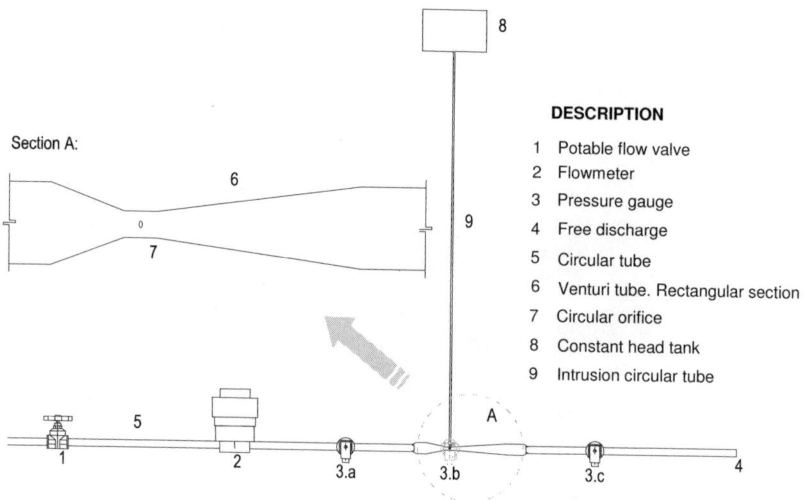

Figure 5. Plane of the physical model

Measuring devices for flow and pressure parameters are described below and shown in Figure 5. The flowmeter is a rotary piston type for residential and commercial consumption applications and shows a nominal flow of 1500 l/h with a range from 15 to 3000 l/h.

There are three pressure transducers: two of them located upstream and downstream of the Venturi tube rectangular, and the third located on the throat of the Venturi tube. The pressure transducers can measure pressure from -40 up to 80 kPa. Intrusion flow was determined by a volumetric procedure. The volume of water coming into the system was counted throughout the time of physical simulation, thus giving the mean flow of intrusion.

Potable water is pumped by a centrifugal pump in a re-circulation system. The flow is conducted by a circular pipe 16 mm diameter. A valve is located at the entrance to the system. According to the power of the pump equipment and its maximum flow, the valve opens at different flows. As a result, nine series of seven tests are conducted. Figure 5 sets out the proposed pilot laboratory device.

The experiment has been designed to measure pressure and velocity at certain points. A suitable interface has been designed by means of LABVIEW. In Figure 6, a computer screen shot shows the points in the laboratory assembly where flow and pressure measurements were taken. The intrusion flow was taken in a volumetric way; in each one of the simulations, a specific time was established in order to obtain this intrusion flow.

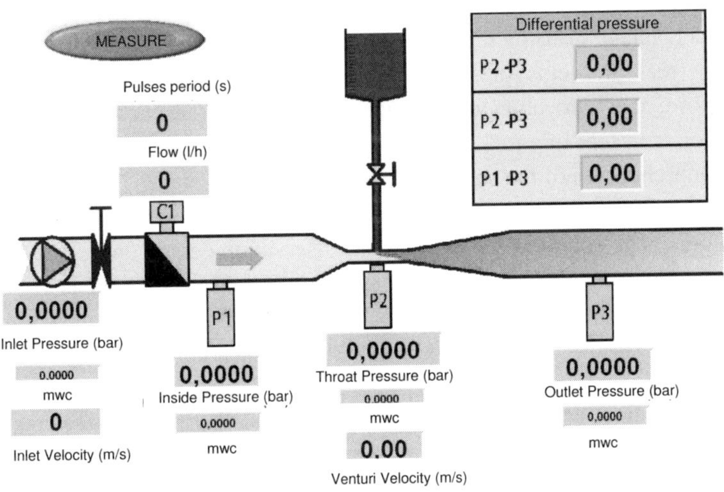

Figure 6. Design of control screen for the measurements in laboratory experiment

2.2. *Computational model*

In order to compare the results of the measurements and to visualize many other aspects, this case has been implemented in FLUENT software. As already indicated, the computational model numerically solves the governing laws of Fluid Dynamics. These equations, taking into account turbulent phenomena, are solved in a geometrical domain, given a number of suitable boundary conditions. In CFD, the relevant magnitudes (like velocity, pressure and temperature) are calculated in a discrete manner at the nodes of a certain mesh or grid and they are represented along the mesh.

The way to think about solving the problem is to consider the total flow that enters or leaves a cell through its boundaries. Continuity implies that the net flow through the boundary of an element is equal to the flow produced by the internal sources. In fact, an equation of balance for the property in as small a volume as desired is solved. However, instead of carrying out a discretization of the differential equation, the process is reversed: each point is associated with a mean value obtained by integrating the differential equation, instead of an approximate value of the property at the point.

FLUENT is CFD code that uses a numerical method based on the discretization of the space by means of finite volumes [3], and it offers a number of CFD options for the steady state solution, which can be used in this case. Specifically, incompressible flow and steady state have been considered, together with water as the current fluid under turbulent regime.

The advantage of using these models resides in the fact that they are able to reproduce real problems of Fluid Mechanics to any degree of complexity. Furthermore, they are able to visualize hydrodynamic aspects impossible to measure or represent in a real case (i.e. stream lines) that have great importance in the comprehension of the studied phenomena.

The conservation equations solved by the code are those of mass and momentum. The continuity or mass conservation equation used in the hydrodynamic study of the present problem and solved by FLUENT is

$$\frac{\partial \rho}{\partial t} + \nabla \rho \vec{v} = S_m \tag{1}$$

where ρ is the fluid density, $\vec{v}$ is the velocity and S_m the mass source contained in the volume of control. For other geometries, suitable coordinates, namely spherical or cylindrical, should be used. Also, the momentum equation is given by the following expression:

$$\frac{\partial(\rho\vec{v})}{\partial t} + \nabla \rho(\vec{v}\vec{v}) = -\nabla p + \nabla \overline{\overline{\tau}} + \rho\vec{g} + \vec{F} \tag{2}$$

where p is the static pressure; $\vec{g}$ and $\vec{F}$ the gravitational and outer forces defined on the control volume, respectively; and $\overline{\overline{\tau}}$ is the stress tensor defined by:

$$\bar{\tau} = \mu\left[\left(\nabla\vec{v} + \nabla\vec{v}^{T}\right) - \frac{2}{3}\nabla\vec{v}I\right] \tag{3}$$

where μ is the eddy viscosity, I is the unit tensor and the third term accounts for the effect of the expansion of volume. The geometry being studied was constructed in AutoCAD® 2006 considering the sections where pressure was measured upstream and downstream in the physical prototype model, and the design was made considering the original scale geometry representing the model in three dimensions (Figure 7).

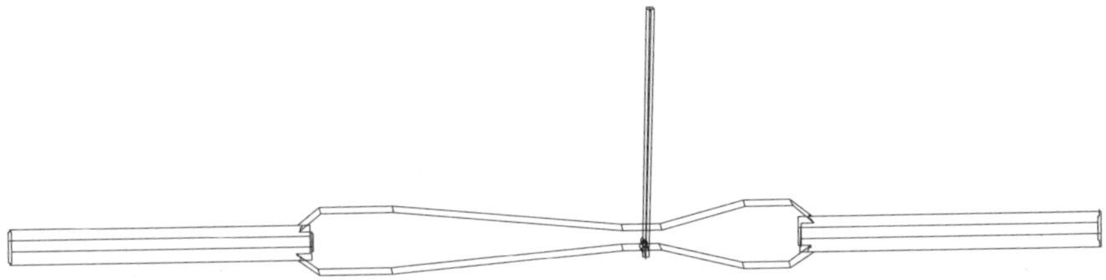

Figure 7. Control volume meshed in the computational model

The mesh was designed on Gambit 2.2.30. In this software, the volume control and the solid and liquid elements established the defining the computational domain and constructed by unstructured tetrahedral and hexahedral elements specially applied to complex and very large surface meshes (Figure 8).

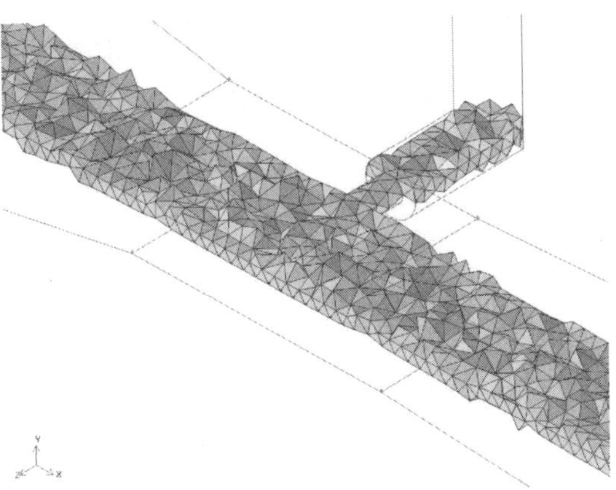

Figure 8. Details of the interior generated mesh

Focusing on the conditions of this model, the RNG k–ε turbulent model (semi-empirical model with a mathematical technique called renormalization group based on model transport equations for the turbulence kinetic energy and its dissipation rate) is the best option. The RNG model has an additional term in its equation that significantly improves the accuracy for rapidly strained flows. The effect of swirl on turbulence is included in the RNG model, enhancing accuracy for swirling flows. The RNG theory provides an analytical formula for turbulent Prandtl numbers, while the standard k-ε model uses user specified, constant values. While the standard k-ε model is a high-Reynolds-number model, the RNG theory provides an analytically-derived differential formula for effective viscosity that accounts for low Reynolds number effects. These features make the RNG k-ε model more accurate and reliable for a wider class of flows than the standard k-ε. In this case, the values for the following constants are C_μ=0.0845; C_1=1.42 and C_2=1.68.

In the FLUENT software, the boundary conditions shown in Figure 9 were defined. The model works with velocity inlet boundaries at the entry of the flow on principal and intrusion sections and a pressure outlet boundary on the final section of the main tube. The hydraulic diameter method and the turbulence intensity were considered for boundaries.

In order to improve the accuracy of the solution, second-order discretization was used for numerical method. The gradient option was modified from cell-based to node-based in order to optimize energy conservation; this option is more suitable for tri-element meshes [3]. One of the main advantages of numerical solvers is that computational models can provide a large amount of information for the magnitudes of interest (in this case pressures and velocities, but the code provides many others related to turbulence or other aspects of momentum, etc.) that cannot be measured with the apparatus arranged in real conditions.

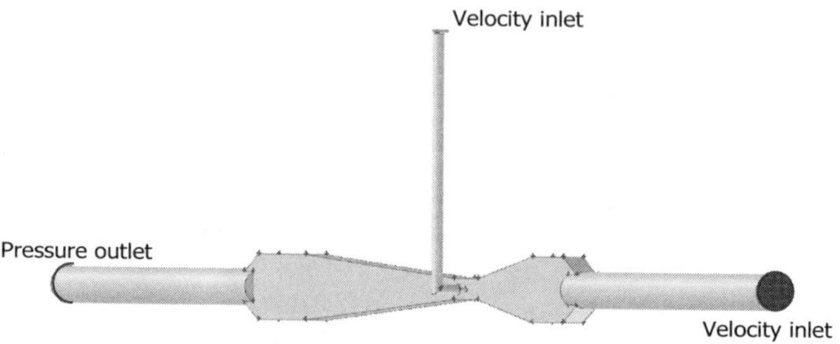

Figure 9. Boundary conditions for the hydrodynamic simulations (reverse angle)

3. Analysis of Results

3.1. *Validation of the numerical simulations for the hydrodynamic model*

Table 1 shows the results obtained from the experimental simulation. In this table, we can observe the flows measured in the experiments inside the main tube (principal) and in the volumetric measurements related to the intruding flow. This is the information needed to establish the boundary conditions for the numerical model.

Table 1. Experimental flows for the different experiments (l/h)

Simulation	Principal	Intrusion	Simulation	Principal	Intrusion
1	341.7	45.4	6	2068.8	50.7
2	540.4	45.4	7	2316.7	53.2
3	1074.4	45.5	8	2548.6	55.9
4	1325.2	46.2	9	2857.1	60.1
5	1828.6	48.1			

In the Venturi throat section, the pressures were negatives as expected in almost all nine different experiments for different flows. The graphical results provided by the CFD modeling are described in Figures 10 and 11.

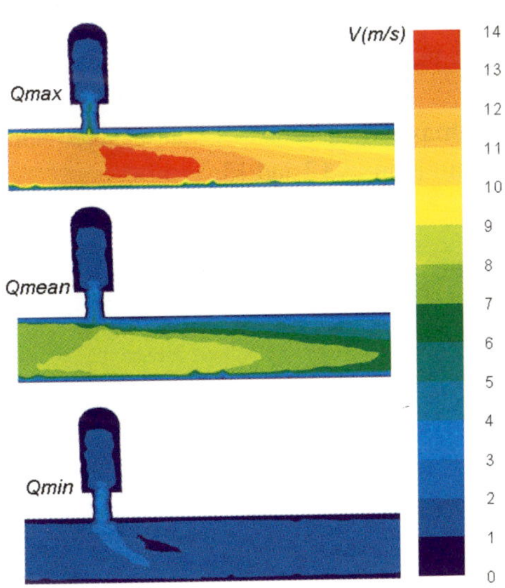

Figure 10. Velocity fields for three representative cases of flow

                          P.A. López-Jiménez et al.

Three representative simulations of flow have been described, as we have considered nine flow situations, but only three of them are shown: maximum, intermediate and minimum flow in the intrusion zone (experiments 1, 5 and 9 in the previous table). Maximum flow occurs after the intrusion section, this maximum velocity registering in the computational model at 13.42 m/s. In Figure 11, the pressure regime of these representative situations is shown.

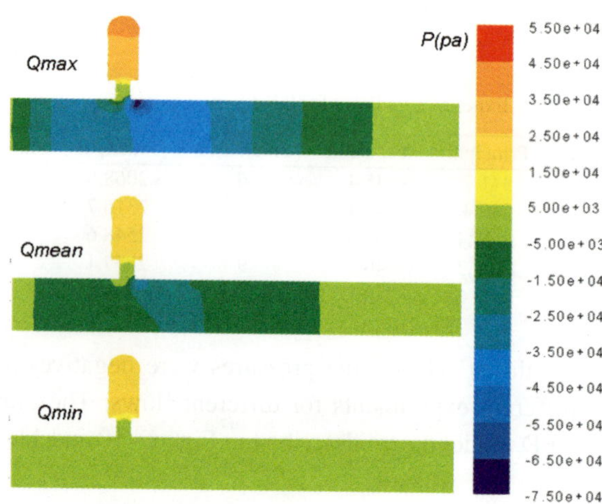

Figure 11. Pressure fields for three representative cases of flow

Physical and computational results, after the computational model calibration, are shown in Figures 12 and 13.

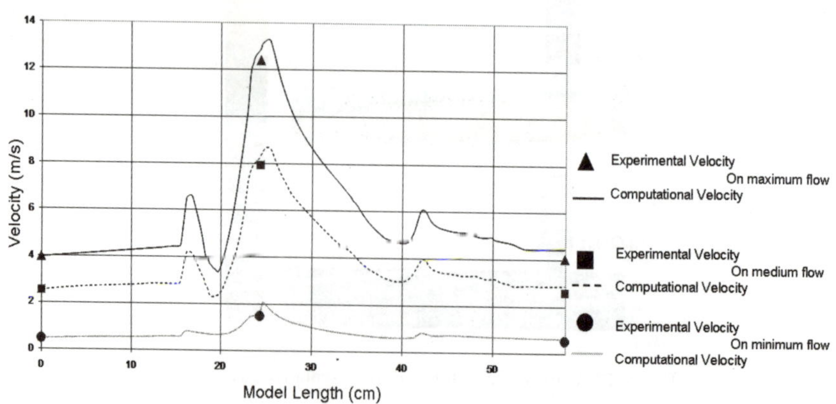

Figure 12. Velocity measurements vs. computational model predictions

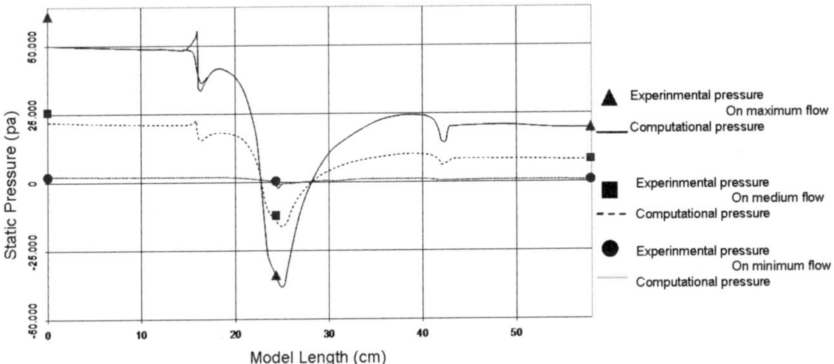

Figure 13. Pressure measurements vs. computational model predictions

As can be seen, agreement of pressure and velocity values between measured and numerical results is satisfactorily achieved in these three representative flow situations. The error was established using the Nash-Sutcliffe efficiency (E), which is one of the most used criteria for hydrologic evaluation between simulated and observed variables [6]. It is defined as one minus the sum of the absolute squared differences between the predicted and observed values normalized by the variance of the observed values during the period under investigation:

$$E = 1 - \frac{\sum_{i=1}^{n}(O_i - P_i)^2}{\sum_{i=1}^{n}(O_i - \overline{O})^2} \tag{4}$$

with O observed and P predicted values. The range of E lies between 1.0 (perfect fit) and $-\infty$. An efficiency of lower than zero indicates that the mean value of the observed would have been a better predictor than the model. In this case, we assess the velocity and pressure of both models in every section of measurement (Table 2). With respect to the velocity configuration that was evaluated in the intrusion and the final sections of the models, the results obtained were 0.988 of similitude from the intrusion section and 0.952 in the final section.

Table 2. Results of Nash-Sutcliffe criteria on hydraulic parameters

Parameter	E Upstream Venturi tube	E Throat Section Venturi tube	E downstream Venturi tube
Velocity	-	0.988	0.952
Pressure	0.926	0.999	-

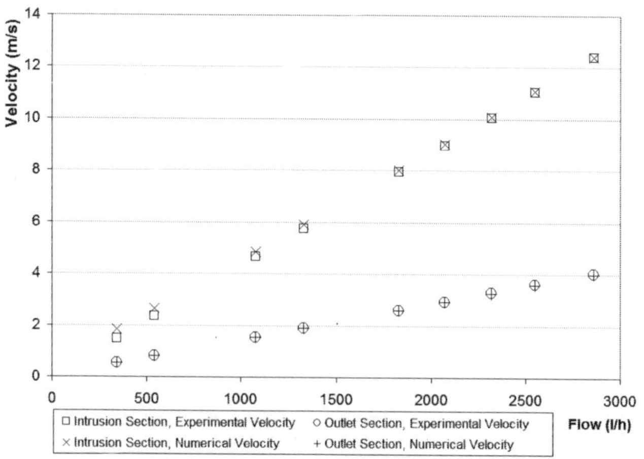

Figure 14. Modeled vs. measured velocities in the modeled flow range

Figure 14 presents the comparison between the velocities obtained in the intrusion region and in the final pressure condition from measurements and computational model. This is a representation of the agreement achieved in the predictions, affirmed by the great mathematical similitude between both results for the range of modeled main flows.

The pressure configuration was evaluated in the same way, in this case by considering the first section and the intrusion sections of the models. The results were 0.926 in the first section, and 0.999 in the intrusion section. All of these results present more than 92% similitude between the computational and experimental models. Thus, the predicted values were adequate for the representation of the experimental model.

Computational modeling is consistent with the physical prototype and adequate in the flow range considered. This conclusion leads us to consider the computational model as an entity that represents the velocity and pressure fields in detail at those points where measurements are not available or at points which require contrast.

3.2. *Quality model*

Based on the hydrodynamic model of the intrusion, some experiments to represent water quality in the intrusion mixture have been carried out. In this case, water quality will be represented by a conservative parameter, such as electrical conductivity. A small volume of water with a huge electrical conductivity is used to represent the pollutant intrusion.

Apart from continuity and momentum equations, FLUENT also predicts the local mass fraction of each species, Y_i, through the solution of a convection-diffusion equation for the ith species in order to solve a conservation equation for chemical species [3]. This conservation equation takes the following general form:

$$\frac{\partial}{\partial t}(\rho Y_i) + \nabla \cdot (\rho \vec{v} Y_i) = -\nabla \cdot \vec{J}_i + R_i + S_i \tag{5}$$

where R_i is the net rate of production of species i by chemical reaction and S_i is the rate of creation by addition from the dispersed phase plus any user-defined sources. In turbulent flows, FLUENT computes the mass diffusion in the following form:

$$J_i = -\left(\rho D_{i,m} + \frac{\mu_t}{Sc_t} \right) \nabla Y_i \tag{6}$$

where Sc_t is the effective Schmidt number for the turbulent flow (which varies with the turbulent viscosity, μ_t, and the turbulent diffusivity), and $D_{i,m}$ the mass diffusion coefficient for species i in the mixture. The input parameters in this case are reflected in Table 3.

Table 3. Input parameters on the numerical model

Simulation	Principal inlet boundary	Intrusion inlet boundary	Outlet boundary
	Mass flow (kg/s)	Mass flow (kg/s)	Velocity (m/s)
1	2.86E-05	1.59E-04	0.309
2	5.97E-05	8.36E-05	0.,625
3	8.58E-05	1.15E-04	0.898
4	1.06E-04	3.59E-04	1.121
5	1.23E-04	3.78E-04	1.307
6	1.44E-04	3.66E-04	1.517
7	1.61E-04	3.66E-04	1.699
8	2.18E-04	3.01E-04	2.629
9	2.51E-04	3.64E-04	2.629
10	2.76E-04	3.71E-04	2.887
11	3.04E-04	3.80E-04	3.177

In this case, to specify boundary conditions, the inlet mass flow condition was established for the principal and intrusion entrances. This boundary condition defines both flow and density of water and also the pressure.

As a result, we present the mixture flow in which it is possible to observe a 6.10% maximum mass flow contaminant in the last section of the model. The configuration of the parameter of contaminant flow can be seen as follows (Figure 15).

                         P.A. López-Jiménez et al.

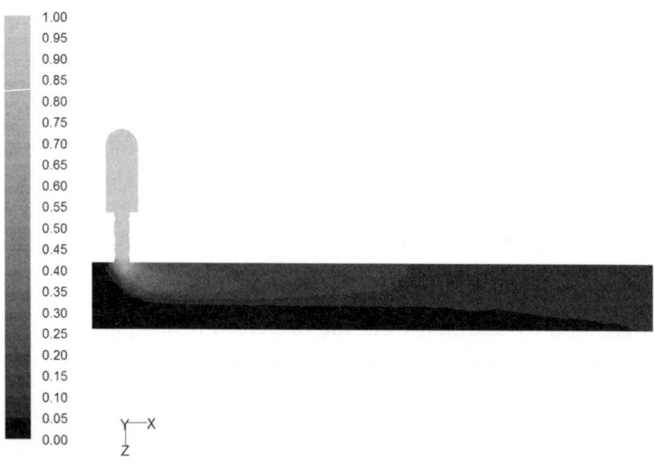

Figure 15. Electrical conductivity configuration when the percent of max flux contaminated is 6.10% on the output of the model

To calibrate the transport model, some comparisons between the theoretical, (CE_R for the final measured mixtured flow in expression 7), experimental and computational modeling have been made.

$$CE_R = \frac{Q_P CE_P + Q_i CE_i}{Q_R} \tag{7}$$

where CE_R is the theoretical resultant electrical conductivity; Q_R is the total flow; Q_P and C_P are, respectively, the flow and electrical conductivity of the main flow in the pipe; and Q_i and C_i are, respectively, the flow and electrical conductivity in the intrusion flow.

The mean absolute error of the computational model when comparing the results of the theoretical and experimental study is 0.090 and 0.082. The certainty of the computational modeling is 91.0% and 91.8%. Again, in this case the agreement between the modeling predictions of the mixture concentration and the theoretical calculations was adequate.

In this case, the viscous model is still considered as the RNG k–ε turbulent model. The constants of turbulent model are the same values: C_μ=0.0845; C_1=1.42 and C_2=1.68. Moreover, the species model was used, specifying the species transport model without considering volumetric reactions. Properties for the mixture material were defined. First, mixture properties were defined in a template form, then they were set to account for the different flows. There were two different kinds of water to represent the two volumetric species. Both of those species were used on the inlet boundaries; the difference between those flows was the electrical conductivity of the water. The physical property for the mixture material was the volume weighted mixing law for the conductivity; in this case it had to be defined for each of the fluid materials that comprise the mixture on the mass flow rate. The total mass fraction species was established for the intrusion inlet.

The computational model allows us to visualize the mixture region along the section of the tube in the conservative transport species as is here described. As can be observed in Figure 16, the agreement between the results in the computational predictions versus measurements again achieves an acceptable value.

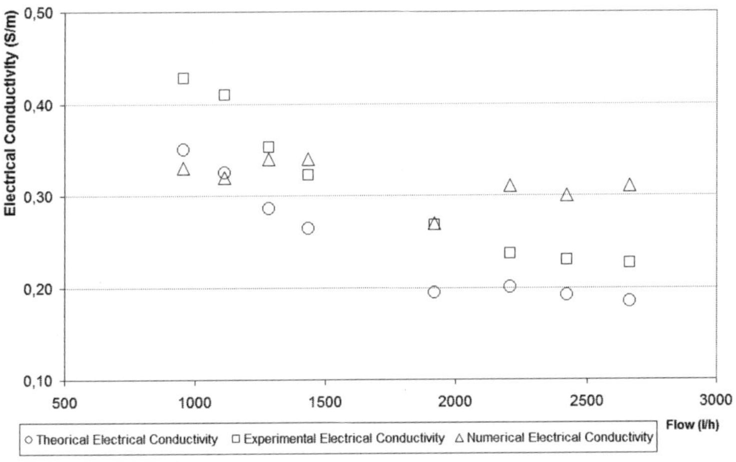

Figure 16. Electrical conductivity measured vs. predicted in the simulated flow range

4. Conclusion

Contamination of drinking water due to exposure to biological and chemical pollutants is a very important cause of diseases. One of the causes of this sort of water pollution is external intrusion of pathogen agents though defects in the pipes. In this contribution, we have studied the phenomenon of external pathogen intrusion in water mains by means of simulation models.

We have developed two ways of solving the problem: by physical prototype and numerical computations with mathematical and physical models. We have attempted to achieve better knowledge of quantities that cannot be measured, such as velocity fields, aspects of turbulence, pressure fields, concentrations, that are present in mixing processes when unusual situations occur in the system.

Through computational modeling, we seek to study conditions under steady state of the leak and the subsequent mixture, entry and diffusion of the pollutant within the pipe to see the phenomenon that occurs in detail and complement the experiments that will be carried out in the laboratory.

The numerical model complements the physical prototype. In the experiments designed in the laboratory, the magnitudes are only monitored at the points at which we have measurements of volume or pressure, whereas in the computational model, we can obtain values of pressure or velocity at all the points considered in the mesh. Therefore, the knowledge of the phenomena obtained by means of mathematical models is much more complete, as long as the values calculated and measured at the same points are reasonably contrasted.

5. Acknowledgements

This contribution has been made possible by the actions of the CMMF researchers involved in the following project: *DANAIDES: Desarrollo de herramientas de simulación para la caracterización hidráulica de redes de abastecimiento a través de indicadores de calidad del agua. REF. DPI2007-63424. Ministerio de Educación y Ciencia de España.* We would like to thank the R&D&I Linguistic Assistance Office, Universidad Politécnica de Valencia (Spain), for granting financial support for the linguistic revision of this paper.

Appendix - list of symbols

List of Symbols		
Symbol	**Definition**	**Dimensions or Units**
C_μ, C_1, C_2	numerical constants	
CE_R, CE_P, CE_i	electrical conductivities	$[L^2{\cdot}M{\cdot}T^{-3}{\cdot}I^{-2}]$
$D_{i,m}$	mass diffusion coefficient in a mixture	$[L^2{\cdot}T^{-1}]$
E	Nash-Sutcliffe efficiency	
$\vec{F}$	force vector	$[M{\cdot}L{\cdot}T^{-2}]$
I	unit tensor	
J	mass flux; diffusion flux	$[M{\cdot}L^{-2}{\cdot}T^{-1}]$
O_i	observed values	

$\overline{O}$	mean of observed values	
P_i	predicted values	
Q_R, Q_P, Q_i	flows	$[L^3 \cdot T^{-1}]$
R_i	net rate of production of species by chemical reaction	$[M \cdot L^{-3} \cdot T^{-1}]$
Sc_t	turbulent Schmidt number	
S_i	rate of species creation	$[M \cdot L^{-3} \cdot T^{-1}]$
S_m	mass source	$[M \cdot L^{-3} \cdot T^{-1}]$
T	temperature	$[K]$
Y	mass fraction	
$\vec{g}$	gravitational acceleration	$[L \cdot T^{-2}]$
p	static pressure	$[M \cdot L^{-1} \cdot T^{-2}]$
t	time	$[T]$
$\vec{v}$	overall velocity vector	$[L \cdot T^{-1}]$
μ	eddy viscosity	$[M \cdot L^{-1} \cdot T^{-1}]$
μ_t	turbulent viscosity	$[M \cdot L^{-1} \cdot T^{-1}]$
ρ	density	$[M \cdot L^{-3}]$
$\overline{\overline{\tau}}$	stress tensor	$[M \cdot L^{-1} \cdot T^{-2}]$
∞	infinitum	

References

1. R.E. Besser, R.B. Moscoso, A.O. Cabanillas, V.L. Gonzalez, L.P. Minaya, P.M. Rodriguez, S.W. Saldana, C.J.L. Seminario, A.K. Highsmith, R.V. Tauxe, *B. Ofic. Sanit. Panam.* **3**, 119 (1995).
2. K. Choe and R. Varley, Conservation and pricing-does raising tariffs to an economic price for water make the people worse off?. *Prepared for the Best Management Practice for Water Conservation Workshop*, South Africa (1997).
3. FLUENT, FLUENT 6.2 *User's Guide* (© Fluent Inc., 2005).
4. V.S. Fuertes, P.L. Iglesias, J. Izquierdo, P.A. López, G. López, F.J. Martínez, R. Pérez, *Ingeniería Hidráulica en los Abastecimientos de Agua*, (Fluid Mechanics Group Ed. Universidad Politécnica de Valencia, Spain 2003).
5. G.J. Kirmeyer, M. Friedman, K. Martel, D. Howie, M. LeChevalier, M. Abbaszadegan, M.R. Karim, J. Funk, J. Harbour, *Pathogen Intrusion Into*

Distribution Systems, (AWWA Research Foundation and the American Water Works Association, USA 2001).

6. P. Krause, D. P. Boyle, F. Bäse, *Advances in Geosciences,* 89-97, **5** (2005).
7. M.W. LeChevallier, R.W. Gullick, M.R. Karim, M. Friedman, J. Funk, *Journal of Water and Health* **1**, 1 (2003).
8. P.A. López, *Metodología para la Calibración de Modelos Matemáticos de Dispersión de Contaminantes Incluyendo Regímenes No Permanentes.* (PhD Thesis Disertation, Universidad Politécnica de Valencia 2001).
9. P.A. López, F.J. Martínez, G. López, B. Lara, *Ing. Hidraul. Mex.* 43-53, **2**, 23 (2007).
10. P.A. López, M. Herrera, J.J. Mora, F.J. García, La necesidad de un adecuado tratamiento de los datos en la modelación hidrodinámica. Caso de estudio: Análisis estadístico de mediciones en modelos físicos. *VII Seminario Iberoamericano de Planificación, Proyecto y Operación de Sistemas de Abastecimiento de Agua.* Morelia, Mexico (2007).
11. D.K. Makepeace, D.W. Smith, S.J. Stanley, *Crit. Rev. Env. Sci. Tec.* **2**, 25 (1995).
12. J. May, *World Water and Environmental Engineering,* **10**, October (1994).
13. M.W. Rosegrant, X. Cai, S.A. Cline, Averting an Impending Crisis. Global Water Outlook to 2025, Food Policy Report. International Water Management Institute (IWMI), Colombo, Sri Lanka (2002).
14. D. Seckler, D. Molden, R. Arker, Water Scarcity in the Twenty First Century. Research Report N. 19. Colombo, Sri Lanka: International Water Management Institute. (1998)
15. A. Speranza, V. Lucarini, *Encyclopedia of Condensed Matter Physics.* Elsevier, Eds. (2005).
16. D.L. Swerdlow, E.D. Mintz, M. Rodriguez, E. Tejada, C. Ocampo, L. Espejo, K.D. Greene, W. Saldana, L. Seminario, R.V. Tauxe, J.G. Wells, N.H. Bean, A.A. Ries, M. Pollack, B. Vertiz, P.A. Blake, *Lancet* **340**, 8810 (1992).
17. K. Vairavamoorthy, Y. Jimin, H.M. Galgale, S. Gorantivar, *Environ. Modell Softw.* 951-965, **22** (2007).

Chapter 12

NUMERICAL SIMULATION OF MASS EXCHANGE PROCESSES IN A DEAD ZONE OF A RIVER

CARLO GUALTIERI

Department of Hydraulic, Geotechnical and Environmental Engineering (DIGA), University of Napoli Federico II, Napoli, Italy

The presence of dead zones in streams and rivers significantly affects the characteristics of mass transport. In a river, dead zones can be due to geometrical irregularities in the riverbanks and riverbed and/or to spur dikes and groyne fields. Dead zones produce a difference between the concentration curves measured and modeled by the classical 1D advection-diffusion equation with sharper front and longer tails. In a dead zone, the mean flow velocity in the main stream direction is essentially zero and the main transport mechanism is transverse turbulent diffusion which controls the exchange processes of solutes with the main stream. This Chapter presents the results of 3D steady-state and time-variable numerical simulations carried out with Multiphysics 3.5™ in a rectangular channel with a lateral square cavity representing a dead zone. This geometry was previously experimentally studied by Muto et al. [18] [19]. The exchange coefficient between the main flow and the dead zones was calculated both from the transverse velocity data along the dead zone-main channel interface and from the temporal decay of the concentration of a tracer that was homogeneously injected in the dead zone.

1. Introduction

Solute transport in streams and rivers is strongly related to river characteristics, such as mean flow velocity, velocity distribution, secondary currents and turbulence features. These parameters are mainly determined by the river morphology and the discharge conditions. Most natural channels are characterized by relevant diversity of morphological conditions. In natural channels, changing river width, curvature, bed form, bed material and vegetation are the reason for this diversity. In rivers which are regulated by man-made constructions, such as spur dikes, groins, stabilized bed and so on, the morphological diversity is often less pronounced and, thus, flow velocities are more homogeneous. In natural channels, some of these morphological irregularities, such as small cavities existing in sand or gravel beds, side arms and embayments, can produce recirculating flows which occur on different scales on both the riverbanks and the riverbed. These irregularities act as dead zones for the current flowing in the main stream direction. In regulated rivers,

249

groyne fields are the most important sort of dead zones. Groyne fields can cover large parts of the river significantly affecting its flow field. Dead-water zones or *dead zones* can be defined as geometrical irregularities existing at the river periphery, within which the mean flow velocity in the main stream direction is approximately equal to zero [33]. Dead zones significantly modify velocity profiles in the main channel as well affect dispersive mass transport within the river. Thus, in recent years exchange processes between the main stream and its dead zones were increasingly studied, mostly using experimental laboratory and field works [3] [6] [10] [11] [13] [14] [18] [19] [25] [26] [28] [32] [33] [36] [37]. Also, computational methods were recently carried out to investigate hydrodynamics in channels with dead zones [6] [8] [9] [15] [16] [27] [35] [36].

The objective of this Chapter is to present the preliminary results of a numerical study undertaken to investigate the fundamental flow phenomena around and inside a simplified geometry, representative of flow in natural rivers with dead zones. Also, the exchange coefficient between the main flow and the dead zones was evaluated. For the present study, 3D steady-state and time-variable numerical simulations were performed with Multiphysics 3.5™, which is a commercial modeling environment, in a rectangular channel with a lateral square cavity representing a dead zone. This geometry was previously experimentally studied by Muto et al. [18] [19].

2. Literature Review

2.1. *Dead zone effect on mass transport in rivers. The Dead-Zone-Model*

Following a tracer cloud from the source until it is spread over the entire river cross-section, three stages of mixing can be distinguished [21]. In the *near-field* mixing is dominated by buoyancy and momentum forces that are determined by the effluent, so transport phenomena must be considered as 3D problems. In the *mid-field*, the tracer mass is already mixed over the river depth, so transport phenomena can be treated as depth-averaged 2D problems. The sum of the above mixing zone is also termed as *advective length*. In the *far-field*, the tracer mass is well-mixed over the entire river cross-section, so transport phenomena are often as cross-sectional 1D problems. Also, in the far-field the skewness of the tracer distribution in the longitudinal direction slowly vanishes. Therefore, in the far-field, solute transport is usually analyzed by using the classical 1D advection-diffusion equation (ADE), where the main problem is to apply a reasonable value of the longitudinal dispersion coefficient [7] [21]. However, many field studies demonstrated that the ADE predictions often are in disagreement with the observed tracer concentrations from instantaneous spills. Two discrepancies were observed [36]. First, tracer concentration curve shows a sharp front and a long tailing. Second, in natural channels, the travel time of the solute is always somewhat smaller than the average velocity of the river. This can be explained by the exchange processes between the main flow in the channel and the dead-water zones existing at its periphery [29]. In natural channels, dead zones can be formed by sand or gravel banks, side cavities, small pockets at the river bed or any other irregularity in the morphology both at the riverbanks and at the riverbed, whereas in regulated rivers, spur dikes and groin fields are typical dead zones. Due to momentum exchange across the interface with the main channel, flow patterns inside the dead zones are characterized by recirculating flows which occur on different scales and exhibit flow velocity in the main stream direction close to zero. Therefore, the most important effect of dead zones on mass transport in rivers is the storage of some amount of the contaminants or nutrients being transported by the main stream inside the dead zones, i.e. some mass of solutes are first trapped in the dead zones and only later, after an average storage time T_{DZ}, are released back into the main flow. The storage time depends on the strength of the exchange processes occurring between the main stream and the dead zones, which are due mostly to turbulent mixing in the lateral or in the vertical direction, if the dead zone is at the riverbanks or at the riverbed, respectively [18] [19] [36].

As discussed below in details, the flow structure in a dead zone consists of a mixing layer, a primary eddy and a core region within this eddy. Depending on the dead zone aspect ratio a smaller corner eddy may exist. Turbulent mixing around a dead zone is mostly governed by two-dimensional coherent structures in the mixing layer between the dead zone and the main channel, even if other mechanisms and 3D structures are believed to be also involved [3] [10] [14]. The effect of these exchange processes on the transport of solutes in a river with dead zones can be quantified considering the *average transport velocity of solutes*, defined as the velocity of the center of mass of the tracer cloud, and the *strength of the longitudinal dispersion process*, which describes the rate of stretching of the tracer cloud [36].

In a channel without dead zones, the average transport velocity of a tracer cloud that is completely mixed over the river cross-section equals the mean flow velocity. If dead zones are present, the part of the tracer cloud trapped in the dead zones is retarded compared to the part of tracer cloud travelling in the main stream with the mean flow velocity. In this case, the transport velocity of the tracer cloud is lower than the average flow velocity. Also, dead zones produce an increased stretching of a passing tracer cloud, which means that the longitudinal dispersion is enhanced. This results in a tail of the contaminants cloud longer than that predicted by the 1D advection-diffusion equation [5] [30] and the length of the tail depends on the exchange between the main flow and the dead zones. This could be explained by two processes [36]. First, dead zones modify the transverse profile of flow velocity in the main channel and increase lateral turbulent mixing, which are the determining parameters for longitudinal dispersion (Fig.1). It is well known that transverse mixing is important in determining the rate of longitudinal mixing because it tends to control the exchange between regions of different longitudinal velocity. Particularly, transverse mixing and longitudinal mixing are inversely proportional. A strong transverse mixing tends to erase the effect of differential longitudinal advection and pollutants particles migrate across the velocity profile so fast that they essentially all move at the mean speed of the flow, causing only a weak longitudinal spreading. On the other hand, a weak transverse mixing implies a long time for differential advection to take effect, so the pollutants patch is highly distorted while it diffuses moderately in the transverse direction and longitudinal mixing is large [4]. Second, tracer cloud is stretched because solute parcels are trapped within the dead zones and only later released back into the main channel. Both processes result in an enhanced longitudinal dispersion, which leads to lower peak levels of tracer concentration but to a longer period during which critical values could be exceeded. Finally, these processes are strongly related to the geomorphological conditions of the dead zones.

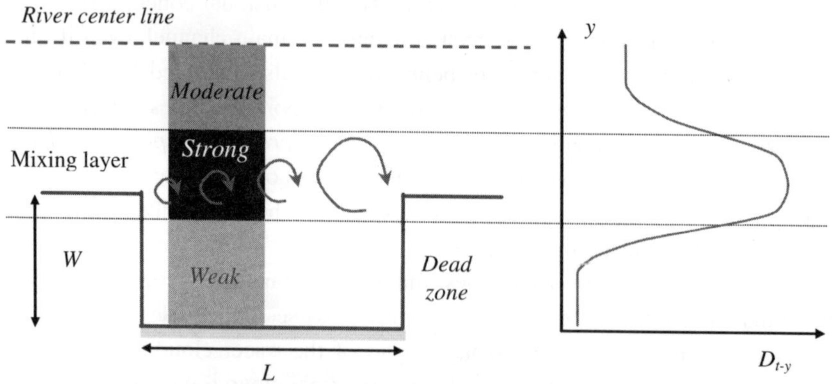

Figure 1. Transverse turbulent diffusivity profile in a river with a dead zone

Therefore, models accounting for the dead zones effects were proposed to be applied in rivers where there is a relevant amount of dead zones. The basic idea of the dead-zone model (DZM) is to distinct two zones within the cross-section of a river, the main stream and the dead-zone. In the main stream the mass transport is governed by advection in the longitudinal direction, longitudinal shear due to the velocity distribution and transverse turbulent diffusion. Thus, in the main stream transport processes can be modeled under well mixed conditions using 1D advection-diffusion equation. In the dead zone, since velocity in the main stream direction is close to zero, transverse turbulent diffusion across the interface between the dead zone and the main stream is the dominant mechanism, which leads to momentum and mass exchange processes. Assuming that in the dead zone the solute concentration is uniform, mass exchange between the dead zone and the main stream is proportional to the difference of the averaged concentration in the dead zone and in the main channel. To set up the DZM, conservation of mass for the main stream and the dead-zone has to be considered, that is [5] [36] [37]:

$$\frac{\partial C_{MS}}{\partial t} = -U\,\frac{\partial C_{MS}}{\partial x} + D_L\,\frac{\partial^2 C_{MS}}{\partial x^2} + \frac{h_E\,E}{h_{MS}\,B}\,\left(C_{DZ} - C_{MS}\right) \tag{1a}$$

$$\frac{\partial C_{DZ}}{\partial t} = -\frac{h_E\,E}{h_{DZ}\,W}\,\left(C_{DZ} - C_{MS}\right) \tag{2a}$$

where C_{MS} and C_{DZ} are the solute concentration in the main stream and in the dead zone, respectively, U is average flow velocity, D_L is longitudinal dispersion coefficient, B and W are channel width and dead zone width, respectively, h_{MS}, h_{DZ} and h_E are the water depths in the main stream, in the dead zone and at their interface, respectively, and E is the exchange velocity across this interface (Fig.1). Usually, the ratios $h_E E/h_{MS}B$ and $h_E E/h_{DZ}W$ are called *exchange coefficients* K_{MS} and K_{DZ} and the above equations yield:

$$\frac{\partial C_{MS}}{\partial t} = -U\,\frac{\partial C_{MS}}{\partial x} + D_L\,\frac{\partial^2 C_{MS}}{\partial x^2} + K_{MS}\,\left(C_{DZ} - C_{MS}\right) \tag{1b}$$

$$\frac{\partial C_{DZ}}{\partial t} = -K_{Dz}\,\left(C_{DZ} - C_{MS}\right) \tag{2b}$$

As previously noted, one important parameter of the dead zone is its *average storage time* T_{DZ}, which could be approximated by calculating the time needed to completely change the water volume in the dead zone. Thus, T_{DZ} could be obtained dividing the dead zone volume by the exchange velocity and the interface area, if L is the dead zone length, as [36]:

$$T_{DZ} = \frac{W\,L\,h_{DZ}}{E\,L\,h_E} = \frac{1}{K_{DZ}}$$ (3)

which demonstrates that T_{DZ} and K_{DZ} are reciprocal, i.e. higher exchange coefficients correspond to lower retention time inside the dead zone. Notably, if the concentration in the main stream is zero, Eq. (2b) corresponds to first-order decay law. Thus, for an instantaneous and homogeneous mass release into the dead zone, the change of concentration inside the dead zone is described by:

$$C_{DZ}\,(t) = C_0\,exp\,(-K_{DZ}\,t) = C_0\,exp\,(-t/T_{DZ})$$ (4)

where C_0 is the initial concentration in the dead zone. Eq. (4) confirms that the exchange coefficient K_{DZ} represents the slope of the decay process occurring inside the dead zone. Usually, K_{DZ} is normalized with the mean velocity in the mean stream and the width of the dead zone giving a dimensionless *exchange coefficient* [30]:

$$k_{DZ} = \frac{K_{DZ}\,W}{U}$$ (5a)

Note that if the water depth at the interface and in the dead zone are equal, i.e. $h_E=h_{DZ}$, the exchange coefficient K_{DZ} becomes:

$$K_{DZ} = \frac{E}{W}$$ (6)

and k_{DZ} is the ratio between the exchange velocity and the main stream velocity:

$$k_{DZ} = \frac{E}{U}$$ (5b)

Typical values of k_{DZ} parameter are in the range from 0.012 to 0.05 [1] [19] [29] [32] [34] [36]. Dimensional analysis leads to conclude that dimensionless exchange coefficient k_{DZ} has the form [36]:

$$k_{DZ} = f\left(Re, Fr, Pe, \frac{k_s}{h_{MS}}, \frac{h_{DZ}}{h_{MS}}, \frac{B}{h_{MS}}, \frac{W}{L}, \alpha, S \right)$$ (7a)

where Re, Fr and Pe are Reynolds, Froude and Peclet number, respectively, k_s is bed roughness and α and S are the inclination to the main flow direction and the shape of the dead zone. However, under typical turbulent flow and low Froude number conditions, a major influence on the mass exchange is given by the aspect ratio of the dead zone *W/L* and by the inclination angle. This leads to discard most of the parameters involved in eq. (7a), which yields:

$$k_{DZ} = f\left(\frac{W}{L}, \alpha\right) \tag{7b}$$

Previous experimental studies demonstrated that the dimensionless exchange coefficient k_{DZ} decreased with the increasing aspect ratio *W/L* of the dead zone [18] [19] [26] [33] [36].

As above outlined, another important parameter in the DZM is the average transport velocity of solutes U_S, which is usually lower than the average flow velocity U [36]. In the DZM, U_S can be derived by a weighted average:

$$U_S = \frac{M_{MS}\,U + M_{DZ}\,U_{DZ}}{M_{MS} + M_{DZ}} \tag{8a}$$

where M_{MS} and M_{DZ} are the solute mass in the main stream and in the dead zone, respectively, and U_{DZ} is the velocity in the dead zone, which is zero. The masses can be replaced by the concentrations by volumes per unit length and eq. (8a) yields [36]:

$$U_S = \frac{h_{MS}\,B\,C_{MS}\,U}{h_{MS}\,B\,C_{MS} + h_{DZ}\,W\,C_{DZ}} = \frac{U}{1 + \dfrac{h_{DZ}\,W\,C_{DZ}}{h_{MS}\,B\,C_{MS}}} \tag{8b}$$

which demonstrates that the transport velocity of a solute in the far-field is determined by the ratios between the cross-sectional areas and the concentrations in the dead zone and in the main stream. Assuming that the concentrations are equal, i.e. $C_{MS}=C_{DZ}$, eq. (8b) yields:

$$U_S = \frac{U}{1 + \dfrac{h_{DZ}\,W}{h_{MS}\,B}} = \frac{U}{1 + \beta} \tag{8c}$$

where the ratio between the cross-sectional areas is termed *dead zone parameter* β [31].

2.2. *Flow patterns in a dead zone*

Literature studies have already pointed out some typical patterns of flow structure around and inside a dead zone which are depending on the aspect ratio W/L [1] [3] [10] [11] [19] [25] [26]. The flow inside a dead zone is a type of shear-driven cavity flow [22] [23], where the flow separates at the upstream

                                    C. Gualtieri

corner of the cavity and forms a wake, like behind a backward facing step, in the lee of the upstream end of the cavity. In turbulent flow, the wake may be bounded by a plane mixing layer (ML), which is the shear flow formed in the region between two co-flowing streams of different velocities [20]. Due to the velocity difference, this region is characterized by a strong inflection in the mean velocity profile which gives rise to hydrodynamic instability processes resulting in the development of the mixing layer. Large scale turbulent structures with the vorticity aligned with the main flow vorticity are continuously fed from the main flow. These structures determine a turbulence length scale in the order of the width of the mixing layer, which develops in self-preserving way with a constant spreading rate, depending only on the relative velocity difference across the mixing layer [2]. Depending on the dead zone aspect ratio, this wake may entirely fill the dead zone as a primary eddy, but for low aspect ratio, secondary or more eddies start to develop. All those eddies are mainly driven by the main stream flow.

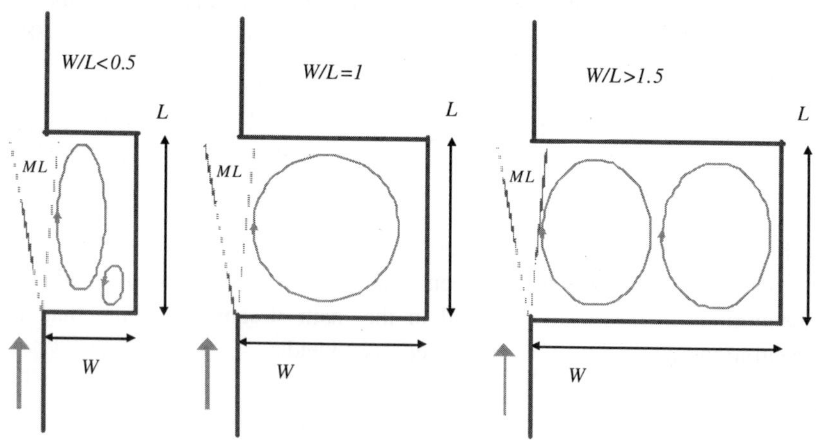

Figure 2. Influence of aspect ratio W/L in flow field in a dead zone

Figure 2 presents typical flow patterns in a dead zone at different aspect ratio. Particularly, if the aspect ratio *W/L* is in the range from 0.5 to 1.5 only the primary recirculating eddy is present inside the dead zone and it covers almost the complete area of the dead zone. If *W/L<0.5*, a second eddy develops in the upstream corner of the dead zone. This secondary gyre has no direct contact with the main stream so it is driven by momentum exchange with the primary eddy. Thus, flow velocity in this area is quite slow. With even smaller aspect ratio, that is *W/L<0.2*, the two eddies remain but the main flow starts to penetrate into the dead zone [25]. On the other hand, if *W/L>1.5*, a second eddy develops behind the primary in transverse direction to the main stream.

Experimental observations with Particle Image Velocimetry (PIV), Laser Induced Fluorescence (LIF) and dye methods of flow dynamics confirmed that large-scale coherent structures were generated at the head of the upstream wall of the dead zone and were growing as they were advected within the mixing layer governing mass exchange across the interface between the dead zone and the main stream [10] [26] [34]. Since a river can be considered to have shallow flow, that is horizontal length scale dominates on vertical length scale, these coherent structures are expected to be mainly two-dimensional. However, some experimental studies pointed out that the exchange process is not uniform over the water depth and the exchange rate increases towards the water surface [3] [33], whereas intermittent vertical structures were observed [10]. Recent experimental studies addressed the role in the exchange process of Kelvin-Helmholtz instabilities occurring at the dead zone-main stream interface [3].

Coming back to the dimensionless exchange coefficient k_{DZ}, it was recognized in some laboratory experimental works that in dead zone with *W/L<0.5* the exchange between the dead zone and the main stream was a two stage process, where two different time scales could be observed [1] [6] [10] [26]. Some authors ascribed this two stage structure to an initial, fast exchange between the primary eddy and the main stream followed by an exchange between the primary eddy and the smaller, secondary eddy in the upstream corner of the dead zone [1] [26]. Other authors suggested a model with a fast exchange for the mixing layer and a slower exchange between the primary eddy and its relatively standing core [6] [10].

2.3. *Computational fluid dynamics (CFD) studies*

Both 2D depth averaged and 3D large-eddy simulation (LES) of the flow past array of submerged and emerged groynes were recently reported in the literature [16] [27]. 2D simulations were more successful for the emerged case when the flow exhibited less three dimensionality. 3D simulations demonstrated that despite that the flow inside the main recirculating eddy in the groyne can be characterized as 2D dimensional, the flow in the mixing layer region was not uniform over the depth [16]. Numerical simulation of solute exchange in a similar geometry demonstrated that the process was lower close to the bottom boundary [15] These numerical results confirmed the above experimental observations [3] [33]. Also, LES simulation [15] confirmed that the mass exchange was characterized by two different time scales, the first one faster and the second one slower, as above highlighted from experiments [1] [6] [10] [26].

Since the approach based on the Reynolds-averaged Navier-Stokes equations (RANS) is largely applied due to its ease in implementation, economy in computation and, most importantly, being able to obtain reasonable accurate solution with the available computer power, standard k-ε model was recently applied to investigate flow patterns around and inside three 2D simplified geometries, representative of flow in natural rivers with dead zones [8]. Exchange rates between the main channel and the dead zone were calculated by using transverse velocity data at the interface according to eq. (5b) [8]. Also, a comparison among the performances of some turbulence models in the analysis of the exchange process in a 2D square dead zone was carried out [9].

2.4. *Lagrangian particle tracking methods*

The effect of a dead zone on solute transport was also investigated by using a 2D Lagrangian Particle Tracking method, which allowed to transfer the local results of laboratory experiments for a single dead zone into the global parameters of a 1D far field model for many dead zones [35] [36]. The method is based on a random walk approach. Under the assumptions of transport dominated by longitudinal shear and transverse diffusion, the movement of a cloud of discrete particles due to longitudinal advection and transverse diffusion was studied in a 2D domain. Advection from a known mean flow profile was superimposed on a random movement in transverse direction, representing turbulent diffusion. The characteristic transport parameters, such as dispersion coefficient, transport velocity and skewness coefficient, were calculated by analyzing the statistics of such a particle cloud at any position of the simulation. In addition to 1D informations, this method yielded concentration distributions in transverse direction, in order to describe near-field phenomena. The method demonstrated that in the presence of large dead water zones at the river banks, an equilibrium between longitudinal dispersion and transverse diffusion can be reached if the morphologic conditions do not change. The simulations resulted in a cross sectional averaged concentration distribution that converged asymptotically to a Gaussian distribution over the longitudinal coordinate. A comparison with field data showed that an agreement of the predicted dispersion coefficient was found within a factor of two [36].

3. Numerical Simulations

As above outlined, the exchange process between a dead zone and the main stream is mainly governed by 2D large-scale coherent structures. Also, the flow structure inside and around the dead zone may be considered as mainly 2D [18]

[26] [34]. However, both experimental and numerical studies demonstrated that the flow in the mixing layer region and the exchange process are not uniform over the water depth [3] [16] [33]. Thus, 3D models should provide a better description of hydrodynamics and mass-transfer processes. These computational fluid dynamics (CFD) models are based on the mass conservation equation and the Navier-Stokes equations of motion. If the flow is turbulent, these equations must be averaged over a small time increment applying Reynolds decomposition, which results in the Reynolds-averaged Navier-Stokes equations (RANS). For an incompressible flow these equations are:

$$\frac{\partial \overline{u}}{\partial x} + \frac{\partial \overline{v}}{\partial y} + \frac{\partial \overline{w}}{\partial z} = 0 \tag{9}$$

$$\overline{\rho}\left(\frac{\partial \overline{u}}{\partial t} + \overline{u}\frac{\partial \overline{u}}{\partial x} + \overline{v}\frac{\partial \overline{u}}{\partial y} + \overline{w}\frac{\partial \overline{u}}{\partial z}\right) = \overline{\rho}\,g_x - \frac{\partial \overline{p}}{\partial x} + \left(\mu + \rho\,v_t\right)\nabla^2\overline{u}$$

$$\overline{\rho}\left(\frac{\partial \overline{v}}{\partial t} + \overline{u}\frac{\partial \overline{v}}{\partial x} + \overline{v}\frac{\partial \overline{v}}{\partial y} + \overline{w}\frac{\partial \overline{v}}{\partial z}\right) = \overline{\rho}\,g_y - \frac{\partial \overline{p}}{\partial y} + \left(\mu + \rho\,v_t\right)\nabla^2\overline{v} \tag{10}$$

$$\overline{\rho}\left(\frac{\partial \overline{w}}{\partial t} + \overline{u}\frac{\partial \overline{w}}{\partial x} + \overline{v}\frac{\partial \overline{w}}{\partial y} + \overline{w}\frac{\partial \overline{w}}{\partial z}\right) = \overline{\rho}\,g_z - \frac{\partial \overline{p}}{\partial z} + \left(\mu + \rho\,v_t\right)\nabla^2\overline{w}$$

where ρ and μ are fluid density and viscosity, p is fluid pressure and u, v and w are velocity components in the x, y and z directions, respectively. The overbar indicates time-averaged quantities. The eddy kinematic viscosity v_t in eq. (10) in the above assumption of isotropic turbulence was estimated as:

$$v_t = \frac{C_\mu\, k^2}{\varepsilon} \tag{11}$$

where k and ε are turbulent kinetic energy per mass unit and its dissipation rate, respectively. These parameters were estimated by using the two-equations of standard k-ε model [17]:

$$\rho\frac{\partial k}{\partial t} + \rho\overline{V}\cdot\nabla k = \nabla\cdot\left[\left(\frac{\mu+\mu_t}{\sigma_k}\right)\nabla k\right] + \frac{1}{2}\mu_t\left(\nabla\overline{V} + \left(\nabla\overline{V}\right)^T\right)^2 - \rho\varepsilon \tag{12}$$

$$\rho\frac{\partial \varepsilon}{\partial t} + \rho\overline{V}\cdot\nabla\varepsilon = \nabla\cdot\left[\left(\frac{\mu+\mu_t}{\sigma_\varepsilon}\right)\nabla\varepsilon\right] + \frac{1}{2}C_{1\varepsilon}\frac{\varepsilon}{k}\mu_t\left(\nabla\overline{V} + \left(\nabla\overline{V}\right)^T\right)^2 - \rho C_{2\varepsilon}\frac{\varepsilon^2}{k} \tag{13}$$

where μ_t is dynamic eddy viscosity, whereas C_μ, σ_k, σ_ε, $C_{1\varepsilon}$ and $C_{2\varepsilon}$ are constants and their values are listed in Table 1.

Table 1. Values of the constants of the standard k-ε model

C_μ	σ_k	σ_ε	$C_{1\varepsilon}$	$C_{2\varepsilon}$
0.09	1.00	1.30	1.44	1.92

For solute transport, the 3D advection-diffusion equation was used [17]:

$$\frac{\partial \overline{C}}{\partial t} + \overline{u}\,\frac{\partial \overline{C}}{\partial x} + \overline{v}\,\frac{\partial \overline{C}}{\partial y} + \overline{w}\,\frac{\partial \overline{C}}{\partial z} = \frac{\partial}{\partial x}\left(D_t\,\frac{\partial \overline{C}}{\partial x}\right) + \frac{\partial}{\partial y}\left(D_t\,\frac{\partial \overline{C}}{\partial y}\right) + \frac{\partial}{\partial z}\left(D_t\,\frac{\partial \overline{C}}{\partial z}\right) \quad (14)$$

where molecular diffusion was neglected, D_t is turbulent diffusivity and $\overline{C}$ is solute concentration. Turbulent diffusivity was derived from turbulent viscosity following Reynolds analogy [4]. Their ratio is the turbulent Schmidt number Sc_t:

$$Sc_t = \frac{v_t}{D_t} \quad (15)$$

which is analogous to the classical Schmidt number $Sc=v/D_m$, ratio of kinematic viscosity to molecular diffusivity. In open channel flows, Sc_t is usually in the range from 0.3 to 1.0 [21]. After some preliminary tests it was assumed that solute transfer was stronger than momentum transfer and $Sc_t=0.75$ was applied.

These equations were solved by using Multiphysics 3.5™ modeling package, which solves for the same flow domain both motion and advection-diffusion equations. The model provides as outputs the velocity components $\overline{u}$, $\overline{v}$ and $\overline{w}$, k-ε model parameters, solute concentration $\overline{C}$ and many others [17]. Multiphysics 3.5™ was applied to the 3D geometry presented in Figure 3, which reproduced the laboratory flume used by Muto et al. [18] [19]. They conducted their experiments in a rectangular flume 2.0 m long and 0.16 m wide. The lateral square cavity was 0.16 m long and 0.16 m wide. The water depth was everywhere 0.038 m, i.e. $h_{MS}=H_{DZ}=h_E=h$. Inlet velocity was $U_{in}=0.37$ m/s, which corresponded to a $Re=12343$. They used two laser Doppler anemometers (LDA) to obtain at the same time 3 components velocity.

Several *boundary conditions* were required. For the k-ε model, boundary conditions were assigned at the inlet, the outlet, at the free surface and at the walls:

- at the inlet, an *inflow* type boundary condition was applied, with uniform velocity profile. Also, inlet turbulent intensity I_T and turbulent length scale L_T were assigned. Usually, I_T can be derived from the Reynolds number as:

$$I_T=0.16\,Re^{-1/8} \quad (16)$$

Eq. (16), for a *Re*=12343, provides a turbulent intensity of about 5%, which corresponds to fully turbulent flows.

The turbulent length scale L_T is a physical quantity related to the size of the large eddies that contain the energy in turbulent flows. Also it is a measure of the size of the turbulent eddies that are not resolved [17]. For fully developed channel flows, this parameter can be approximately derived as 0.1×B, where B is the channel width. Thus, L_T=0.016 m;

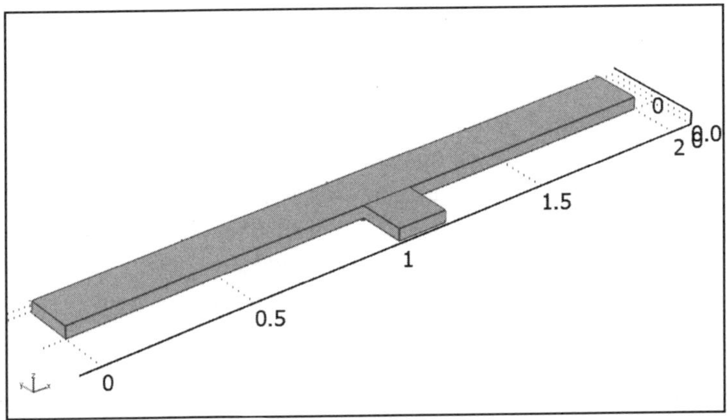

Figure 3. Dead zone geometry investigated in this study

- at the outlet, a *zero pressure* type condition was assigned;
- for the free surface, the simplest approach is to assume a symmetry plane condition, which will result in the velocity contours at the free surface appearing normal to the boundary and in the maximum velocity occurring at the free surface [12]. This is known as the *fixed lid* approach. This approach holds only in first approximation, since, secondary currents lead to the maximum velocity occurring beneath the free surface. This is the *velocity dip* phenomenon [12];
- at the walls, *logarithmic law of the wall* boundary condition was applied. It is well-known that turbulent flows are significantly affected by the presence of walls. First, no-slip condition must be satisfied at the walls for the mean velocity field. Very close to the wall, viscous damping reduces the tangential velocity fluctuations, while kinematic blocking reduces the normal fluctuations. Turbulent eddies are distorted and constrained in size, being compressed in the wall-normal direction and elongated in the streamwise direction. Moreover, as at the wall turbulent

                           C. Gualtieri

energy is damped and dissipated into heat through molecular viscosity, toward the outer part of the near-wall region, however, the turbulence is rapidly increasing by the production of turbulence kinetic energy due to the large gradients in the mean velocity and in the other parameters of the flow field. Also, in this region it should be expected that momentum and scalar transport occur most vigorously. Hence, accurate flow modeling in the near-wall region significantly affects successful prediction of turbulent wall-bounded flows. The classical k-ε model due to its basic hypothesis of isotropy needs to be modified to account for the effect of the walls on the local structure of turbulence. Different approaches were proposed in the literature [24]. Herein to account for solid walls the approach based on the so-called *wall functions* was applied to bridge the viscosity-affected region between the wall and the fully-turbulent region. This approach is the most commonly because it eliminates the need for very fine computational meshes that are required for direct integration of the governing equations all the way to the wall. Also, it reduces numerical stiffness due to the stiff source terms in the turbulence closure equations and large aspect ratio meshes. Basically a wall function is a semi-empirical relation between the value of velocity and wall friction which replaces the thin boundary layer near the wall [17]. This approach is expected to be accurate for high Reynolds numbers and situations where pressure variations along the walls are not very large. However, it can often be used outside its frame of validity with reasonable success [17]. In particular, logarithmic wall functions applied to finite elements assume that the computational domain begins a distance δ_w from the real wall. They also assume that the flow is parallel to the wall and that the velocity can be described by:

$$u^+ = \frac{u}{u*} = \frac{1}{\kappa} ln\left(\frac{\delta_w \, u*}{v}\right) + C_l \qquad (17)$$

where κ is Von Kármán constant, which is equal to 0.42, and C_l is universal constant for smooth walls equal to 5.5 . Note that the ratio of the kinematic viscosity to the friction velocity, $v/u*$ is the viscous length scale. Moreover, the term in the round brackets is $\delta_w{}^+$ and the logarithmic wall functions are formally valid for values of between 30 and 300. This is the usual practice in standard turbulence models with wall functions [12]. Multiphysics 3.5™ allows to select $\delta_w{}^+$ value, which was set up to 150.

For the advection-diffusion equation, boundary conditions were assigned at the inlet, the outlet, at the free surface and at the walls:

- at the inlet, a *concentration* type boundary condition was applied, assuming zero concentration entering the domain;
- at the outlet, an *advective flux* type condition was assigned;
- at the free surface walls, an *insulation* type condition was applied. This conditions means that the solute cannot cross the free surface;
- at the walls, again, an *insulation* type condition was applied.

As initial condition, a constant, homogeneous concentration of a solute was assigned in the dead zone domain. The initial value is 100.

For the simulation water with density $\rho=1000$ Kg/m³ and molecular viscosity $\mu=1.14\cdot10^{-3}$ Kg/m×s, was selected as fluid.

Different mesh characteristics were tested. After that, the mesh generation process was made assuming, among the others, as *element growth rate* and *resolution of narrow regions* 1.2 and 1.00, respectively. The *element growth rate* determines the maximum rate at which the element size can grow from a region with small elements to a region with larger elements [17]. The value must be greater or equal to 1. In the *resolution of narrow regions* field the number of layer of elements created in narrow regions is controlled [17]. Moreover, the maximum element sizes were 0.04 m, 0.02 m and 0.005 m in the main stream, in the dead zone and at the interface between them, respectively. Thus, the mesh had 19086 tetrahedral elements, with a minimum *element quality* of 0.249. The *element quality* measure is related to its aspect ratio [17]. It is a scalar from 0 to 1. Mesh quality visualization demonstrated a quite uniform quality of the elements of the mesh. Moreover, it is well-known that by using a mesh, the dependent variables are approximated with a function that can be described with a finite number of parameters, the degrees of freedom [17], which was 182883.

About the solver settings, stationary segregated solver with non-linear system solver was used, where the relative tolerance and the maximum number of segregated iterations were set to $1.0\cdot10^{-3}$ and 100, respectively. The segregated solver allows to split the solution steps into substeps. These are defined by grouping solution components together. This procedure can save both memory and assembly time. Three groups were considered, namely velocity components $\overline{u}$, $\overline{v}$ and $\overline{w}$ and pressure $\overline{p}$, turbulence model parameters k and ε, and the solute concentration $\overline{C}$. For time-variable analysis, a time step of 1 second was selected extending the simulation until 200 seconds.

4. Results. Discussion

Numerical simulations provided as outputs both velocity and concentration field, pressure, turbulent viscosity v_t, k and ε values throughout the flow domain. Figure 4 presents the flow field in the considered geometry. As expected, in the dead zone velocities lower (blue) than in the main stream were predicted by Multiphysics 3.5™.

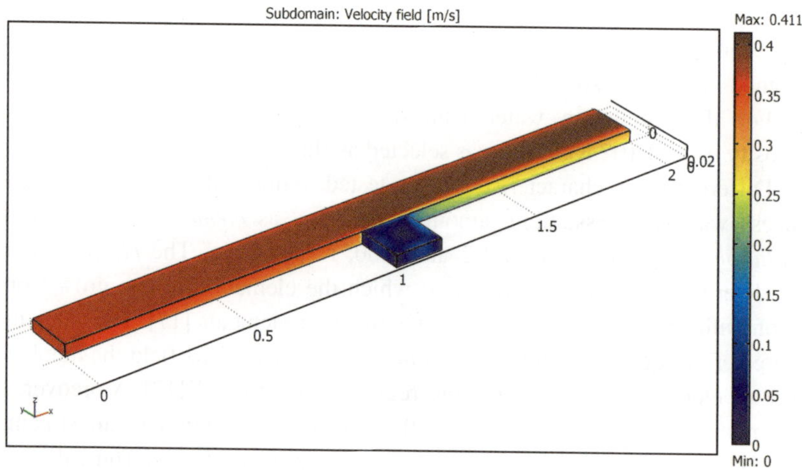

Figure 4. Flow field in the considered geometry

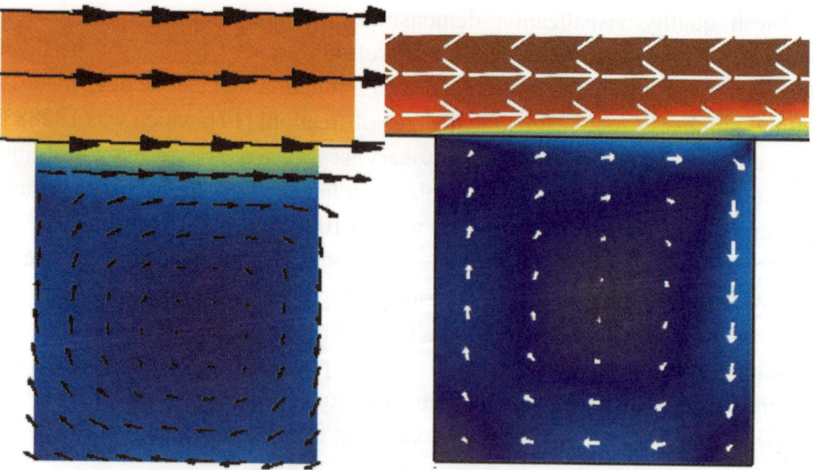

Figure 5. Measured (left) and simulated (right) velocity values and velocity vectors in the square cavity

Figure 5 presents a view in the plan *xy* of measured (left) and simulated (right) velocity values (surface color) and velocity vectors (arrows) in the square cavity experimentally studied by Muto et al. [18 [19], where the flow in the main stream is from the left. Inside the cavity a recirculating primary eddy was observed and its core with very low velocities (blue) was located around the midpoint of the cavity. Higher velocities can be observed in the mixing layer developing at the interface between the cavity and the main channel.

Figure 6 shows a comparison between the transverse velocity along the interface between the dead zone and the main stream predicted by Multiphysics 3.5™ and the experimental values collected at mid-depth, i.e. z/h_E=0.50. Numerical values refer to different dimensionless water depths, z/h_E=0, 0.25, 0.50, 0.75 and 1. Positive values are directed toward the main stream.

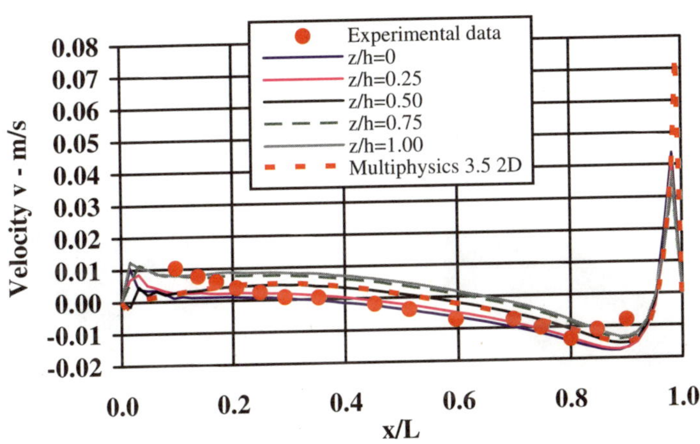

Figure 6. Transverse velocities along the dead zone-main stream interface

Negative and positive values represented the exchange processes between the dead zone and the main stream. At the bottom, the simulated velocity profile was mainly negative, whereas at the free surface it is mainly positive. The best agreement between experimental data and numerical results was at z/h=0.25. Note that in Figure 6 are also presented the results from a 2D numerical simulations which are quite similar to those from the 3D model at z/h=0.50.

Further comparison between experimental observations and numerical simulations was carried out by using the values of dimensionless exchange coefficient k_{DZ}, which as above stated represents the rate of momentum and mass exchange between the dead zone and the main stream. This parameter was

                    C. Gualtieri

determined using the transverse velocity data along the dead zone-main channel interface and the procedure proposed by Weitbrecht and Jirka [34]. That is, the norm of transverse velocity v_i was integrated over the interface, resulting in the total specific volume flux through the exchange area. Division by 2 yielded the total specific out flux into the main stream:

$$\overline{E} = \frac{1}{2\,A_{int}} \int_{A_{int}} |v|$$

(17)

where A_{int} is the area of the interface, which provides the exchange coefficient K_{DZ} and the dimensionless exchange coefficient k_{DZ} through eqs. (6) and (5a), respectively. Note that Muto et al. determined two different exchange coefficients, for mass and for momentum [19]. The former is mainly related to the lateral component of velocity, whereas the latter is mostly governed by the lateral shear stress and by the development of the mixing layer at the interface between the dead zone and the main stream. Their values were 0.0148 and 0.0116 [19].

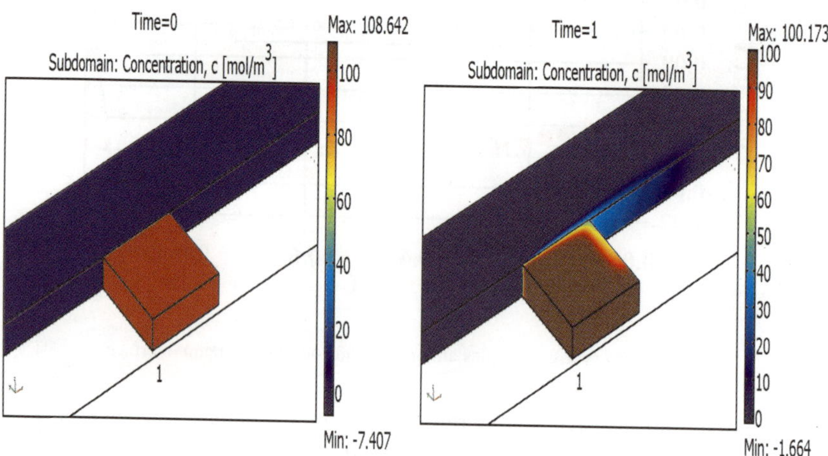

Figure 7. Concentration field in the dead zone at t=0 s (left) and t=1 s (right)

From eq. (17), numerical dimensionless exchange coefficient k_{DZ} was 0.0090, which means that the numerical model underestimated the exchange occurring between the dead zone and the main stream. This was not unexpected because the k-ε model assumes that the turbulence is isotropic, whereas turbulence in open channel flows is known to be anisotropic. The anisotropic behavior of turbulence as it approaches the walls and free surface, namely the imbalance in the normal Reynolds stresses, creates secondary circulation even if in a straight channel [12]. Moreover, in a shallow river reach with a lateral dead zone, shallow flow typical properties in combination with large horizontal structures leads to a high degree of anisotropy of flow turbulence [25]. That is, as above outlined, the transverse turbulent exchange of momentum is very strong. Therefore, the k-ε model either fails to predict any evidence of secondary flow in the case prismatic channels and tends to underestimate the lateral exchange of momentum across the interface between the dead zone and the main stream.

Second, the exchange rate was not constant over the depth, but was maximum at the free surface, where $k_{DZ}=0.0097$. This trend confirmed previous literature experimental observations [3] [32].

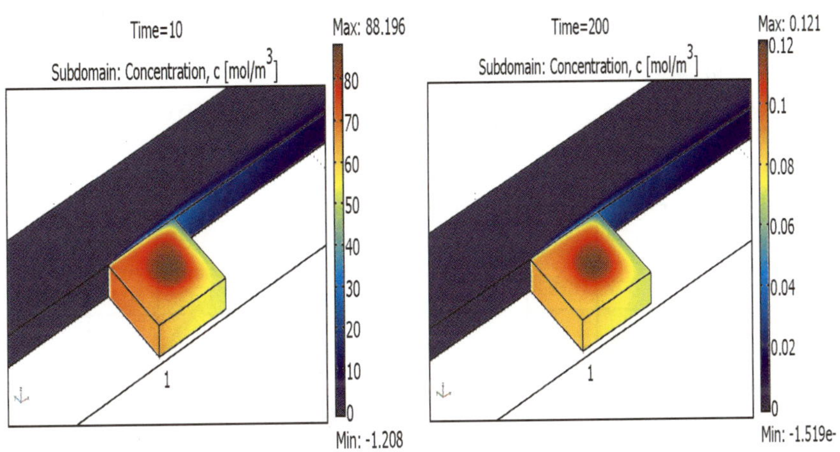

Figure 8. Concentration field in the dead zone at t=10 s (left) and t=200 s (right)

A second estimation of the exchange rate between the dead zone and the main stream was carried out using the data of temporal concentration decay of a solute that was instantaneously and homogeneously released at *t=0* into the volume of the dead zone. Figures 7 and 8 present the concentration field around and inside the dead zone at different times, namely *t*=0, 1, 10 and 200 seconds. Note that *t*=200 seconds is the end of the simulation. It can be observed that solute purging process from the dead zone followed closely flow patterns. Solute is first removed in the downstream part of the dead zone (right in Fig.5), where *clean* fluid enters from the main stream. After that, following the pattern of the primary eddy, the remaining part of the dead zone was cleaned. At the end of the time-variable simulation, the concentration in the dead zone was at maximum equal to 0.12% of the initial value. This concentration was located in the core region, around the center of the primary eddy. In this region, higher values were observed at mid-depth, i.e. z=0.019 m. Moreover, in the interfacial region, solute fluxes were higher at the free surface than at the bottom confirming previous literature experimental observations [3] [32].

Finally, dimensionless exchange rate k_{DZ} was evaluated by assuming that the change in concentration inside the dead zone was described by eq. (4). Thus, the average storage time T_{DZ} and the dimensionless exchange rate k_{DZ} were obtained. By integrating the concentration to the entire volume of the dead zone, an averaged decay of the concentration over the time was calculated (Fig.9), which corresponded to a dimensionless exchange rate k_{DZ}=0.0151. Note that concentration decay was also studied in several different points of the dead zone and k_{DZ} was in the range from 0.0150 to 0.0152. In the core region, k_{DZ} was 0.150. This value is not far from that determined by Muto et al. [19], that is 0.0148. Note that by assuming Sc_t=1.0, the dimensionless exchange rate was k_{DZ}=0.0126, so the numerical results proved to be quite sensitive to the turbulent Schmidt number values.

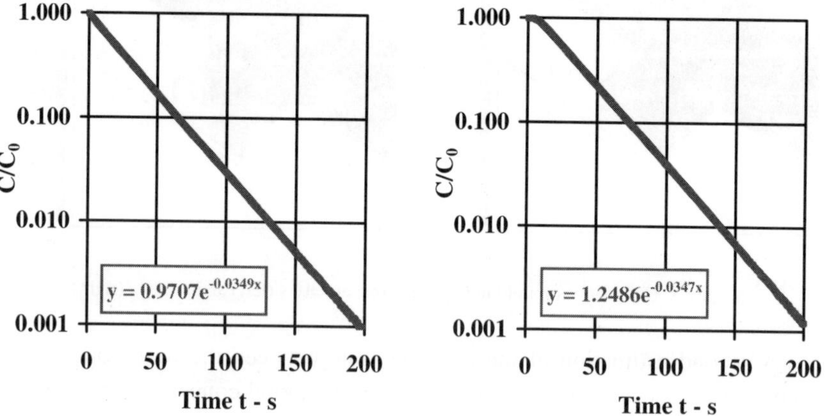

Figure 9. Temporal decay of the concentration in the dead zone. Volume averaged (left) and in the core region (right)

The difference in the numerical k_{DZ} values obtained from velocity and concentration data can be first explained considering that across the interfacial area between the dead zone and the main stream simulated concentration gradients were generally larger than velocity gradients. Concentration gradients were the *driving force* of large diffusive fluxes from the dead zone to the main stream. These fluxes added to those of advective nature due to the transverse velocity. Second, the flow patterns around the interface between the dead zone and the main stream (Fig.5, right) further contributed to diffusive fluxes from the dead zone. That is, part of the solute flux escaping from the dead zone around the upstream corner was again introduced into the dead zone at the downstream corner. This enhanced concentration gradients and, thus, diffusive fluxes toward the main stream. Third, by assuming that $Sc_t=0.75$, it was expected that solute mass transfer was stronger than momentum transfer.

5. Conclusions

The presence of dead zones in streams and rivers significantly affect the characteristics of mass transport. In a river, dead zones can be due to geometrical irregularities in the riverbanks and the riverbed and/or to spur dikes and groyne fields. Dead zones produce a difference between the concentration curves measured and modeled by the classical 1D advection-diffusion equation with sharper front and longer tails. In a dead zone, the mean flow velocity in the main stream direction is essentially zero and the main transport mechanism is transverse turbulent diffusion which controls the exchange processes of solutes with the main stream.

The chapter presented some results of a numerical 3D study undertaken to investigate the fundamental flow phenomena around and inside a simplified geometry, representative of the flow in a natural channel with a lateral square dead zone. In the numerical study an approach based on the Reynolds Averaged Navier-Stokes (RANS) equations was applied, where the closure problem was solved by using a turbulent viscosity concept. Particularly, the classical two-equations k-ε model was used. Two methods were applied to estimate the exchange rate between the dead zone and the main stream. The first one was based upon the analysis of transverse velocity at the interface between the dead zone and the main stream. The second used concentration data over the time after that a tracer was instantaneously and homogeneously released into the dead zone.

First, the analysis of flow field demonstrated that the numerical simulations qualitatively reproduced the observed flow patterns but underestimated the exchange rate between the dead zone and the main stream. This is due to the assumption of isotropic turbulence which is done in the k-ε model. In a reach of a shallow river with a lateral dead zone, an higher degree of anisotropy of flow turbulence is expected. Therefore, this model fails to reproduce the strong lateral exchange of momentum existing across the interface between the dead zone and the main stream.

Second, the analysis of concentration data determined a predicted dimensionless exchange rate k_{DZ} close to the experimental value. The difference between k_{DZ} values obtained with the two methods can be explained considering that the flow patterns around the interface between the dead zone and the main stream contributed to increase diffusive fluxes from the dead zone. Moreover, numerical results were quite sensitive to the assumption of a solute transfer stronger than the momentum transfer, i.e. Sc_t=0.75.

Future research will be addressed to apply different turbulent models to the study of the exchange process in channels with lateral dead zones and to investigate the effects of other parameters of the dead zone, such as the aspect ratio W/L and the inclination to the main flow direction α.

Appendix – List of Symbols

List of Symbols		
Symbol	**Definition**	**Dimensions or Units**
A_{int}	area of the interface between the dead zone and the main stream	$[L^2]$
B	channel or main stream width	$[L]$
C	concentration	$[M{\cdot}L^{-3}]$
C_{DZ}	concentration in the dead zone	$[M{\cdot}L^{-3}]$
C_{MS}	concentration in the main stream	$[M{\cdot}L^{-3}]$
C_0	concentration in the dead zone at t=0	$[M{\cdot}L^{-3}]$
C_1	universal constant for smooth walls	
$C_{1\varepsilon}, C_{2\varepsilon}$	numerical constants of k-ε model	
C_μ	numerical constant of k-ε model	
$\overline{C}$	temporal mean concentration	$[M{\cdot}L^{-3}]$
D_L	longitudinal dispersion coefficient	$[L^2\,T^{-1}]$
D_t	turbulent diffusion coefficient	$[L^2\,T^{-1}]$
E	exchange velocity	$[L\,T^{-1}]$
Fr	Froude number	
J_b	channel bed slope	
I_T	turbulent intensity	
K_{DZ}	dead zone exchange coefficient	$[T^{-1}]$
K_{MS}	main stream exchange coefficient	$[T^{-1}]$
L	dead zone length	$[L]$
L_T	turbulent length scale	$[L]$

M_{DZ}	solute mass in the dead zone	$[M]$
M_{MS}	solute mass in the main stream	$[M]$
Pe	Peclet number	
Re	Reynolds number	
S	dead zone shape	
Sc	Schmidt number	
Sc_t	turbulent Schmidt number	
T_{DZ}	average storage time in the dead zone	$[T]$
U	average flow velocity	$[L{\cdot}T^{-1}]$
U_{in}	inlet velocity	$[L{\cdot}T^{-1}]$
U_{DZ}	average velocity in the dead zone	$[L{\cdot}T^{-1}]$
U_S	transport velocity of a solute	$[L{\cdot}T^{-1}]$
W	dead zone width	$[L]$
g	gravitational acceleration constant	$[L\,T^{-2}]$
h	water depth	$[L]$
h_{DZ}	water depth in the dead zone	$[L]$
h_E	water depth at the interface	$[L]$
h_{MZ}	water depth in the main stream	$[L]$
k	turbulent kinetic energy per unit mass	$[L^2\,T^{-2}]$
k_{DZ}	dimensionless exchange coefficient	
k_s	bed roughness	$[L]$
t	time	$[T]$
u^*	shear velocity	$[L{\cdot}T^{-1}]$
u^+	dimensionless velocity	
$\bar{u}$	time-averaged component velocity in the x direction	$[L{\cdot}T^{-1}]$
$\bar{v}$	time-averaged component velocity in the y direction	$[L{\cdot}T^{-1}]$
$\bar{w}$	time-averaged component velocity in the z direction	$[L{\cdot}T^{-1}]$
X	longitudinal coordinate	$[L]$
Y	transverse coordinate	$[L]$
Z	vertical coordinate	$[L]$
α	inclination of the dead zone to the main stream direction	
β	dead zone parameter	

δ_w	wall offset	[L]
δ_w^+	wall offset in viscous unit	
ε	dissipation rate of turbulent kinetic energy per unit mass	$[L^2 T^{-3}]$
κ	Von Karman constant	
μ	water dynamic viscosity	$[M L^{-1} T^{-1}]$
ν	water kinematic viscosity	$[L^2 T^{-1}]$
ν_t	water turbulent kinematic viscosity	$[L^2 T^{-1}]$
ρ	water density	$[M L^{-3}]$
σ_ε , σ_k	numerical constants of k-ε model	

References

1. Booij, R., Exchange of mass in harbors, *Proceedings of the XXIII IAHR Congress*, Ottawa, Canada, (1989).
2. Booij, R., and Tukker. J., Integral model of shallow mixing layers, *Journal of Hydraulic Research,* **39** (2), 169-179, (2001).
3. Brevis, W., Niño, Y., and Vargas, J., Experimental characterization and visualization of mass exchange process in dead zones in rivers, *Proceedings of 3rd International Conference on Fluvial Hydraulics (RiverFlow 2006)*, 235-242, (2006).
4. Cushman-Roisin, B., Environmental fate and transport, Lecture Notes, Thayer School of Engineering, Dartmouth College, NH, USA, (2007).
5. Czernuszenko, W., Rowiński, P.M., and Sukhodolov, A., Experimental and numerical validation of the dead-zone model for longitudinal dispersion in rivers, *Journal of Hydraulic Research,* **36** (2), 269-280, (1998).
6. Engelhardt, C., Kruger, A., Sukhodolov, A., and Nicklisch, A., A study of phytoplankton spatial distributions, flow structure and characteristics of mixing in a river reach with groynes, *Journal of Plankton Research*, 26 (11), 1351-1366, (2004).
7. Gualtieri, C., Ciaravino, G., Pulci Doria, G., Analysis of longitudinal dispersion equations in streams and rivers, *Proceedings of the 7th International Conference on HydroScience & Engineering (ICHE 2006)*, Philadelphia, USA, (2006).
8. Gualtieri C., Numerical simulation of flow patterns and mass exchange processes in dead zones, *Proceedings of the iEMSs Fourth Biennial Meeting: International Congress on Environmental Modelling and Software (iEMSs 2008)*, Barcelona, Spain, (2008).
9. Gualtieri, C., López Jiménez, P.A., and Mora Rodríguez J.J., (2009). A comparison among turbulence modelling approaches in the simulation of a square dead zone. *XXXIII IAHR Congress*, Vancouver, Canada, (2009).

10. Jamieson, E., and Gaskin, S.J., Laboratory study of 3 dimensional characteristics of recirculating flow in a river embayment, *Proceedings of the XXXII IAHR Congress*, Venice, Italy, (2007).

11. Kimura, I., and Hosoda, T., Fundamental properties of flows in open channels with dead zone, *Journal of Hydraulic Engineering, ASCE,* **123** (2), 322-347, (1997).

12. Knight, D.W., Wright, N.G., and Morvan, H.P., *Guidelines for Applying Commercial CFD Software to Open Channel Flow*, Report based on research work conducted under EPSRC Grants GR/R43716/01 and GR/R43723/01, pp.31, (2005).

13. Kurzke, M., Weitbrecht, V., and Jirka G.H.J., Laboratory concentration measurements for determination of mass exchange between groin fields and main stream. *Proceedings of 1st International Conference on Fluvial Hydraulics (River Flow 2002)*, vol.1, 369-376, (2002).

14. Le Coz, J., Brevis, W., Niño, Y., Paquier, A., and Riviére, N., Open-channel side-cavities: A comparison of field and flume experiments, *Proceedings of 3rd International Conference on Fluvial Hydraulics (RiverFlow 2006)*, 235-242, (2006).

15. McCoy, A., Constantinescu, G., and Weber, L., Exchange processes in a channel with two vertical emerged obstructions, *Flow, Turbulence and Combustion,* **77** (1-4), 97-126, (2006).

16. McCoy, A., Numerical investigation of flow hydrodynamics in a channel with a series of groynes. *Journal of Hydraulic Engineering,* **134** (2), 157-172, (2008).

17. Multiphysics 3.5 *User's Guide*, ComSol AB. Sweden, (2008).

18. Muto, Y., Imamoto, H., and Ishigaki, T., Turbulence characteristics of a shear flow in an embayment attached to a straight open channel, *Proceedings of the 4th International Conference on HydroScience & Engineering (ICHE 2000)*, Seoul, Korea, (2000).

19. Muto, Y., Imamoto, H., and Ishigaki, T., Velocity measurements in a straight open channel with a rectangular embayment, *Proceedings of the 12th APD-IAHR, Bangkok, Thailand*, (2000).

20. Raupach. M.R., Finnigan, J.J., and Brunet, Y., Coherent eddies and turbulence in vegetation canopies: the mixing-layer analogy, *Boundary-Layer Meteorology,* **78**, 351-382, (1996).

21. Rutherford, J.C., *River Mixing*, John Wiley & Sons, Chichester, U.K., pp.348, (1994).

22. Shankar, P.N., and Deshpande, M.D., Fluid mechanics in the driven cavity, *Annual Review in Fluid Mechanics,* **32**, 93-136, (2000).

23. Shen, C., and Floryan, J.M., Low Reynolds number flow over cavities, *Physics of Fluids A,* **28**, n.11, November 1985, 3191-3202, (1985).

24. Sotiropoulos, F., Introduction to statistical turbulence modeling for hydraulic engineering flows, in Bates, P.D., Lane, S.N., and Ferguson, R.I.,

(Eds) *Computational Fluid Dynamics. Applications in Environmental Hydraulics*, John Wiley & Sons, Chichester, England, pp.534, (2005).

25. Uijttewaal, W.S.J., Groyne field velocity patterns determined with particle tracking velocimetry, *Proceedings of the XXVIII IAHR Congress*, Graz, Austria, (1999).

26. Uijttewaal, W.S.J., Lehmann, D., and Mazijk, A. van, Exchange processes between a river and its groyne fields: model experiments, *Journal of Hydraulic Engineering, ASCE,* **127** (11), 928-936, (2001).

27. Uijttewaal, W.S.J., and Van Schijndel, S.A.H., The complex flow in groyne fields: numerical modeling compared with experiments, *Proceedings of 2nd International Conference on Fluvial Hydraulics (River Flow 2004)*, vol.2, 1331-1338, (2004).

28. Uijttewaal, W.S.J., Effects of groyne layout on the flow in groyne fields: laboratory experiments, *Journal of Hydraulic Engineering, ASCE,* **131** (9), 782-791, (2005).

29. Valentine, E.M., and Wood, I.R., Longitudinal dispersion with dead zones, *Journal of Hydraulics Division, ASCE,* **103** (9), 975-990, (1977).

30. Valentine E.M., and Wood I.R., Experiments in longitudinal dispersion with dead zones, *Journal of Hydraulics Division, ASCE,* **105** (8), 999-1016, (1979).

31. Van Mazijk, A., *One-dimensional approach of transport phenomena of dissolved matter in rivers*, PhD Thesis, TU-DELFT, Faculty of Civil Engineering, (1996).

32. Wallast, I., Uijttewaal, W., and Van Mazijk, A., Exchange processes between groyne field and main stream, *Proceedings of the XXVIII IAHR Congress*, Graz, Austria, (1999).

33. Weitbrecht, V., and Jirka, G.H.J., Flow patterns and exchange processes in dead zones of rivers, *Proceedings of the XXIX IAHR Congress*, Seoul, Korea, (2001).

34. Weitbrecht, V., and Jirka, G.H.J., Flow patterns in dead zones of rivers and their effect on exchange processes, *Proceedings of the 2001 International Symposium on Environmental Hydraulics*, Tempe, USA, (2001).

35. Weitbrecht, V., Uijttewaal, W., and Jirka G.H.J., A random walk approach for investigating near- and far-field transport phenomena in rivers with groin fields, *Proceedings of 2nd International Conference on Fluvial Hydraulics (River Flow 2004)*, vol.2, 1157-1166, (2004).

36. Weitbrecht, V., *Influence of dead-water zones on the dispersive mass-transport in rivers*, Dissertation, (2004).

37. Weitbrecht, V., Socolofsky, S.A, Jirka, G.H., Experiments on mass exchange between groin fields and the main stream in rivers, *Journal of Hydraulic Engineering*, **134** (2), 173-183, (2008).

Chapter 13

MODELING MERCURY FATE AND TRANSPORT IN AQUATIC SYSTEMS

ARASH MASSOUDIEH[1], DUŠAN ŽAGAR[2], PETER G. GREEN[3], CARLOS CABRERA-TOLEDO[3], MILENA HORVAT[4], TIMOTHY R. GINN[3], TAMMER BARKOUKI[3], TESS WEATHERS[3], FABIAN A. BOMBARDELLI[3, 5]

[1]*Department of Civil Engineering, The Catholic University of America, USA*
[2]*Faculty of Civil and Geodetic Engineering, University of Ljubljana, Slovenia*
[3]*Department of Civil and Environmental Engineering, University of California, USA*
[4]*Department of Environmental Sciences, Jožef Stefan Institute, Slovenia*

Mercury in the aquatic environment is a neurotoxin with several known adverse effects on the natural ecosystem and the human health. Mathematical modeling is a cost-effective way for assessing the risk associated with mercury to aquatic organisms and for developing management plans for the reduction of mercury exposure in such systems. However, the analysis of mercury fate and transport in the aquatic environment requires multiple disciplines of science ranging from sediment transport and hydraulics, to geochemistry and microbiology. Also, it involves the knowledge of some less understood processes such as the microbial and diagenetic processes affecting the chemical speciation of mercury and various mechanisms involved in the mass-exchange of mercury species between the benthic sediments and the overlying water. Due to these complexities, there are many challenges involved in developing an integrated mercury fate and transport model in aquatic systems. This paper identifies the various processes that are potentially important in mercury fate and transport as well as the knowns and unknowns about these processes. Also, an integrated multi-component reactive transport modeling approach is suggested to capture several of those processes. This integrated modeling framework includes the coupled advective-dispersive transport of mercury species in the water body, both in dissolved phase and as associated to mobile suspended sediments. The flux of mercury in the benthic sediments as a result of diffusive mass exchange, bio-dispersion, and hyporheic flow, and the flow generated due to consolidation of newly deposited sediments is also addressed. The model considers in addition the deposition and resuspension of sediments and their effect on the mass exchange of mercury species between the top water and the benthic sediments. As for the biogeochemical processes, the effect of redox stratification and activities of sulfate and iron-reducing bacteria on the methylation of mercury is discussed, and the modeling approach is described. Some results for the application of the model to the Colusa Basin Drain in California are presented. At the end of the paper, the shortcomings of our current knowledge in predicting the fate of mercury in water-sediment systems, the potential improvements, and additional complexities required to make the model more realistic, are discussed.

Keywords: aquatic, mercury, methylation, resuspension, shear stress, non-cohesive, mass-exchange, diagenesis, sorption

1. Introduction

Mercury in the aquatic environment is a neurotoxin with several known adverse effects on the natural ecosystem and on human health. A large number of water bodies around the world are contaminated with mercury as a result of direct anthropogenic activities such as mining, or indirectly via dry and wet deposition of mercury. An essential component of any conceptual or mathematical model of mercury fate and transport involves the types and rates of mercury transformations. Important among the different chemical forms or "species" is its methylated form (MeHg) which is the most bioavailable. Mercury is believed to be methylated in anoxic waters and sediments [85] and then transported (upward usually) to oxic zones through advection and turbulent dispersion, and in case of benthic methylation, also through molecular diffusion in porous media. There are also speculations that the sometimes large concentrations of methylmercury in surface waters of deep lakes and oceans cannot arise solely as a result of transport from deeper layers, and so there should be some MeHg production in oxic layers [85]. In any case, the most accepted theory is that methylation of mercury is a bacterial-mediated process mainly done by sulfate reducing bacteria, [8] [9] [43] [44]. However, it was recently found that iron reducing bacteria can also mediate mercury methylation [37]. Further, [29] noted in their measurements in sediments from eight sites including lakes, and freshwater and brackish estuaries that methylation rates, %MeHg (MeHg to total Hg ratio), and absolute MeHg concentrations were not significantly correlated to the availability of sulfate as a terminal electron acceptor for SRB. In shallower waters and also waters with larger mixing, the anoxic and in particular sulfate and iron reducing conditions more often occur in benthic sediments and at certain depths where oxygen and nitrate are depleted.

[53] noted apparent higher %MeHg downstream of reservoirs in river systems suggesting that the higher organic content due to settling in these systems promoted microbially mediated processes. Monperrus et al. (2007) [84] measured methylation in the oxic waters of the euphotic zone of the Mediterranean Sea and suggested a microbially mediated pathway due to the higher methylation rates observed during periods of higher temperatures and high biological turnover. Hg reduction and de-methylation were also studied and both were mainly attributed to photochemical processes near the surface. Thus, while current conceptual models treat anoxic biotic transformation to MeHg as important, this is not certain at all and much more remains to be done in order to fix the understanding.

Mathematical modeling is one cost-effective approach to evaluate the associated risk and the utility of remediation options. There have been many efforts to develop realistic mercury cycling models for aquatic environments including lakes, wetlands, rivers and coastal waters. These efforts can be categorized into a) models that have treated the air-water-sediment system as respective batch reactors exchanging mercury species through rate-limited processes (e.g., [26] [33] [56]) without explicit consideration of transport of mercury species as dissolved or associated with sediments; b) models that have treated mercury as a non-reactive conservative metal mainly being transported as bound to sediments [102]; and c) models that have emphasized the detailed geo-chemical cycling of mercury [47] [48] [138]. Also there are bioaccumulation models focused on the uptake of different forms of mercury by living species in the aquatic system [52]. More recently some researchers have coupled a more sophisticated representation of physical processes including the hydrodynamics, sediment transport and the biogeochemical transformation of mercury in the water body using one-dimensional [19] [20] [135] [76] [77], two-dimensional [107] and three-dimensional [94] [93] [136] approaches.

In the following review we first consider sediment transport processes and modeling, because of the fundamental role of sediments as vehicles for mercury and other metal species. Then we turn to models developed specifically for mercury in surface aquatic systems.

1.1. *Sediment transport*

From a historical point of view, the interest in predicting sediment transport (and the resulting changes in riverbed morphology) has been driven mainly by the need to enable undisturbed navigation of vessels in rivers and channels [44]. Several empirical and semi-empirical predictors have been developed to describe the changes in river beds due to bed-load and suspended sediment movement, and to estimate the decrease in reservoir storage capacity. In the last twenty years, the attention on sediment transport has been enhanced by the important role that sediment plays in transporting adsorbed pollutants, considering all the environmental and social implications of the transport of contaminants.

When the water velocity increases close to the bottom of a water body, the particles of the bed start to move. The conditions for this incipient motion have been quantified since the seminal experimental work of [101], and details can be found in books, book chapters, and e-books such as those of [60] [40] and [88], respectively. Further increase in velocity results in entrainment of particles into suspension. Several expressions for predicting the entrainment rate of sediment into suspension (mostly for non-cohesive sediment) have been proposed by diverse authors (see, [54] [114] [115] [117] [41] for instance). In those equations, the rate of sediment entrainment into suspension is described by the difference between the bed shear stress due to currents and waves, and the critical bed shear stress, which is a function of sediment grain size and density. In some formulations, the wall-friction (shear) velocity is used as a surrogate of the bed shear stress. The main deposition parameter, the particle settling velocity, is also determined from grain and liquid properties: sediment density, grain shape and the viscosity of liquid. The transport parameter is then calculated, which further determines the conditions and areas of resuspension, transport and deposition of moving sediment. The processes and equations which can be used to determine onshore – offshore and longshore transport due to currents and waves in coastal areas are described in the literature (e.g., [74]).

Numerous sediment transport models have been constructed from these principles for river and estuary simulation, mostly for non-cohesive sediments (e.g., [116] [120] [86] [34] [73] [68] [93] [123] [6] [87] [30]). However, these models are valid only for uniform grain size and non-cohesive sediment, conditions rarely met in natural waters. With increasing discharge, finer non-cohesive fractions (silt and fine sands) usually enter into suspension, while other, coarser fractions require higher shear stresses to initiate their movement, first as bed-load and later as suspended particles. Such phenomena can often be observed in the coastal zone, where sediment composition is heterogeneous due to spatially and temporally varying hydrodynamic conditions [110]. Multi-fractional techniques (e.g. [97]) and models (e.g., [128]) have therefore been developed to increase the accuracy of non-cohesive sediment transport modeling. One such approach is dividing suspended sediment into washload and coarse suspended sediment, as described in [19].

Cohesive sediments, on the other hand, are more difficult to describe and particularly to simulate with a numerical model. The mechanisms and equations of cohesive sediment erosion and resuspension are well described in the literature (e.g., [126]); however, they have not been adopted for use in mathematical models of sediment transport that deal with transport of particle-bound pollutants. Few models for estuarine waters include the simulation of cohesive sediments; moreover, their authors report relatively poor agreement with measurements of sediment concentration in the water column [50] [123] [70]. Even more complicated is the modeling of transport of colloidal particles and pollutants bound to colloids as the particle movement cannot be described in accordance to available formulae of sediment transport (valid for non-cohesive sediment). On the other hand, the role of colloidal particles in transporting the contaminants to/from the deeper sediment layers has been overlooked in the literature. A few models are capable of simulating colloid-facilitated transport in saturated and unsaturated porous media (e.g., [78] [32] [59] [105]), while at present no known sediment transport model for surface waters accounts for colloids.

With some exceptions, only qualitative agreement among model predictions and measurements in natural environments can be expected; often, model results are in accordance with measurements by an order of magnitude or at best a factor of three (e.g., [109] [132] [93] [94]). Using the one-dimensional coupled US-EPA models, the same level of agreement was achieved in the Idrijca and Soča Rivers [135], while spatiotemporal moments showed significantly better agreement for the Carson River [19], mostly due to a very large quantity of measured data used for the calibration of the model. The studies of sediment transport in coastal areas [55] [93] as well as studies of riverine sediment transport [21] [135] have confirmed that the transport of particulate-bound mercury represents a major part of its overall transport in rivers, lakes and estuaries. Moreover, a single major flood event can account for more than 90% of the total annual mercury transport. Therefore, it is very important to calibrate the sediment transport models in accordance with high discharges, although it is very difficult to ensure enough measurements for calibration and validation of models in such conditions. Furthermore, it has been established that concentrations of suspended matter are different during the increasing and decreasing flow at the same discharge [46] [125] [66]. The hysteresis phenomenon has often been observed, but rarely measured and/or modeled [2]. The ratio between discharge and sediment discharge is a function of the characteristics of each stream and catchment and as such cannot be generalized. Moreover, sediment transport also depends on the availability of sediment, which is again a function of the stream itself. For the Carson River, an unlimited quantity of material was reported [19], while the situation in the Idrijca and Soča is different due to dams and the geology of the stream and the catchment [135]. In the case of rivers having dams, the transport capacity in the basin is heavily decreased and sediment is stored in the reservoir. Furthermore, during floods, large quantities of fine sediment deposited during low flows can be released from the upstream accumulations.

Mass balances of sediment and sediment-bound pollutants are often established, either solely from measured data or from combined modeling results and measurements [108] [93] [133] [94] [75] [127] [96] [63]. They are a valuable tool, which is often helpful in deciding which type of model to apply and which parameters need to be measured in the future. The simulation results, on the other hand, are often used to improve the mass balances.

1.2. *Model tools for understanding mercury fate and transport*

Early models have used relatively simplistic assumptions to predict the dispersion of mercury in water bodies. For example, Turner and Lindberg (1978) [111] used a simple dilution model to describe the decline of mercury concentration in a river-reservoir system downstream of a chemical plant. Clearly, since the effect of sedimentation was not taken into account, the model over-predicted the mercury concentration downstream of the release point.

Not all earlier models have adopted such a simplistic approach. Herrick et al. (1982) [52] developed a model for mercury cycling in woodland streams that considers mercury and methyl-mercury to be in five phases (sediment, water, invertebrates, detritus and fish), and the detritus phase is discretized into several bins representing various compositions and sizes of organic particles. However, this model does not have a transport component and the deposition and resuspension of particles were not accounted for. Fontaine (1984) [38] developed a fully non-equilibrium one-dimensional reactive transport model for the fate of metals in aquatic systems. Three phases of metals (i.e., dissolved, particulate and associated with organic substrates) were considered in the model.

The mass-exchange between the phases and also between the water column and sediments were all considered to be kinetically controlled. A relatively sophisticated model was used for sorption and geochemical transformation of various mercury species.

[71] developed a mass balance model based on the concept of aquivalence and fugacity called QWASI (Quantitative Water-Air-Sediment Interaction) for lake systems. They considered the effects of inflow and outflow of chemicals (dissolved or particle-bound), diffusive mass exchange from/to sediments, resuspension/deposition and burial, and dry and wet deposition, and solved the system assuming a steady-state condition overall. In each compartment (water and sediments), concentrations of sorbed and dissolved species were assumed to be in equilibrium. No transport process was considered in this model. [26] used a modified version of QWASI to simulate cycling of mercury in a hypothetical lake. Ethier et al. (2008) [33] further simplified QWASI by assuming an instant equilibrium between different species of mercury and applied the resulting model to Big Dam West, Canada.

Using a similar approach, [56] developed a lake dynamic mercury cycling model called MCM. They considered three vertical compartments of epilimnion, hypolimnion, and sediments and four biotic compartments including phytoplankton, zooplankton, a prey and a predator fish species. Mercury species include elemental, reactive and methylmercury. The distinct feature of MCM compared to even more recently developed models is its more sophisticated treatment of the rates of methylation and de-methylation. While methylation was considered to be a function of reactive mercury concentration, sulfate ion, dissolved organic content and temperature through a Monod type relationship, the latter was considered as a function of methylmercury concentration, H+ and dissolved organic carbon. Leonard et al. (1995) [67] used MCM to predict the fate of mercury in the Great Lakes. Harris et al. (1996) [49] upgraded MCM to simulate mercury cycling in a group of lakes. Knightes and Ambrose (2007) [62] applied this model to 91 lakes in Vermont and New Hampshire and found that the performance of the model is inadequate as a prediction tool. Kotnik et al. (2002) [65] applied the same reaction rate functional forms suggested by Hudson et al. (1994) [56] to simulate mercury cycling in Lake Velenje, Slovenia. The approach of considering lake systems with well mixed layers has been adopted by some other researchers recently. [61] developed a spreadsheet-based steady-state mass balance model (SERAFM) for mercury cycling and evaluation of wildlife exposure risk. Similar to other models, SERAFM categorizes mercury into three species (Hg0, HgII, MeHg) and in five phases (abiotic solids, phytoplanktonic, zooplanktonic, detrital and associated with dissolved organic matter). Also, the aquatic system was modeled with three compartments (epilimnion, hypolimnion, and sediments) in addition to an equilibrium sorption assumption. [15] applied SERAFM to a constructed wetland and a stream in Nevada, USA.

In most of the models used for circulation of mercury in lake systems including QWASI [71] and MCM [49] the system is considered as horizontally well mixed, sometimes with single-layer and sometimes with multi-layer water columns with a uniform and time invariant mass exchanges between the layers. Although this approach may be appropriate for small lakes with mainly depositing suspended particles, it is inadequate for more complicated systems including rivers, coastal areas as well as larger lakes where the circulation of water and resuspension of sediments plays an important role in mercury cycling. Assumptions of net deposition and of equilibrium between particulate and dissolved mercury phases artificially restrict re-entrainment of mercury. On the other hand, most of the surface water metal cycling models developed in the past have relied on the partitioning coefficient (K_D) concept. As opposed to organic chemicals, the value of K_D for metals is sensitive to the chemical conditions such as pH, occurrence of complexing ligands, and redox potential among others [45] [92]. Thus, the constant K_D approach may lead to erroneous results in some applications.

Faced with these shortcomings, researchers have used two distinct pathways to improve the modeling approaches for mercury. One group (mainly hydrologists) have improved the treatment of flow and sediment transport and the other group (mainly geochemists) have focused on using more accurate and complicated biogeochemical reaction networks. [121] briefly reviewed mercury models applied to rivers, lakes and coastal zones.

Among the group of researchers focusing on the hydrological processes are [103], who used the cohesive sediment and pollutant transport model TSEDH [103] coupled with the hydrodynamic circulation model RMA-2V to model mercury cycling in Clear Lake, California. They considered multiple layers of bed sediments with different critical shear stresses but did not consider the different distribution of mercury in each layer. Bale (2000) [5] suggested using an advection-dispersion transport model and as well as incorporating a more mechanistic approach to calculate rates of deposition and resuspension to simulate the fate of mercury in Clear Lake. He used a slightly modified version of the reaction-rate functional forms proposed by [49]. Another way to consider heterogeneity and transport without explicitly modeling the hydrodynamics and sediment transport is to use compartmental models where the flow and mass exchange between the compartments are given as model parameters. AScI (1992) [1] incorporated the kinetic transformation subroutine from MCM into the water quality model WASP4 and called the resulting model MERC4. [51] coupled MERC4 with both a fish bioenergetics/bio-uptake model and a lake eutrophication model to predict the transport and fate of mercury in Onondaga Lake, NY. The lake eutrophication model was used to estimate the planktonic populations and their settling rates.

In systems where the transport is more important such as rivers, hydrodynamic forces affecting plays a more important role. In [19] Carroll et al. used MERC4 along with WASP and the RIVMOD hydrodynamic model to simulate mercury transport in the Carson River, Nevada. They modeled sorption and desorption as kinetically controlled processes. This model was also applied to the Idrijca and Soca River systems in Slovenia [135].

STATRIM, a two-dimensional model of mercury transport and fate, was developed first for coastal environments [107], where mercury in its particulate form was considered for the first time in marine environments. The model used annually averaged input data combined with depth-averaged parameters and was subsequently upgraded to a three-dimensional non-steady state model (PCFLOW3D). The PCFLOW3D model together with a new sediment transport module was further used to simulate resuspension and other sediment and mercury processes in coastal environments [132] [93] [94] and applied to the Mediterranean Sea with simplified sediment transport and some improvement in the biogeochemical transformation and atmospheric mass exchange modules [136]. This model incorporates mercury methylation/de-methylation and other Hg transformations such as reduction-oxidation. Different forcing factors were considered, the most important being extreme (flood-wave) river inflow, extreme wind-induced currents and waves.

Among the works focusing on the biogeochemical cycling are models mainly devoted to the geochemistry of mercury in the water column, including the full reaction network affecting mercury speciation in batch systems (i.e., [47] [138]). An extensive review of the biogeochemical cycling of mercury in aquatic systems can be found in [85] and, for marine environments, [35-36]. [11] developed a steady state fate and transport model called TRANSPEC by incorporating the geochemical speciation model MINTEQ+ with the aquivalence/fugacity approach introduced by [72], but including a more sophisticated geochemical reaction network. The species were assumed to be in chemical equilibrium; however, the mass transfer between various compartments (water, air, and sediments) was considered as a dynamic process. In this model, flow between various water body compartments and also resuspension/deposition were considered as steady pre-specified processes. The model was applied to predict the fate of Zn in several lakes in Canada [11-12].

Although many reversible abiotic geochemical reactions affecting the speciation of metals can be treated as equilibrium reactions, their microbe-mediated transformations are typically slow, irreversible in typical modeling time-scales and kinetically controlled. In case of mercury, both methylation and de-methylation are known to be bacteria-mediated. [39] modified TRANSPEC to handle kinetically controlled reactions and called the resulting model BIOTRANSPEC. They also included the effect of uptake by fish in their model. They applied the model to mercury cycling in Lahontan reservoir in Nevada, USA. The methylation and de-methylation rates were considered constant but site-specific rates.

Because sulfate reducing bacteria, and possibly iron-reducing bacteria, are the most important microorganisms in the formation of methylmercury in temperate locations, it is important that attempts to quantify rates of mercury methylation consider redox-specific and kinetically-controlled microbially-driven degradation reactions, within the context of equilibrium multiphase and multicomponent chemical conditions. Modeling platforms available for coupling microbe-mediated kinetics with multicomponent geochemical equilibria are available as subroutines that can be linked with other models. Tools such as PHREEQC2 [89] can be used for exploration of the relationship between mercury methylation and ambient chemical conditions as controlled by ambient microbial communities. PHREEQC2 is a computer program that can perform speciation, batch-reaction, one-dimensional transport, and inverse geochemical calculations, as well as arbitrarily defined kinetics associated with microbially-mediated reactions.

The rate of release of the methylmercury produced in sediment layers is a function of the depth of the layers. Sediment diagenesis models focusing on modeling redox stratification have been developed in the past [10] [13] [14] [113] and have been used to predict the cycling of contaminants and nutrients in lakes [16] [17] [100], deep seas [69], as well as in coastal zones and estuarine sediments. These models however have not been integrated with fate and transport models representing the overlying waters. Particularly in the case of mercury cycling, it is important to integrate redox zonation in the sediments and its effect on mercury methylation with (a) the release mechanisms of the produced MeHg to the overlying water and (b) with the transport processes in the overlying water. In the present work a framework for coupling the sediment processes with the cycling of mercury in the overlying water is introduced.

2. Model Development

The objective of this section is to introduce a modeling framework applicable to a wide variety of surface water bodies including lakes, estuaries, rivers and coastal zones. The formulation is developed considering a two-dimensional representation of the water column; however, at the end of the chapter demonstration results are generated assuming a one-dimensional case. To preserve the generality of the modeling framework, an unlimited number of suspended solid phases are considered in the model. Depending on the aquatic system being studied, these phases can be assigned to mineral abiotic solids, phytoplanktonic, detrital, or various classes or size categories of suspended solids. The model is also capable of predicting the fate and transport of an unlimited number of major components and ligands influencing the speciation of mercury and the redox zonation in the sediments. The choice of these major components are also problem dependent. The computational domain of the model is considered to be the overlying water (assumed to be totally mixed in the vertical direction) and the active benthic sediment layer. The depth of the benthic sediment layer should be chosen based on the problem. Also, the two-dimensional representation of the water body can be easily replaced by a three-dimensional one by a generalization of the governing equations. This step may be necessary in problems involving deep water bodies where the anoxic layer can be present in the water column or in cases where vertical mixing does not occur due to temperature stratification. As it was explained in the Introduction, the aim here is to track the vertical distribution of mercury and the rest of major components involved in the active bed sediment layer. This is necessary since, first, the rate of release of contaminants from the sediments to the overlying water as a result of both diffusive mass exchange and resuspension depends on the distribution of contaminants in the sediment layers, and, second, the rate of methylmercury production depends on the magnitude of overlap between the sulfate or iron reducing layer and the layers containing reactive mercury.

2.1. *Fate and transport of mercury in the overlying water*

The depth-averaged fate and transport governing equations for the species in the overlying water considering a kinetic mass transfer between the phases can be written as [76-77]:

$$\frac{\partial HC_i}{\partial t} + \frac{\partial UHC_i}{\partial x} + \frac{\partial VHC_i}{\partial y} = \frac{\partial}{\partial x}\left(D_{hx}H\frac{\partial C_i}{\partial x}\right) + \frac{\partial}{\partial y}\left(D_{hy}H\frac{\partial C_i}{\partial y}\right) + k_b\left[c_i(0) - C_i\right]$$
$$-H\sum_{j=1}^{n}\phi_j k_j\left(K_{Dij}C_i - S_{ij}\right) + qC_{in} + k_{at,i}\left(C_{i,at} - C_i\right) - u_f\theta_0 c_i(0) + HR_i + \psi_i \tag{1}$$

$$\frac{\partial H\phi_j S_{ij}}{\partial t} + \frac{\partial UH\phi_j S_{ij}}{\partial x} + \frac{\partial VH\phi_j S_{ij}}{\partial y} = \frac{\partial}{\partial x}\left(D_{xj}H\frac{\partial \phi_j S_{ij}}{\partial x}\right) + \frac{\partial}{\partial y}\left(D_{yj}H\frac{\partial \phi_j S_{ij}}{\partial y}\right)$$
$$+Er_j s_{ij}(0) - w_{pj}\phi_j S_{ij} + q\phi_{j,in}S_{ij,in} + Hk_j\phi_j\left(K_{Dij}C_i - S_{ij}\right) + HR_{s,ij} + \psi_{ij} \tag{2}$$

in which t[T] is time; x *and* y [L] are spatial coordinates; C_i[M_c/L^3] is the dissolved concentration of chemical i in bulk water; U *and* V [L/T] are the depth averaged velocity components; D_{hx} *and* D_{hy} [L^2/T] are the mechanical dispersion coefficients for dissolved species in the x and y directions; k_b is the sediment-water mass exchange coefficient for the dissolved species [L/T]; $c_i(0)$ is the pore-water concentration of species i at the topmost layer of sediments [M_c/L^3]; k_j [T^{-1}] is the mass exchange coefficient between the suspended solid phase j and water; ϕ_j [M/L^3] is the concentration of the suspended solid phase category j; K_{Dij} [L^3/M] is the water-solid distribution coefficient for solid phase j and species i; S_{ij}[M_c/M] is the sorbed phase concentration of species i to solid phase j; R_i and R_{sij} are the sum of rates of elimination or production of species i at phase j due to reactions for dissolved and sorbed phases, respectively; q [$L^3 T^{-1}L^{-2}$] is the amount of inflow/outflow per surface area; $\phi_{j,in}$ [M/L^3] is the concentration of solid phase j in the inflow; $S_{ij,in}$ [M_c/M] is the concentration of species i sorbed to solid phase class j in the lateral influx; $k_{at,i}$[L/T] is the atmospheric exchange rate coefficient for species i; $C_{i,at}$ [M_c/L^3] is the saturation concentration for species i calculated using Henry's law; D_{xj} *and* D_{yj} $[L^2/T]$ are the dispersion coefficient for suspended particles in the x and y directions respectively; $Er_j[ML^{-2} T^{-1}]$ is the sediment resuspension rate for solid phase class j; $w_{pj}[L/T]$ is the deposition rate parameter for solid phase class j; $s_{ij}(0)$ $[M_c/M]$ is the sorbed concentration of chemical species i to solid phase j at the topmost layer of the bed sediments; u_f $[L/T]$ is the pore water velocity in bed sediments due to consolidation or hyporheic flow (downward defined as positive); θ_0 is the bed sediment porosity at the sediment-water interface and ψ_{ij} $[LT^{-1}M_cL^{-3}]$ is a source term representing for example the effect of plant uptake and decomposition.

2.2. *Sediment Resuspension/Deposition*

To solve Eq. (2), the transport of the suspended solid phase needs to be modeled. Many commercial and open-source software packages are available for modeling the transport of cohesive or non-cohesive sediments in various systems including rivers, estuaries or coastal areas [e.g., MIKE 21(DHI), HEC-6 (USACE), STAND [137] and GSTARS [131]. Many of these models assume quasi-equilibrium conditions to calculate the suspended and bed loads (e.g., HEC-6, GSTARS). Although this approach is appropriate for engineering purposes such as studying the morphological changes of the water body and sedimentation in reservoirs and similar applications, it is insufficient for tracking contaminants associated with sediments. The general formulation used in sediment transport models is represented by an advection-dispersion governing equation:

$$\frac{\partial H\phi_j}{\partial t} + \frac{\partial UH\phi_j}{\partial x} + \frac{\partial VH\phi_j}{\partial y} =$$

$$\frac{\partial}{\partial x}\left(D_{xj}H\frac{\partial \phi_j}{\partial x}\right) + \frac{\partial}{\partial y}\left(D_{yj}H\frac{\partial \phi_j}{\partial y}\right) + Er_j - w_{pj}\phi_j + q\phi_{j,in}$$

(3)

The main difference between various sediment transport models is essentially the way they calculate Er and deposition terms. [40] lists all different formulas used to calculate the sediment entrainment for non-cohesive sediments. [41] suggested a formulation for mixed size sediments. The fall velocity estimated using the Stokes equation or empirical relationships (e.g., [27]) is often used to model the deposition of non-cohesive sediment. Since the main mechanism for settling of cohesive sediments is coagulation and flocculation, their rate of deposition depends on the concentration of sediments. [80] suggested the following relationship for the rate of deposition:

$$w_p = pv_s$$

(4)

in which p is a factor to incorporate the effect of shear stress, calculated as:

$$p = 1 - \tau_b / \tau_{cd} \tag{5}$$

So no deposition occurs when the shear stress τ_b is above the critical shear stress τ_{cd} in Eq. (5). v_s is the fall velocity which is considered as a function of the concentration of particles, as suggested by [80]:

$$v_s = \alpha_8 C_s^{4/3} \tag{6}$$

Suggested equations for the rate of resuspension of cohesive sediments typically have the following form:

$$Er = M \left(\tau_b - \tau_c \right) \tag{7}$$

where M is the erodibility coefficient that can be a function of the degree of consolidation of sediments, their size distribution and factors affecting their stability such as biofilm formation or armoring due to the existence of mixtures of large and fine particles. Some researchers have proposed more complicated power-law or exponential relationships between the erosion rate and the bed shear stress [23].

Some models have addressed the effect of the depth of erosion into the erodibility rate [98] and only a few of them have incorporated the effect of consolidation into the model directly by defining distinct layers of bed sediments [99].

2.3. *The fate of mercury in sediments (burial, resuspension, diffusive mass exchange)*

Several benthic sediments contaminant fate and transport models have been developed in the past mainly to address sediment diagenesis and nutrient cycling in the sediments [10] [13] [112]. In these models the sediment layer is represented by a one-dimensional column sometimes with a dispersion coefficient and porosity changing with depth as a result of consolidation (e.g., [13]). The top boundary condition (the overlying water) is typically considered to have a fixed concentration of compounds and the burial is modeled using a transformation of coordinates. Here we propose a 3-D model for the transport of contaminants in the sediments in order to incorporate the spatial heterogeneities in resuspension and deposition. The governing equation for multi-phase transport in the sediments can be expressed as:

$$\frac{\partial (\theta c_i)}{\partial t} + \frac{\partial (\theta u_{fx} c_i)}{\partial x} + \frac{\partial (\theta u_{fy} c_i)}{\partial y} + \frac{\partial (\theta u_{fz} c_i)}{\partial z^*} = \frac{\partial}{\partial z^*} \left((D_m + D_B) \theta \frac{\partial c_i}{\partial z^*} \right)$$
$$- \sum_j f_j B_d k_j \left(K_{Dij} c_i - s_{ij} \right) + R_i + \sum_{j'} k_{j'0} f_{j'} B_d s_{ij'} - \psi_i \tag{8}$$

$$\frac{\partial\left(f_{j}B_{d}s_{ij}\right)}{\partial t}+\frac{\partial\left(f_{j}B_{d}u_{s}s_{ij}\right)}{\partial z^{*}}=\frac{\partial}{\partial z^{*}}\left(D_{B}f_{j}B_{d}\frac{\partial c_{i}}{\partial z^{*}}\right)$$

$$+B_{d}k_{j}\left(K_{Dij}c_{i}-s_{ij}\right)+R_{s,i}-\sum_{j'}k_{jj'}f_{j}B_{d}s_{ij}+\sum_{j'}k_{j'j}f_{j'}B_{d}s_{ij'}$$

$$\tag{9}$$

The first terms on the right hand side of both equations account for the effect of consolidation. The boundary conditions are as follows:

$$\left(D_{m}+D_{B}\right)\theta\frac{\partial c_{i}}{\partial z^{*}}=k_{b}\left(c_{i}-C_{i}\right)\ \text{at } z^{*}=z_{0} \tag{10}$$

$$\frac{\partial c_{i}}{\partial z}=0\ \text{ at } z^{*}=-\infty \tag{11}$$

and for Eq. (9):

$$s_{ij}\Big|_{z^{*}=-\infty}=0\ \text{ for } J_{0}<0\ \text{ erosion} \tag{12}$$

$$s_{ij}\Big|_{z^{*}=z_{0}}=S_{ij}\ \text{ for } J_{0}>0\ \text{ deposition} \tag{13}$$

In Eqs. (8) to (13) z^{*} [L] is the vertical coordinate based on a fixed datum which increases increasing upward; $c_i=c_i(x,z^{*},t)$ [M_c/L^{3}] is the dissolved concentration of species i in pore water in the bed sediments; s_{ij} [M_c/M] is the mass concentration of sorbed species i on solid phase j; u_{fx}, u_{fy}, u_{fz} [L/T] are the pore water velocity components; u_s [L/T] is the advective velocity of sediments due to sediment consolidation; B_d [M/L^{3}] is the total bulk density of the sediment materials and f_j is the fraction of sediments consisting of phase j (i.e., planktonic, detrital, minerals); D_B [L^{2}/T] is the mechanical diffusion coefficient due to the inhomogeneity of bed materials and the mixing due to the activities of benthic organisms; D_m [L^{2}/T] is the molecular diffusion coefficient and θ is the porosity of the bed material. It is assumed that the effect of diffusion in the horizontal directions is negligible due to the large scales in the two directions. ψ_i is a sink term that can represent the rate of root uptake of species i; $k_{jj'}$ [1/T] is the rate of transformation of sediment phase j to phase j' and $j=0$ indicates dissolved phase. Including the x and y directions of pore water velocity allows for inclusion of hyporheic flow. In the derivation of Eqs. (8) and (9) it is assumed that none of the solid phases is mobile in the pore-water in benthic sediments, i.e., no colloid-facilitated transport in the benthic sediments is taken into consideration. This phenomenon can be incorporated into the model by adding advective transport terms to the solid phase.

Imposing a transformed coordinate system in order to attach the origin of the coordinate system to the sediment-water interface $z(t)=z_{0}(t)-z^{*}$ on Eqs. (8) and (9), where $z_0(t)$ is the elevation of sediment-water interface, yields:

$$\frac{\partial(\theta c_i)}{\partial t} + \frac{\partial(\theta u_{fx} c_i)}{\partial x} + \frac{\partial(\theta u_{fy} c_i)}{\partial y} + \frac{\partial\left[\theta(u_{fz} + J_0)c_i\right]}{\partial z} = \frac{\partial}{\partial z}\left((D_m + D_B)\theta\frac{\partial c_i}{\partial z}\right)$$
$$-\sum_j f_j B_d k_j \left(K_{Dij} c_i - s_{ij}\right) + R_i + \sum_{j'} k_{j'0} f_{j'} B_d s_{ij'} - \psi_i \tag{14}$$

$$\frac{\partial(f_j B_d s_i)}{\partial t} + \frac{\partial\left[f_j B_d (u_s + J_0)s_{ij}\right]}{\partial z} = \frac{\partial}{\partial z}\left(D_B B_d \frac{\partial f_j s_{ij}}{\partial z}\right) + B_d k_j \left(K_{Dij} c_i - s_{ij}\right)$$
$$+R_{s,i} - \sum_{j'} k_{jj'} f_j B_d s_{ij} + \sum_{j'} k_{j'j} f_{j'} B_d s_{ij'} \tag{15}$$

The elevation of the sediment-water interface, $z_0(t)$, is variable with time due to erosion and deposition of sediments. The positive direction for u_f and u_s is changed according to the new coordinate system for the sake of simplicity. Therefore, henceforth positive value indicates downward flow.

The terms u_f and u_s depend on the rate of deposition or erosion at the surface and the consolidation of sediments. Here it is assumed that the porosity of sediments decreases with depth as indicated in [13] via:

$$\theta(z) = (\theta_0 - \theta_\infty)e^{-k_\theta z} + \theta_\infty \tag{16}$$

where θ_0 is the porosity of the topmost layer of sediments and θ_∞ is the porosity of deep sediments. In this case, u_s and u_f can be calculated as follows, using a simple mass balance

$$u_s = J_0\left(\frac{\theta - \theta_\infty}{1 - \theta}\right) \tag{17}$$

$$u_f = -J_0\left(\frac{\theta - \theta_\infty}{\theta}\right) \tag{18}$$

where u_s and u_f are functions of z through θ, and J_0 is the rate of change in the elevation of the sediment-water interface. In turn, $J_0 = \partial z_0/\partial t$, which is calculated as:

$$J_0 = \frac{1}{(1 - \theta_\infty)\rho_s}\sum_j (w_{pj}\phi_j - Er_j) \tag{19}$$

where ρ_s [M/L^3] is the density of sediment particles. The governing equation controlling the mass balance for various solid phases in the benthic sediments can be written as:

$$\frac{\partial f_j}{\partial t} + \frac{\partial\left[f_j(u_s + J_0)\right]}{\partial z} = \frac{\partial}{\partial z}\left(D_B \frac{\partial f_j}{\partial z}\right) - \sum_{j'} k_{jj'} f_j + \sum_{j'} k_{j'j} f_{j'} \tag{20}$$

2.4. *Modeling mercury bio-geochemistry*

Most mercury-cycling models have lumped all inorganic species of mercury into a single species identified by reactive mercury (HgII) and, therefore, mercury species in the models include HgII, methylmercury (MeHg) and elemental mercury (Hg0). In addition, in many models the rate coefficients defining the transformation of these species have been considered constant or solely a function of temperature [19] [26] [38] [94] [93] [135] [136]. There have also been modeling studies concentrating on the detailed geochemical speciation of mercury, but these models have been applied only to batch systems [47]. Based on the effect of sulfate- and iron-reducing bacteria, some later models have considered the mercury methylation rate as a function of the concentration of sulfate ion [56] [49]. However, a high concentration of sulfate ion does not automatically lead to a higher methylation rate since, first, the activity of sulfate reducing bacteria can be inhibited by the presence of terminal elector acceptor with higher energy yield (e.g., oxygen, nitrate, and ironIII) and, second, sulfate reduction leads to production of S^{-2} ions that can react with dissolved Hg and lead it to precipitate (e.g., [100]). So here we consider the rate of methylation to be proportional to the activity of sulfate reducing biomass. To do this, mercury species are categorized into four groups of HgII, MeHg, Hg0 and HgS (cinnabar). In addition, to predict the sulfate reduction rate at each layer of sediments the cycling of the major terminal electron acceptors including oxygen, nitrate, iron, and sulfate should be modeled. The reaction network suggested by [113] for sediment diagenesis is adopted with some simplifications, with primary redox reactions listed in Table 1, speciation reactions in Table 2, and secondary redox reactions in Table 3. The main difference between the reaction network used herein and previous models such as MCM is that the methylation rate is considered as a function of the sulfate reduction rate and not of the concentration of sulfate ion. Due to the small concentration of mercury compared to the major components affecting its speciation, it is not necessary to consider the effect of mercury speciation reactions on the concentration of major species such as oxygen and organic matter.

Table 1: Primary redox reaction network for major components affecting mercury cycling

Reaction	Rate Expression
$OM + O2 \xrightarrow{\;R1\;} TIC + \dfrac{1}{(C:N)} NH_4^+$	$R_1 = k_{OM}\,[OM]\,\dfrac{[O_2]}{[O_2]+K_{O_2}}$
$OM + 0.8NO_3^- \xrightarrow{\;R2\;}$ $TIC + \dfrac{1}{(C:N)} NH_4^+ + 0.4N_2$	$R_2 = k_{OM}\,[OM]\,\dfrac{[NO_3^-]}{[NO_3^-]+K_{NO_3^-}}\,\dfrac{K'_{O_2}}{[O_2]+K'_{O_2}}$
$OM + 0.5FeOOH \xrightarrow{\;R3\;}$ $TIC + \dfrac{1}{(C:N)} NH_4^+ + 0.5Fe^{2+}$	$R_3 = k_{OM}\,[OM]\,\dfrac{K'_{O_2}}{[O_2]+K'_{O_2}}\,\dfrac{K'_{NO_3^-}}{[NO_3^-]+K'_{NO_3^-}}\,\dfrac{[Fe^{3+}]}{[Fe^{3+}]+K_{Fe^{3+}}}$
$OM + 0.5SO_4^{2-} \xrightarrow{\;R4\;}$ $TIC + \dfrac{1}{(C:N)} NH_4^+ + 0.5H_2S$	$R_4 = k_{OM}\,[OM]\,\dfrac{K'_{O_2}}{[O_2]+K'_{O_2}}\,\dfrac{K'_{NO_3^-}}{[NO_3^-]+K'_{NO_3^-}}\,\dfrac{K'_{Fe^{3+}}}{[Fe^{3+}]+K'_{Fe^{3+}}}\,\dfrac{[SO_4^{3-}]}{[SO_4^{2-}]+K_{SO_4^{2-}}}$
$OM \xrightarrow{\;R_3\;}$ $0.5TIC + 0.5CH_4 + \dfrac{1}{(C:N)} NH_4^+$	$R_5 = k_{OM}\,[OM]\,\dfrac{K'_{O_2}}{[O_2]+K'_{O_2}}\,\dfrac{K'_{NO_3^-}}{[NO_3^-]+K'_{NO_3^-}}\,\dfrac{K'_{Fe^{3+}}}{[Fe^{3+}]+K'_{Fe^{3+}}}\,\dfrac{K'_{SO_4^{2-}}}{[SO_4^{2-}]+K'_{SO_4^{2-}}}$

Table 2: Mercury speciation/methylation reactions

Mercury Speciation Reactions	
$HgS \xleftrightarrow{\;k_{Hg,1},k_{Hg,2}\;} Hg^{2+} + S^{2-}$	Equilibrium
$Hg^{2+} + OM \xrightarrow{\;R_{me}\;} MeHg$	$R_{me} = k_{me}\,\theta^{T-T_0}\,R_4\left[Hg^{2+}\right]$
$MeHg \xrightarrow{\;R_{dm}\;} Hg^{2+}$	$R_{me} = k_{dm}\left[MeHg\right]$
$Hg^{2+} \xrightarrow{\;R_r\;} Hg^0$	$R_r = k_{red}\left[Hg^0\right]$
$Hg^0 \xrightarrow{\;k_o\;} Hg^{2+}$	$R_o = k_o\left[Hg^{2+}\right]$

Table 3: Primary redox reaction network for major components affecting mercury cycling

Secondary Redox Reactions	
$CH_4 + 2O_2 \xrightarrow{R6} TIC$	$R_6 = k_6 [CH_4][O_2]$
$4Fe^{2+} + O_2 \xrightarrow{R7} 4FeOOH + 8H^+$	$R_7 = k_7 [Fe^{2+}][O_2]$
$2Fe^{2+} + H_2S \xrightarrow{R8} 2FeS + 2H^+$	$R_8 = k_8 [Fe^{2+}][H_2S]$
$NH_4^+ + 2O_2 + TIC \xrightarrow{R9} NO_3^- + TIC$	$R_9 = k_9 [NH_4^+][O_2]$
$H_2S + 2O_2 \xrightarrow{R10} SO_4^{2-} + 2H^+$	$R_{10} = k_{10} [H_2S][O_2]$
$H_2S + 2FeOOH + 4H^+ \xrightarrow{R11}$ $S_0 + 2Fe^{2+}$	$R_{11} = k_{11} [FeOOH][H_2S]$
$2O_2 + 2FeS \xrightarrow{R12} Fe^{2+} + SO_4^{2-}$	$R_{12} = k_{12} [FeS][O_2]$
$FeS + S_0 \xrightarrow{R13} FeS_2$	$R_{13} = k_{13} [FeS][S_0]$
$7O_2 + 2FeS_2 \xrightarrow{R14} 4SO_4^{2-} + 2Fe^{2+}$	$R_{14} = k_{14} [FeS_2][O_2]$
$4S_0 \xrightarrow{R15} 3H_2S + SO_4^{2-} + 2H^+$	$R_{15} = k_{15} [S_0]$

3. Numerical Implementation

Equations (1) and (2) are solved using the finite differences method. Centered differences were employed for the dispersion terms, weighted equally at the old and new time step. The advection terms were treated using upwinding. The time derivative was discretized at t and $t+1$ time steps using the Crank-Nicholson scheme. The equations controlling the fate of species in the benthic sediments (8) and (9) form a set of one-dimensional equations each linked to one grid cell of the overlying water body. Due to the possibility of sharp fronts of concentrations in the sediments, especially in the sorbed phase, a higher-order method, QUICKEST, was utilized to solve these equations with Crank-Nicholson time-weighting. The coupling between the sediments and the water body was conducted using a non-sequential explicit method.

4. Demonstration Simulation (Colusa Basin Drain)

For demonstration purposes the model described above was applied to simulate the mercury cycling in the Colusa Basin Drain in Northern California. The Colusa Basin Drain transfers surface runoff and irrigation return flow from agricultural lands in the northern central valley to the Sacramento River (Figure 1). The sorption properties of the mercury were obtained from [3]. Sediment characteristics and concentrations in the Colusa Basin Drain were measured by [81] [82] Mirbagheri et al. (1988a, 1988b). In this study, a 30 km reach of the drain is considered (Figure 1). The water body is considered one-dimensional and therefore one-dimensional versions of Eqs. (1) and (2) are considered. Also to simulate the water flow in the system, which dictate the velocities appearing in Eqs. (1) and (2), a kinematic wave model is utilized [106] (Singh, 1997). In addition, all the hyporheic flows are neglected, and therefore Eqs. (8) and (9) actually behave as a series of one-dimensional PDEs.

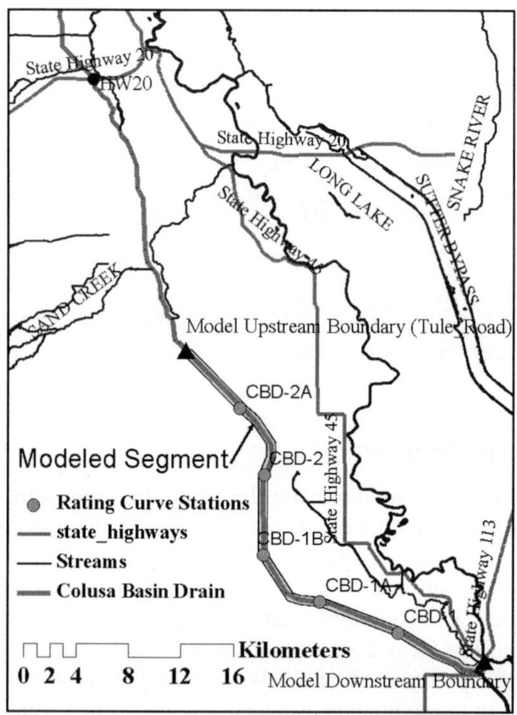

Figure 1: Location map of model domain and sampling stations in the Colusa Basin Drain.

4.1. *Flow and sediment transport*

Flow and sediment transport were modeled using the governing equations described in the previous section for a three-year simulation period (1996-98). The rating curve parameters for the river segments between stations with specified rating curves were obtained by interpolation. The upstream boundary condition for the kinematic wave model was obtained from the observed flow hydrograph provided by the U.S. Geological Survey (USGS 2000).

Figure 2 shows predicted versus measured total suspended solids at the CBD-1 station close to the downstream end of the modeling domain. Considering the uncertainties in the inflow concentrations of suspended sediment and the variations in the sediment lateral inflow with time due to agricultural return flows, the agreement can be considered acceptable. Further, the agreement is of the same nature than comparisons reported elsewhere (see [28]). The model simulation reflects high-frequency variations that are not captured by the lower-frequency sampling, indicating that data collection techniques might benefit by either cumulative (averaged) sampling and/or higher-frequency sampling. Figure 3 shows the net deposition rate (deposition-erosion) over the reach for the period of modeling. The reason for larger deposition rates compared to resuspension is the amount of sediments carried by the lateral inflow.

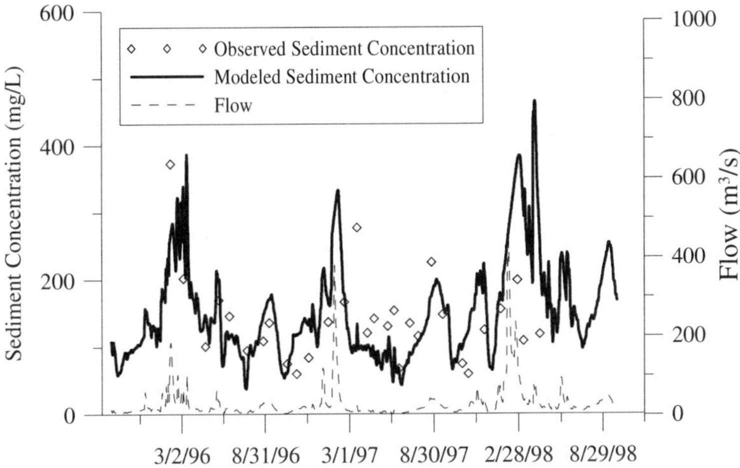

Figure 2: Measured and modeled suspended sediment concentration at CBD-1 Station.

Deposition mainly takes place at kilometers 20 to 30 in the river due to smaller velocities in that region, and to larger erosion rates in the upstream reach that provide source material for the deposition zone. Higher erosion rates during high flow conditions, followed immediately by a higher rate of deposition in lower flow velocity regions (kilometers 20-30) after each high flow event can be noticed. Figure 3 shows the accumulation of sediments due to deposition and erosion processes during the modeling period. The largest deposition rate occurs in river kilometers 20-30 due to the lower flow velocity in that region.

4.2. *Multi-component reactive transport*

The forward and reverse solid-water mass exchange coefficients are assumed to be large enough to mimic equilibrium sorption conditions. It should be noted that all the organic matter in this demonstration study was assumed to be easily mineralizable and therefore only one organic carbon pool was considered for the sake of simplicity. Other recent studies have considered up to three pools of organic matter with different mineralizability degrees [10] [113]. The effect of temperature on the methylation reaction rate was ignored in this simulation.

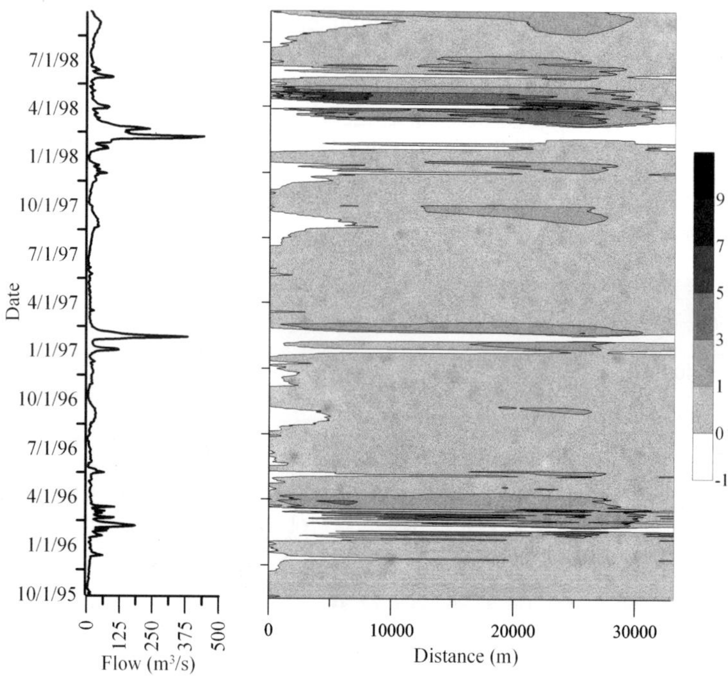

Figure 3: Net deposition (deposition-erosion) rate (g/m²day) versus time along the reach during the simulation period. Positive values indicate deposition whereas negative values represent erosion.

Figure 4 shows the measured vs. modeled total and methylmercury in the water column, close to the downstream end of the domain. It can be seen that the model nicely predicts the total mercury concentration considering its ability to capture the concentration of suspended sediments. There is a large correlation between the total mercury and suspended solids. The majority of the total mercury is carried by the sediments. This fraction increases during the high flow conditions due to the resuspension of sediments containing mercury. Although the model does predict the magnitude of the methylmercury concentration relatively well, it sometimes misses the trends, as is common with the current state-of-the-art (see [28]). This can be attributed to ignoring the temperature effects and also the margin of error in measuring the concentration of methylmercury. The other factor can be the small scale heterogeneities affecting the methylmercury production considering the fact that its production requires certain chemical conditions including an anoxic region in the sediments. These regions, involving dead zones and the areas where vegetation can slow down the flow, have scales smaller than the grid size and therefore cannot be captured in the model. On the other hand it should be noted that the spatial resolution determines the resolution of the gradients in sediment and its fluxes. Sampling schemes are typically designed on much smaller (e.g., bucket) scales than the scale of the numerical grid.

Table 4: Boundary conditions and other parameters used in the modeling study

Parameter		Value	Reference
Boundary Conditions	OM-particle associated (mMols/g)	0.603	(a)
	OM-dissolved(mM)	0.853	(a)
	O_2(mM)	0.390	(a)
	NO_3^-(mM)	0.103	(a)
	NH_4^+(mM)	0.011	(a), calc'd
	Fe^{3+}(mMols/g)	0.003	(a)
	$SO4^{2-}$(mM)	2.1	(a)
$K_D(OM)$ (L/kg)		707	(a), calc'd
$K_D(NH_4^+)$(L/kg)		691	(b)
k_{OM} (Yr^{-1})		25	(b)
k_{NH4+} ($mM^{-1}Yr^{-1}$)		20	(b)
K_{O2}(mM)		0.02	(c)
K_{NO2}(mM)		0.002	(c)
K_{SO4}(mM)		0.02	(c)
K_{Fe}(mM)		0.002	(c)
Bio/mechanical-dispersion coeff. (cm^2/Yr)		$14500e^{-0.25z}$	(b) modified
D_{O2} (cm^2/Yr)		369	(c)
D_{NH4} (cm^2/Yr)		309	(c)
D_{NO3} (cm^2/Yr)		309	(c)
D_{OM} (cm^2/Yr)		298	(c)
Organic Matter (C:N) ratio		0.13	(b)
O_2 atmospheric exchange coeff. (Yr^{-1})		8000	(d)
O_2 saturation concentration (mM)		0.9	calc'd from (d)
$K_D(Hg^{2+})$(L/kg)		125800	(e)
$K_D(MeHg)$(L/kg)		125800	(e)
k_{me}		186	(f)
k_{dm}		0.365	(f)

(a) USGS, 2000; (b) Canavan et al., 2007; (c) Berg et al., 2003; (d) Chapra, 1996; (e) Allison and Allison, 2005; (f) Calibration.

Table 5: Parameters used to model cohesive sediment Transport in the reach.

TSS Concentration Lateral Inflows (mg/L)	1370 mg/L [(assumed)]
Erosion parameter $\tau_c / \rho C_f$ (m²/s²)	0.25 [(calibrated)]
Deposition parameter $\tau_{bc} / \rho C_f$ (m²/s²)	0.30 [(calibrated)]
Deposition rate coefficient α_8	1.0 [(a)]
Uniform lateral inflow rate (m³/day.m)	$3 \times 10^{-6} Q_{upst}$ [(estimated from flow data)]
Erosion rate coefficient E (gr/m².day)	43.2 [(calibrated)]
Spatial horizontal grid size (m)	1072
Number of horizontal grids (river reach)	30
Vertical grid size in sediments(m)	0.03
Number of vertical grid points	12
Minimum dry density of bed material (kg/L)	0.8 [(b)]
Maximum porosity θ_0	0.631 [(b)]
Minimum porosity θ_∞	0.3 [(b)]
Porosity decrease rate k_θ (1/m)	10 [(b)]

(a) Partheniades, 2007; (b) within the reasonable range from the literature

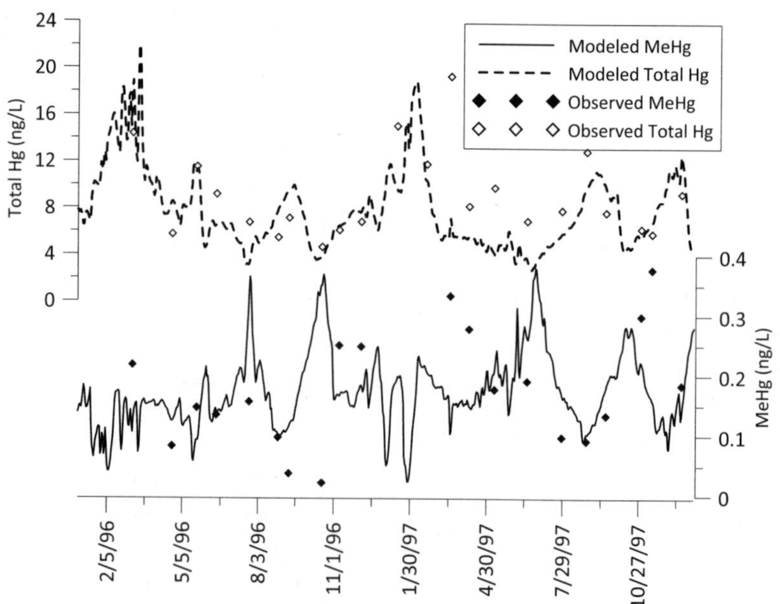

Figure 4: Measured total and methylated mercury at CBD-1 Station.

5. Future Directions

5.1. *Mercury bioavailability and methylation*

The development of a more comprehensive understanding of the biotic and abiotic processes that control Hg transformations are noted by a majority of authors as a key point for further research. It is thus important to study these processes and to quantify the roles of the environmental factors that govern Hg transformation. Research has shown that Hg methylation and de-methylation occur in both the water column and sediments and can occur from a variety of biotic and abiotic pathways [84] [7].

It is clear that Hg reduction, oxidation, methylation, and de-methylation are governed by complex processes still to be determined and quantified. A detailed study of the effect of environmental factors such as temperature, bioavailable organic matter, total Hg (and Hg++) concentrations, and microbial activity on Hg transformation rates will contribute to model development and to MeHg remediation strategies continued worldwide.

The primary missing links in current mathematical model formulations of reaction networks include a) de-methylation in the water column and benthic zones; b) complete mercury speciation; and c) transience in amount and type of available organic matter and its interaction with the cycle. Other phenomena of importance may include phytoplankton cycling, which can increase organic matter concentrations within the water column and thus increase methylation rates. Phytoplankton may also provide a residence for sulfate reducing bacteria or other cells of interest, and deposition of phytoplankton from the water column to the benthos provides another source of organic matter at the interface. Although data about mercury methylation by phytoplankton are scarce [24], hypothesized relationships can be explored within reaction network models by adjusting solution concentrations or by developing and testing kinetic reactions.

5.2. *Sediment particle size heterogeneity, compaction, erodibility and mercury transport*

The size of particles significantly controls the mass exchange of contaminants between sediments and the bulk water. Smaller particles have a larger surface area per unit mass and therefore can absorb larger amounts of metals on their surfaces. The rate of mass transfer between the particles and the water body is also a function of particle size due to the fact that intra-particle diffusion controls the mass exchange [78]. On the other hand, the deposition/resuspension and the transport of particles are largely influenced by their size. These make it important for future models to consider multiple size mixtures of sediments, instead of assuming uniformly-sized particles. This is especially important in cases where high floods capable of resuspending a large range of particle sizes play a major role in the transport of sediments and, therefore, in the transport of the mercury bound to them. Sediment resuspension predictors capable of handling multi-disperse sediment particles have been developed in the past (e.g., [41]) but not many efforts have been made to develop coupled sediment-contaminant models based on multi-disperse sediment mixtures. One reason can be the large computation expense involved in solving the mass exchange with a range of particle sizes and the lack of experimental data on the effect of sediment particle sizes on the sediment-water mass-exchange rates.

Although much progress has been made in modeling sediment dynamics in water bodies in terms of quantifying sediment fluxes, not much research has been conducted on the processes controlling the interactions among different size categories of suspended and bed sediments. This can especially be important in the fate of solid-bound contaminants that often exhibit variation in concentration with particle size. The mixing and sorting of the different fractions of top sediments is a controlling factor in the burial rates of mercury and also in how the resuspension of sediments releases mercury back to the overlying water. Also consolidation or cementation as a result of biomass growth can influence the stability or erodibility of sediments, this is an important factor in the fate of buried mercury contaminants in water bodies. These processes have not been studied in the context of classical sediment transport models, although they play a large role in the fate, transport, and attenuation of solid-bound contaminants.

The knowledge gained in the recent years regarding transport of cohesive sediment is comparatively less massive than that of non-cohesive particles. Cohesiveness is the result of small-scale mechanisms which promote interactions of generally chemical nature among particles. Formulas for resuspension of cohesive/non-cohesive sediments in large water bodies are lacking. In fact, most formulas have been developed from laboratory experiments, and there are reasonable questions as to whether these formulas are applicable to larger scales. Recent work [23] has shown that laboratory formulas for sediment resuspension can be adapted to predict resuspension in shallow lakes.

An almost completely unexplored field is the case of non-dilute mixtures (say, larger than 2-5% in volume concentration). Under these conditions, the transport of mercury could vary significantly due to the changes in the diffusivity of sediment occurring under non-dilute conditions. Very recent work by [58] suggests that while the Schmidt number is smaller than one for dilute mixtures, it is larger than one for non-dilute cases. (We recall herein that the Schmidt number is defined as the ratio between the eddy viscosity of the carrier phase and the diffusivity of the disperse phase.) More research is needed, of experimental and numerical nature, to clarify these issues.

Another unaddressed issue is the nature of rainfall events. Initial rainfalls tend to mobilize more source sediment from watersheds than later serial rainfalls, with an extreme case being the rain-on-snow event in which only the river-bed and river bank erosion contributes to sediment transport [19]. In the case of serial flood events in a relatively short time period, the sediment transport can be by an order of magnitude lower than during the previous flood event. Therefore, the "history" of flood events needs to be recorded and considered, which is seldom done in sediment transport modeling. A possible way of increasing the reliability of the relation between flow and sediment transport is to use artificial intelligence tools, such as the "model-tree" technique [91] [119]. These techniques and models have successfully been applied in ecological modeling (e.g., [68] [31] [4]).

5.3. *Mercury fate and transport in wetlands*

5.3.1. *Plant uptake and release of mercury*

Lacking motility, plants control their surroundings by chemical means. Besides providing the structure for wetlands, trapping and holding particles, plants deliver oxygen and exude organic matter to the sub-surface. In addition, as they carry nutrient metals upward, they bring along mercury, to an unknown degree, in the methylated form. (Alternately, mercury may be methylated inside the plant.) The example in the literature shows a dramatic seasonal rise of percent methyl vs. total Hg in late spring [124]. In either case, these forms are bound to soluble organic matter or on colloidal particles and enter the plant with water being taken up for the normal process of transpiration. In one salt marsh example, the seasonal cycle of total Hg in the plant shoot (especially at the tips) has been studied [57]. The annual flux of mercury released as detritus was estimated to be comparable to atmospheric deposition, and may be in a more bioavailable form.

5.3.2. *Wetting/drying cycles and mercury transformation/mobilization*

Since mercury transformation is driven by oxidation/reduction conditions, the wetting/drying cycle due to weather (or irrigation), as well as tides in coastal areas, must be included in any modeling effort. Furthermore, wet/dry cycling produces net local transport, not merely mixing. As oxygen and subsequent electron acceptors are depleted in the sub-surface, the drying cycle delivers the former. Reduced species produced in lower layers are transported to the surface during the wetting half of the cycle.

Assuming fairly small particle concentrations, fluxes of mercury in an undisturbed sediment may be relatively low. However, upon seasonal wet/dry cycles, or irrigation-related flow changes, strong chemical gradients will move through the sediment strata as the aeration/inundation condition reverses. These intermittent episodes may be a stronger driver of microbial metabolism (and therefore mercury methylation) than the longer periods of diffusion-limited, steady-state flux. Additionally, where drying may change the cohesion of sediments, and high flows transport the majority of the time-averaged particle

                    A. Massoudieh et al.

flux, the average condition may be less important than the departures from it, given the low solubility of both organic and inorganic mercury.

5.4. *Numerical approaches*

Numerical sediment transport models usually use the Eulerian principle, with regular or irregular grids, although it is well known that biogeochemical transformation processes are easier to describe using particle-based Langrangean models. In the particle-based methods, each particle can carry a lot of information on various parameters (e.g., concentration of different Hg species, environmental variables, etc.). However, Lagrangean methods such as particle-tracking and SPH (smoothed particle hydrodynamics) can only be used to describe point- or small-scale pollution domains, such as oil spills [95] [118]. The SPH method (e.g., [83]) is particularly promising; it has so far been successfully applied to simulations of two- or even multi-phase flows on local scale [25], mud-flows [18], and flushing of sediment from retention basins or through bottom outlets of the dams [42]. By coupling the environmental and pollutant information to sediment particles and using the ever increasing processor-power, Lagrangean models will more effectively compete with Eulerian approaches in the simulation of particle-bound pollutants in aquatic environments.

Acknowledgements

We thank very much Dr. Sanjeev Jha and Mr. Kaveh Zamani for their help with the manuscript.

Appendix – List of Symbols

List of Symbols		
Symbol	**Definition**	**Dimensions or Units**
B_d	bulk density of the sediment materials	$[M/L^3]$
c_i	dissolved concentration of chemical species i in the benthic sediments	$[M/L^3]$
C_i	dissolved concentration of chemical species i in the water column	$[M/L^3]$
$C_{i,at}$	equilibrium concentration of species i in the atmosphere	$[M/L^3]$
D_{hx}	dispersion coefficient in x direction	$[L^2/T]$
D_{hy}	dispersion coefficient in y direction	$[L^2/T]$
D_m	molecular diffusion coefficient	$[L^2/T]$

Er_j	sediment resuspension rate for solid phase class j	$[ML^{-2}T^{-1}]$
f_j	fraction of sediments consisting of phase j	
H	depth of water	$[L]$
J_0	rate of change of the elevation of the sediment-water interface with time	$[L/T]$
$k_{at,i}$	atmospheric mass exchange coefficient for chemical species i	$[L/T]$
k_b	diffusive mass exchange coefficient with bed sediments	$[L/T]$
K_{Dij}	water-solid distribution coefficient for solid phase j and species i	$[L^3/M]$
$k_{jj'}$	rate of transformation of solid phase j to phase j'	$[1/T]$
k_θ	rate of porosity decrease with depth	$[1/L]$
M	erosion rate coefficient	$[MT^{-1}/F]$
P	a factor to incorporate the effect of shear stress on settling rate	
Q	lateral inflow per unit area	$[L/T]$
R_i	reaction rate for dissolved species i	$[M/L^3T]$
R_{sij}	reaction rate for species i sorbed to solid phase j	$[M/L^3T]$
s_{ij}	concentration of chemical species i sorbed to solid phase category j in the benthic sediments	$[M/M]$
S_{ij}	concentration of chemical species i sorbed to solid phase category j in the water column	$[M/M]$
$S_{ij,in}$	concentration of species i sorbed to solid phase class j in the lateral influx	$[M_c/M]$
u_f	vertical pore water velocity at the water-sediment interface	$[L/T]$
u_s	advective velocity of sediments due to sediment consolidation	$[L/T]$
U	depth averaged velocity in x direction	$[L/T]$
v_s	particle fall velocity	$[L/T]$
V	depth averaged velocity in y direction	$[L/T]$
w_{pj}	deposition rate parameter for solid phase class j	$[L/T]$
z^*	vertical coordinate based on a fixed datum which increases upward	$[L]$
z_0	elevation of the sediment-water interface	$[L]$

α_8	coefficient for calculating the fall velocity	
ϕ_j	concentration of suspended solid phase category j	$[M/L^3]$
θ	porosity of the bed material	
θ_0	porosity at the water-sediment interface	
ψ_{ij}	source term representing for example the effect of plant uptake and decomposition	$[LT^{-1}M_cL^{-3}]$
ρ_s	density of sediment particles	$[M/L^3]$
τ_b	bed shear stress	$[F/L^2]$
τ_{cd}	critical shear stress for sedimentation	$[F/L^2]$

References

1. A Mercury Transport and Kinetics Model (beta test version 1.0), Model Theory and User's Guide, AScI Corporation, Georgia, USA (1992).
2. M.A. Ahanger, G.L. Asawa and M.A. Lone, *J. Hydraul. Res.*, IAHR **46**, 628 – 635 (2008).
3. J.D. Allison and T.L. Allison, *Soil and Waste*, (2005).
4. N. Atanasova, L. Todorovski, S. Džeroski and B. Kompare, *Ecol. Model.* **212**, 92-98 (2008).
5. A.E. Bale, *J. Environ. Eng.*, ASCE **126**,153-163 (2000).
6. B.D. Barkdoll and J.G. Duan, *J. Hydraul. Eng.*, ASCE **134(3)**, 285-285.
7. K.M.Batten and K.M. Scow, *Microbial Ecology* **46**, 429-441, 2003.
8. J.M. Benoit, C.C. Gilmour, R.P. Mason and A. Heyes, *Environ. Sci. & Technol.* **33**, 951-957 (1999a).
9. J.M. Benoit, R.P. Mason and C.C. Gilmour, *Environ. Toxicol. Chem.* **18**, 2138-2141 (1999b).
10. P. Berg, S. Rysgaard and B. Thamdrup, *Am. J. Sci.* **303**, 905-955 (2003).
11. S.P. Bhavsar, et al., *Environ. Toxicol. Chem.* **23**, 1376-1385 (2004a).
12. S.P. Bhavsar, M.L. Diamond, N. Gandhi and J. Nilsen, *Environ. Toxicol. Chem.* **23**, 2410-2420 (2004b).
13. B.P. Boudreau, *Computers & Geosciences* **22**, 479-496 (1996).
14. B.P. Boudreau, *Diagenetic Models and Their Implementation: Modelling Transport and Reactions in Aquatic Sediments* (Springer Verlag, 1997).
15. S. Brown, L. Saito, C. Knightes and M. Gustin, *Water Air Soil Pollut.* **182**, 275-290 (2007).
16. R.W. Canavan et al., *Geochimica et Cosmochimica Acta* **70**, 2836-2855 (2006).
17. R.W. Canavan, P. Van Cappellen, J.J.G. Zwolsman, G.A. van den Berg and C.P. Slomp, *Sci Total Environ.* **381**, 263-279 (2007).

18. T. Capone and A. Panizzo, SPH Simulation of non-Newtonian Mud Flows, *3rd Ercoftag SPHERIC International Workshop on SPH applications*, Lausanne, Switzerland, 134 – 138 (2008).

19. R.W.H. Carroll et al., *Ecol. Model.* **125**, 255-278 (2000).

20. R.W.H. Carroll and J.J. Warwick, *Ecol. Model.* **137**, 211–224 (2001).

21. R.W.H. Carroll, J.J. Warwick, A.I. James and J.R. Miller, *J. Hydrol.* **297**, 1-21 (2004).

22. S. Chapra, *Surface water quality modeling* (McGraw-Hill, 1996).

23. E.G. Chung, F.A. Bombardelli and G.S. Schladow, *Water Resour. Res.* available online (2009).

24. S.A. Coelho-Souza et al., *Sci. Tot. Environ.* **364**, 188-199 (2006).

25. A. Colagrossi, M. Antuono, N. Grenier and D. Le Touze, Simulation of Interfacial and Free-Surface Flows Using a New SPH Formulation, *3rd Ercoftag SPHERIC International Workshop on SPH applications*, Lausanne, Switzerland , 17 -24 (2008).

26. M.L. Diamond, *Water Air Soil Pollut.* **111**, 337-357 (1999).

27. W.E. Dietrich, *Water Resour. Res.* **18**, 1615-1626 (1982).

28. D.M. DiToro, *Sediment Flux Modeling*, John Wiley and Sons, Inc. (2001).

29. A. Drott et al., *Environ Sci Technol.* **42**(1):153-8 (2008).

30. B.M. Duc and W. Rodi, *J. Hydraul. Eng.*, ASCE **134(4)**, 367-377.

31. S. Džeroski, L. Todorovski, I. Bratko, B. Kompare and V. Križman, Kluwer *Equation discovery with ecological applicatins. In: Fielding, A.H. (Ed.), Machine Learning Methods for Ecological Applications* (Academic Publishers, Boston 1999).

32. C. G. Enfield and G. Bengtsson, *Ground Water* **26(1)**, 64-70. (1988).

33. A.L.M. Ethier et al., *Appl. Geochem.* **23**, 467-481 (2008).

34. R.A. Falconer and P.H. Owens Estuarine, *Coast. Shelf Sci.* **31**, 745-762 (1990).

35. W.F. Fitzgerald and R.P. Mason, Biogeochemical Cycling of Mercury in the Marine Environment, *Mercury and its effects on environment and biology-metal ions in biological systems*, New York, USA (1997).

36. W.F. Fitzgerald, C.H. Lamborg and C.R. Hammerschmidt, *Chem. Rev.*, **107**, 641-662 (2007).

37. E.J. Fleming, E.E. Mack, P.G. Green and D.C. Nelson, *Appl. Environ. Microbiol.* **72**, 457-464 (2006).

38. T.D. Fontaine, *Ecol. Model.* **22**, 85-100 (1984).

39. N. Gandhi, S.P. Bhavsar, M.L. Diamond and J.S. Kuwabara, *Environ. Toxicol. Chem.* **26**, 2260-2273 (2007).

40. M.H. Garcia, *Sedimentation and Erosion Hydraulics*, Chapter 6, Mays, L.W. Ed., Mcgraw-Hill (1999).

41. M.H. Garcia and G. Parker, *J. Hydraul. Eng.*, ASCE **117**, 414-435 (1991).

42. D. Gatti, A. Maffio, D. Zuccalà and A. Di Monaco, SPH simulation of hydrodynamics problems related to dam safety, *XXXII biennial IAHR Congress*, Venice, Italy (2007).

43. C.C. Gilmour and E.A. Henry, *Abstracts of Papers of the Am. Chem. Soc.* **203**, 140 (1992).

44. C.C. Gilmour, E.A. Henry and R. Mitchell, *Environ. Sci. Technol.* **26**, 2281-2287 (1992).

45. M.L. Goyette and B.A.G. Lewis, *J. Environ. Eng.*, ASCE **121**, 537-541 (1995).

46. W.H. Graf and L. Suszka, Unsteady Flow and its Effect on Sediment Transport, *21st IAHR Congress*, Melbourne, Australia, 539–544 (1985).

47. L. Gunneriusson and S. Sjoberg, *Nordic Hydrology* **22**, 67-80 (1991).

48. S. Han, R.D. Lehman, K.Y. Choe and G.A. Gill, *Limnology and Oceanography.* **52**, 1380-1392 (2007).

49. R.C. Harris, S. Gherini and R.J. Hudson, Regional Mercury Cycling Model: A Model for Mercury Cycling in Lakes, *Electric Power Research Institute and Washington Department of Natural Resources*, California, USA (1996).

50. E.J. Hayter and A.J. Mehta, *Appl. Math. Model.* **10**, 294-303 (1986).

51. E.A. Henry, L.J. Dodgemurphy, G.N. Bigham and S.M. Klein, *Water Air Soil Pollut.* **80**, 489-498 (1995).

52. C. J. Herrick et al., *Ecological Modelling* **15(1)**, 1-28 (1982).

53. M.E. Hines et al., *Environ. Res. Sect. A* **83**: 129-139 (2000).

54. A.T. Hjelmfelt and C.W. Lenau, *J. Hydraul. Eng.*, ASCE **96**, 1567-1586 (1970).

55. M. Horvat, S. Covelli, J. Faganeli, M. Logar, V. Mandić, R Rajar, A. Širca and D. Žagar, *Sci. Total. Environ.* **237–238**, (1999).

56. R.J. Hudson, S.A. Gherini, C.J. Watras and D.B. Porcella, *Modeling the Biogeochemical Cycle of Mercury in Lakes: the Mercury Cycling Model (MCM) and its Application to the MTL study lakes. Mercury as a Global Pollutant: Toward Integration and Synthesis* (Lewis Pub. 1994).

57. J. Jensen, Mercury uptake by a dominant plant species in San Francisco Bay tidal marshes, Saliconia virginica, MS Thesis, University of California, Davis (2007).

58. S.K. Jha, and F.A. Bombardelli, in preparation (2009).

59. S.Y. Jiang and M.Y. Corapcioglu, *Colloids and Surfaces a-Physicochemical and Engineering Aspects* **73**, 275-286. (1993).

60. P.Y. Julien, *Erosion and Sedimentation* (Cambridge University Press, 1998).

61. C.D. Knightes, *Environmental Modelling & Software* **23**, 495-510 (2008).

62. C.D. Knightes and R.B. Ambrose, *Environ. Toxicol. Chem.* **26**, 807-815 (2007).

63. D. Kocman, Mass Balance of Mercury in the Idrijca River Catchment. Doctoral dissertation. *Jožef Stefan Institute International Postgraduate School*, Ljubljana, 152 (2008).

64. B. Kompare, *Water Science and Technology* **37**, 9–18 (1998).

65. J. Kotnik, M. Horvat and V. Jereb, *Environ. Model. & Software* **17**, 593-611 (2002).

66. K.T. Lee, Y.L. Liu and K.H. Cheng, *Hydrol. Process.* **18**, 2439–2454 (2004).

67. D. Leonard et al., *Water Air Soil Pollut.* **80,** 519-528 (1995).

68. B. Lin, R.A. Falconer, *J. Hydraul. Res.*, IAHR **34**, 435-455 (1996).

69. R. Luff, K. Wallmann, S. Grandel and M. Schluter, *Topical Studies in Oceanography* **47**, 3039-3072 (2000).

70. U. Lumborg and A. Windelin, *J. Mar. Syst.* **38**, 287-303 (2003).

71. D. Mackay and M. Diamond, *Chemosphere* **18**, 1343-1365 (1989).

72. D. Mackay, S. Paterson and M. Joy,. *Acs Symposium Series* **225**, 175-196 (1983).

73. A. Malcherek, C. LeNormant, E. Peltier, C. Teisson, M. Markofsky and W. Zielke, *Est. Coast. Model.* **21**, 43-57 (1993).

74. P.A. Martinez and J.W. Harbaugh, *Simulating Nearshore Processes. Computer Methods in the Geosciences* (Pergamon Press, Oxford 1993).

75. R.P Mason and G.A. Gill, *Mineral. Soc. Canada* **34**, 179 – 216 (2005).

76. A. Massoudieh, Mathematical and numerical modeling of contaminant transport in aqueous systems involving mobile solid phases, Ph.D. Thesis, University of California, Davis (2006).

77. A. Massoudieh, F.A. Bombardelli, T.R. Ginn and P.G. Green, Mathematical modeling of the biogeochemistry of dissolved and sediment associated trace metal contaminants in riverine systems, *River Flow 2006, Proc. Inst. Conf. on Fluvial Hydraulics*, Lisbon, Portugal, 1993 (2006).

78. A. Massoudieh and T.R. Ginn, *J. Contam. Hydrol.* **92**, 162-183 (2007).

79. A. Massoudieh, D. Ju, T.M. Young and T.R. Ginn, *J. Contaminant Hyd.* **97(1-2)**, 55-66.

80. A. J. Mehta, E. J. Hayter, W. R. Parker, R. B. Krone, and A.M. Teeter, *J. Hydraul. Eng.*, ASCE **115(8)**: 1076-1093. (1989).

81. S.A. Mirbagheri, K.K. Tanji and R.B. Krone, *J. Environ. Eng.*, ASCE **114**, 1257-1273 (1988a).

82. S.A. Mirbagheri, K.K. Tanji and R.B. Krone, *J. Environ. Eng.*, ASCE **114,** 1275-1294 (1988b).

83. J.J. Monaghan, *J. Computational Physics* **110**, 399-406 (1994).

84. M. Monperrus et al., *Marine Chemistry* **107**, 49-63 (2007).

85. F.M.M. Morel, A.M.L. Kraepiel and M. Amyot, *Annu. Rev. Ecol. Syst.* **29**, 543-566 (1998).

86. B.A. O'Connor and J. Nicholson, *Coastal Engineering* **12**, 157-174 (1988).

87. A.N. Papanicolau, M. Elhakeem, G. Krallis, S. Prakash, and J. Edinger, *J. Hydraul. Eng., ASCE* **134(1)**, 1-14.

88. G. Parker, *1D Sediment Transport Morphodynamics*, e-book, http://vtchl.uiuc.edu/people/morphodynamics_e-book.htm.

89. D. Parkhurst and C. Appelo. Water-Resources Investigations Report 99-4259 (1999).

90. E. Partheniades, *Engineering Properties and Hydraulic Behavior of Cohesive Sediments* (CRC Press, Boca Raton 2007).

91. J. Quinlan, Learning with Continuous Classes, *5th Australian Joint Conference on Artificial Intelligence*, Singapore (1992).

92. H. Radovanovic and A.A. Koelmans, *Environ. Sci. & Technol.* **32**, 753-759 (1998).

93. R. Rajar, D. Žagar, A. Širca, and M. Horvat, *Sci. Total Environ.* **260**, 109-123 (2000).

94. R. Rajar, D. Žagar, M. Četina, H. Akagi, S. Yano, T. Tomiyasu and M. Horvat, *Ecol. Model.* **171**, 139 – 155 (2004).

95. R. Rajar, M. Četina and D. Žagar, Three-dimensional Modelling of Oil Spill in the Adriatic. Computer modelling of seas and coastal regions II. Wrobel, L.C., Brebbia, C.A., Traversoni, L. Eds., *Comput. Mech. Publ.* (1995).

96. R. Rajar, M. Četina, M. Horvat and D. Žagar, *Mar. Chem.* **107**, 89 – 102 (2007).

97. J.S. Ribberink, A. Blom and P. van der Scheer, Multi-fraction Techniques for Sediment Transport and Mophological Modelling in Sand-Gravel Rivers, *International Conference on Fluvial Hydraulics*, Lisse, Netherlands (2002).

98. L.P. Sanford and J.P.Y. Maa, *Marine Geology* **179**, 9-23. xxx

99. L.P. Sanford, *Computers and Geosciences* **34(10)**, 1263-1283. xxx

100. S.S. Sengör, N.F. Spycher, T.R. Ginn, R.K. Sani and B. Peyton, *Appl. Geochem.* **22**, 2569-2594 (2007).

101. A.F. Shields, Anwendung der achnlichkeitsmechanik und der turbulenz forschung auf die gescheiekbewegung, Ph.D. Dissertation, Mitt. Press Ver. – Anst., Berlin, Germany.

102. P.L. Shrestha, *Adv. Eng. Softw.*, **27**, 201-212 (1996).

103. P.L. Shrestha, and G.T. Orlob, *J. Envirn. Eng., ASCE* **122**, 730-740 (1996).

104. D.B. Simons and F. Sentürk, *Sediment Transport Technology: Water and Sediment Dynamics* (Water Resources Publication 1992).

105. J. Simunek, C.M. He, L.P. Pang, and S.A. Bradford, *Vadose Zone Journal,* **5(3)**, 1035-1047 (2006).

106. V.P. Singh, *Kinematic Wave Modeling in Water Resources, Environmental Hydrology.* (Wiley-Interscience 1997).

107. A. Širca, R. Rajar, R.C. Harris and M. Horvat, *Environ. Model. Software* **14**, 645-655 (1999a).

108. A. Širca, M. Horvat, R. Rajar, S. Covelli, D. Žagar and J. Faganeli, *Acta Adriatica* **40,** 75 –85 (1999b).

109. A. Širca and R. Rajar, Calibration of a 2D Mercury Transport and Fate Model of the Gulf of Trieste, *4th International Conference on Water Pollution,* Bled, Slovenia, 503-512 (1997).

110. P.G.J. Sistermans, Multi-Fraction Net Sediment Transport by Irregular Waves and a Current, *4th Conference on Coastal Dynamics*, Lund, Sweden, 588 – 597 (2001).

111. R.R. Turner and S.E. Lindberg, *Environ. Sci. and Techn.* **12**, 918-923 (1978).

112. P. VanCappellen, J.F. Gaillard and C. Rabouille, Biogeochemical Transformation in Sediments: Kinetic Models of Early Diagenesis. Interactions of C, N, P, and S Biogeochemical Cycles and Global Change, NATO ASI Series, 14 (1993).

113. P. VanCappellen and Y.F. Wang, *Am. J. Sci.* **296**, 197-243 (1996).

114. L.C. van Rijn, *J. Hydraul. Eng.*, ASCE **110**, 1613–1641 (1984a)

115. L.C. van Rijn, *J. Hydraul. Eng.*, ASCE **110**, 1431–1456 (1984b).

116. L.C. van Rijn, *J. Hydraul. Eng.*, ASCE **112**, 433-455 (1986).

117. L.C. van Rijn, *Principles of sediment transport in rivers, estuaries and coastal sea* (Aqua Publications 1993).

118. D. Violeau, C. Buvat, K. Abed-Meraim and E. de Nanteuil, *Coast. Eng.* **54**, 895-913 (2007).

119. Y. Wang and I.H. Witten, Induction of model trees for predicting continuous classes. *Poster Papers of the European Conference on Machine Learning*, University of Economics, Prague, Czech Republic, 128–137 (1997).

120. S.S.Y. Wang and S.E. Adeff, *J. Hydraul. Res.*, IAHR **24**, 53-67 (1986).

121. Q.R. Wang, D. Kim, D.D. Dionysiou, G.A. Sorial and D. Timberlake, *Environ. Pollut.* **131**, 323-336 (2004).

122. Water-Quality Assessment of the Sacramento River Basin, California: Water-Quality, Sediment and Tissue Chemistry, and Biological Data, 1995-1998: USGS Project (2000).

123. Y. Wu and R.A. Falconer, *Adv. Water Res.* **23**, 531-543 (2000).

124. J. S. Weis and P. Weis, *Environment International* **30**, 685– 700 (2004).

125. G.P. Williams, *J. Hydrol.* **111**, 89–106 (1989).

126. J.C.Winterwerp and W.G.M. van Kersteren, *Dev. in Sedimentology* **56** (2004).

127. S. Wright and D. H. Schoellhamer, *Water Resour. Res.* **41**, W09428.

128. W. Wu and S.S.Y. Wang, *Development and Application of NCCHE's Sediment Transport Models. US-China workshop on advanced Computational Modelling in Hydroscience & Engineering* (Oxford 2006).

129. W. Wu, S.S.Y. Wang and Y. Jia, *J. Hydr. Res.*, IAHR **38**, 427-434 (2000).

130. User's Manual for General Stream Tube model for Alluvial River Simulation version 2.0 (GSTARS 2.0), U.S. Dept. Interior, Bureau of Reclamation, Technical Service Center (1998).

131. C.T. Yang, and F.J.M. Simoes, *Int. J. Sed. Res.* **23(3)**, 197-211. (2008).

132. D. Žagar, *Acta Hydrotechnica* **17**, 68 (1999).

133. D. Žagar and A. Širca, *RMZ Mat Geoenviron* **48**, 179– 85 (2001).

134. D. Žagar, A. Knap, J.J. Warwick, R. Rajar, M. Horvat and M. Četina, *Sci. Total Environ.* **368**, 149–163 (2006).

135. D. Žagar et al., *Mar. Chem.* **10**, 64-88 (2006).

136. D. Zagar et al., *Sci. Total Environ.* **11**, 149-163 (2007).

137. W. Zeng and M.B. Beck, *J. Hydrol.* **277**, 125-133 (2003).

138. J.Z. Zhang, F.Y. Wang, J.D. House and B. Page, *Limnology and Oceanography* **49**, 2276-2286 (2004).

Chapter 14

3D ECOLOGICAL MODELLING OF THE AVEIRO COAST (PORTUGAL)

J.F. LOPES and ANA C. CARDOSO

CESAM, Department of Physics, University of Aveiro, Portugal

The present work is aimed to study the distribution of the temperature and the phytoplankton biomass in the near-shore Aveiro coastal zone (Portugal) using a three-dimensional ecological model. The study area is located in the western coast of the Iberian Peninsula, characterized by meteorological conditions of strong north/northwest prevailing winds, which favours the upwelling of nutrient enriched waters resulting from the divergences associated to the Ekman transport and, therefore, generating high nutrients availability. The results show that the model is able to reproduce the horizontal and the vertical temperatures and chlorophyll-a (*Chl-a*) patterns. They show, as well, the setup of a layer of cold water along the coastal side and the increasing of the declivity of the thermocline and the nutricline toward the coast. The model predicts satisfactorily the values for the maximum *Chl-a* concentration and the depth of the inshore subsurface chlorophyll maximum. The chlorophyll-*a* concentration shows maximum concentration values higher than 6 mg m^{-3}, during summer, near the coast. The subsurface chlorophyll-a maximum concentration is located within the euphotic zone, about 10 m bellow the surface, near the coast, deepening to value close to 40 m offshore.

1. Introduction

The development of three-dimensional ecological models has followed the availability of circulation models for the oceans and the shelf seas that could be used to force the ecological models. These models were setup, validated and extensively applied to the North Sea, the best studied shelf regimes, with a number of 3-D modelling efforts performed in the past decade in order to describe its ecosystem dynamics [13], and are now being used worldwide as tools to better understanding the ecological dynamics of coastal ecosystems. The models of the Greater North Sea have provided consistent regional and annual distributions of the dynamics of state variables representing the lower trophic levels, results which cannot be derived to this degree of coverage by observations. They have provided an understanding of the quantitative dynamics of primary production, especially about its spreading from the coasts to the northwest over the open North Sea [13]. Tett [23], Tett and Grenz [24] and Tett and Walne [25] have applied an ecological model to the southern North Sea and showed that the model realistically simulated the nitrate concentrations and the

309

mid-water chlorophyll maximum concentrations during the upwelling period. They evidenced the importance of the light availability in the bloom development, the influence of the net irradiance, which depends on the light extinction within the water column and on the thickness of the surface layer, in the upper mixed layer. Moll and Radach [13] have shown that although those models realistically simulate observations they do not reproduce accurately the nitrogen nutrients as well as the chlorophyll-a spatial distribution. Burchard et al. [3] applied two different biogeochemical models, a conservative nitrogen-based model to the North Sea, and a more complex model including an oxygen equation to the Baltic Sea, allowing for the reproduction of chemical processes under anoxic conditions. Although the models reproduce qualitatively observations, they attributed the lack of better quantitative agreement, for the maximum standing stock of phytoplankton for the North Sea and the average concentrations or the interannual variability of chlorophyll and nutrients for the Baltic Sea, to the strong dependence of biogeochemical parameters on the applied turbulence closure scheme. Their results show significant differences in all biogeochemical parameters when changing from the Mellor and Yamada [11] to the k–ε model for the North Sea, whereas for the Baltic Sea, using the k–ε model the diatom growth rate was increased as the mixing layer becomes deeper. The Iberian Atlantic coastal waters and, in particular, the Aveiro coastal ecosystem is a demanding area to model realistically [22], therefore, numerical models, namely ecological ones, are needed to be implemented. The main purpose of this paper is to apply a coupled three-dimensional physical and ecological model in the study of the temperature and chlorophyll-a spatial distribution at the Aveiro coastal zone and investigate the impact of the upwelling event and the nutrient availability in the distribution of the phytoplankton along the coast.

The study area corresponds to the western coast of the Aveiro ecosystem (Figure 1) and extends between 8°27' and 9°W W and 37°N and 42°N along the western side of the Portuguese Atlantic coast. The continental shelf is relatively wide (~60km) and gently sloping with an edge defined by the 200-m isobath, where the Aveiro Canyon (40°42'N) is the most significant topographic feature due to the fact that the slope gets very steep in just a few kilometres [14]. Climatologically the study area is located at the northern part of the northern-hemisphere subtropical high-pressure belt and its climate depends on the Azores anticyclone location. It is the northernmost limit of the Eastern North Atlantic Upwelling System [14]. The atmospheric current carried into the circulation of the Azores anticyclone corresponds, during winter, to small westerly winds and, during summer, to relatively strong north and north-westerly winds, with mean intensity of about 5-6 m/s [16]. In March and April the winds over the whole region turn north-westerly and by May they are generally northerly with wind stresses near the coast at 41°N of the order of 0.03 Pa. This pattern of northerlies remains unchanged through June, July and August, with maximum wind stresses of the order of 0.1 Pa in September. By October the winds relax in strength and a pattern similar to that of winter appears. There is a predominance of the southerly winds, during autumn and winter along the western coast of the Iberian Peninsula.

2. Materials and methods

2.1. *The physical model*

The physical model can be represented by a set of 3D hydrodynamic equations associated to the transport equation (which includes the heat and the radiative exchanges with the atmosphere). The hydrodynamic model is similar to the *Blumberg and Mellor* primitive equation models, which use the sigma coordinates and an embedded turbulent closure scheme [2]. The hydrodynamic model uses a time split technique, in order to ensure numerical stability, and calculates separately the external barotropic equations (2-D depth integrated equations), from the 3-D internal baroclinic equations. The transport model is solved using a *TVD* scheme, a weighted average between the *Upwind-* and *Lax-Wendroff* fluxes in the horizontal and between the upwind and the central fluxes in the vertical. The turbulence closure scheme used in this work is the widely known level 2.5 turbulence closure of Mellor and Yamada. The computational domain, with the origin situated at 40°38'N and 8°27'W, corresponds to an area centred at the Aveiro station, extending 40 km offshore and 130 km alongshore (Figure 1). It is represented by a uniform horizontal grid composed by 40x20 cells and a uniform 22 sigma levels distributed in vertical direction. The time steps for the model integration have to be small in order to guarantee the CFL criterion [2], based on stability analysis for linear surface-gravity waves [5, 10]. This criterion implies small time step for the 2-D mode (3 s), whereas for the 3-D mode the time steps may be greater (90 s), as in the last case the wave speed of the surface-gravity waves depends on the reduced gravity. The model was initialised with realistic data concerning the temperature as well as the chemical and the biological variables [10]. The wind intensity and direction were taken from the WRF model (Weather Research and Forecast Model), a nesting mesoscale and assimilation forecast model [16], which allows a mesh refinement enabling a high resolution output focused over the Aveiro region.

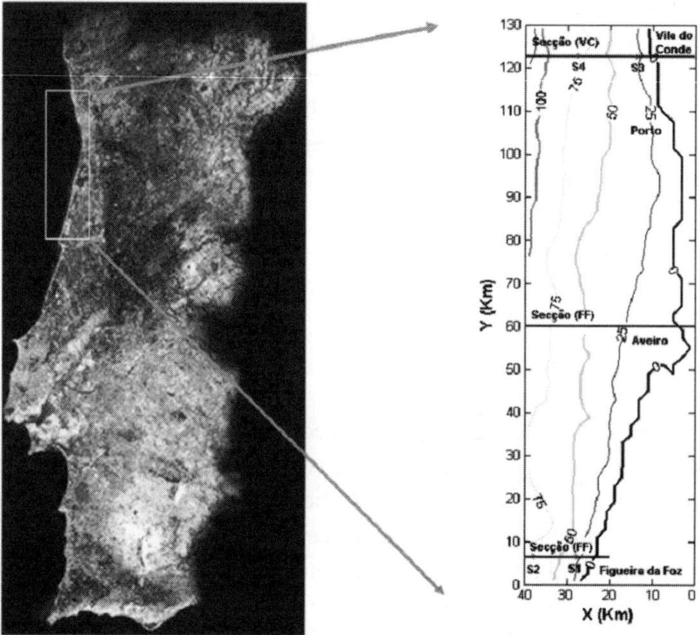

Figure 1. The study area at the Portuguese coast and the bathymetry (depth in m) between Vila do Conde (VC) and Figueira da Foz (FF), centred in Aveiro (Av).

 The von Newmann boundary conditions, relating gradients of the horizontal currents or of a general scalar quantity to the respective fluxes are applied at both surface and bottom boundaries. At the open sea, specific boundary conditions must be supplied for both 2-D and 3-D hydrodynamic modes. For the 2-D mode a radiation condition based on the method of characteristic is applied, using depth-averaged value of currents [10, 17, 19]. The boundary conditions for the 3-D mode are then calculated in the form of a prescription of the deviation of the currents from their depth-averaged value [10]. The momentum exchanges between the atmosphere and the ocean are solved by calculating the wind stress, using several standard formulations, as Charnock [4], Geernaert [7] and Smith and Banke [21]. The incident solar radiation flux, Q_{rad} , used in this model follows the Rosati and Miyakoda formulation [18] whereas the radiative exchanges between the atmosphere and the ocean are solved by calculating the total incoming short-wave radiation flux at the ocean surface, using an empirical formula derived from Reed [15]. The solar irradiance within the water column contributes as source term of the temperature equation. It is expressed as the sum of an infrared and a short-wave component. The non-solar heat flux (long wave emission, latent and sensible heats) lost by the surface is calculated using the well known empirical bulk formulations [1, 7, 8].

2.2. *The ecological model*

The biological sub-model is based on Tett conceptual model [23, 24] and has the following structure: a) internal non-conservative biological or chemical processes; b) photosynthesis by the absorption of PAR; c) physical transport by advection and diffusion; d) vertical sinking; e) deposition and erosion via a "fluff" layer; f) exchanges between the water column and the sediment layer (Figure 2). Its distinguishing features are the use: (1) of a microplankton compartment to include the biomasses and the microbial loop organisms activity as well as those for the phytoplankton, and (2) of variations in the chemical composition (especially, the nitrogen:carbon ratio) and of the microplankton and the detrital components to control many of the biological processes.

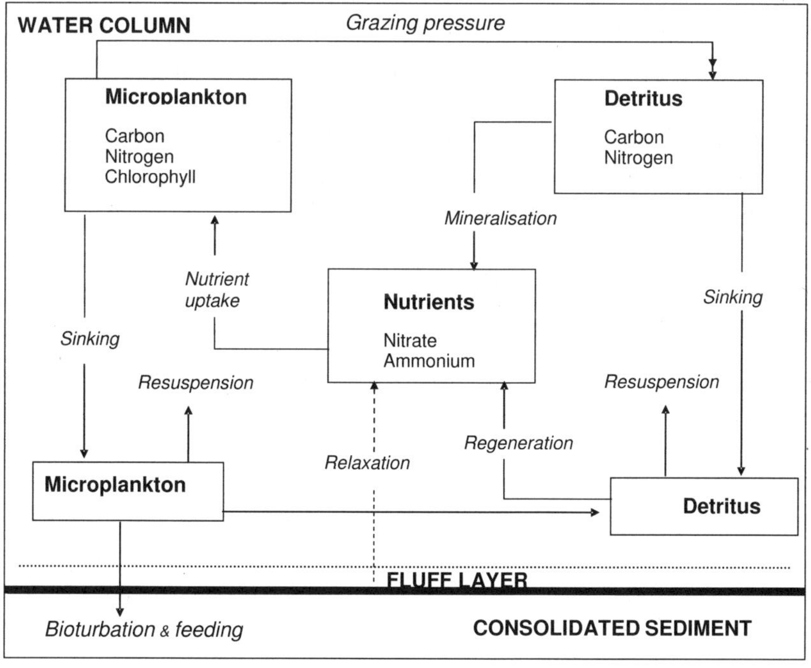

Figure 2. General feature of the biological model [10].

The present version of the biological model considers only nitrogen as a potentially limiting nutrient, and simulates changes in the concentration of ammonium (resulting directly from the zooplankton metabolic activity) and nitrate (deemed to include nitrite) formed by oxidation from ammonium. It contains eight state variables: micro-plankton carbon and nitrogen, detrital carbon and nitrogen, dissolved nutrients (nitrate, ammonium), dissolved oxygen and accumulated zooplankton nitrogen. The chlorophyll is not included as a state variable; it is derived from the microplankton carbon and nitrogen. One state variable for the inorganic particulate material is included in the sediment sub-model.

The micro-plankton growth rate calculates the minimum threshold limitation of both light intensity and nutrients:

$$\mu = \min\left[\mu(I_p), \mu(Q),\right] \tag{1}$$

where $Q=N/B$ is the nitrogen quota defined as the ratio of the organic nitrogen N (mmol N) and the nitrogen microplankton carbon B (mmol C).

I_p is the photosynthetically active radiation (PAR) defined at the sea surface as:

$$I_p = (1-R_1)Q_{rad} \tag{2}$$

where R_1 is defined as the infrared fraction of the solar irradiance and Q_{rad} the incident radiation at the sea surface.

Within the water column I_p satisfies an exponential law:

$$\frac{dI_p}{dx_3} = (k_2 + k_3)I_p \tag{3}$$

with k_2 the diffuse attenuation coefficient for the monochromatic light and k_3 defined as function of R_2, the exponential fraction of hyper-exponential decay, and of Δ_{opt} the thickness of the absorption layer for hyper-exponential PAR decay:

$$k_3 = -\frac{\ln(R_2)}{\Delta_{opt}} \tag{4}$$

For a mixed population of autotrophs and heterotrophs, growth rate is the result of net growth less net respiration:

$$\mu = \mu_a(1-\eta) - r_h\eta \tag{5}$$

where r_h is the heterotrophic respiration losses:

$$r_h = r_{0h} + b_h \mu_1(I_p) \tag{6}$$

with r_{0h} corresponding to the basal respiration rate for heterothrops and b_h the heterotrophic respiration-growth slope.

The heterotrophic fraction η is defined as:

$$\eta = \frac{B_h}{B_h + B_a} \tag{7}$$

with B_a and B_h, respectively, the autotrophs and the heterotrophs carbon biomass.

Phytoplankton also needs light for growth, but under some conditions (e.g. during winter), light rather than nitrogen supply, limits the microplankton growth. The light intensity controlled growth is calculated as the sum of the photosynthetic production and respiration loss, by a formalism described by Droop [6] and Tett [23], which takes into account the photosynthetical efficiency and the ratio of chlorophyll to autotrophic carbon:

$$\mu_1(I_p) = (\alpha \chi_a I_p - r_a)(1 - \eta) - r_h \eta \tag{8}$$

where $\alpha = k \ \varepsilon^X \ \Phi$ ((mmol C)(mg Chl)$^{-1}$(day^{-1} W^{-1} m^2)) is the photosynthetic efficiency, ε^X (m^2 mg^{-1} Chl^{-1}) the phytoplankton attenuation cross-section, Φ (nmol C μE^{-1}) the photosynthetic quantum yield and k a conversion factor. The chlorophyll to autotroph carbon ratio, χ_a, (mg Chl)(mmol C)$^{-1}$ can be expressed as:

$$\chi_a = \chi_\alpha (Q - q_h \eta) \tag{9}$$

where χ_α is the chlorophyll to autotroph nitrogen ratio ((mg Chl)(mmol N)$^{-1}$ q_h the heterotroph nitrogen to carbon quota and r_a the autotrophic respiration losses:

$$r_a = r_{0a} + b_a \mu_1(I_p) \tag{10}$$

with r_{0a} corresponding to the basal respiration rate for autotrophs and b_a the autotrophic respiration-growth slope.

The nutrients controlled growth is calculated as a function of the nutrients cell quota and the respiration loss. The nutrient uptake is calculated by a Michaelis-Menten type kinetics function. Under nitrogen-limiting conditions, growth rate should, ideally, be calculated as a fraction of the maximum specific growth rate depending on the cell nutrient quota, less respiration:

$$\mu_2(Q) = \mu_{max,a} f(T)(1 - \frac{Q_{min,a}}{Q_a})(1 - \eta) - r_h \eta \qquad (11)$$

where Q_a is the (variable) autotrophic cell quota, $Q_{min,a}$ the minimum cell nitrogen content, $\mu_{max,a}$ the maximum nutrient controlled growth and $f(T)$ the temperature growth factor given by :

$$f(T) = e^{q_T(T-T_r)} \qquad (12)$$

With q_T the growth rate coefficient of the temperature growth factor and T_r the reference temperature of the temperature growth factor.

The heterotrophic respiration is defined as:

$$r_h = \frac{\mu_h(1+b_h) - r_{oh}}{q^*} \qquad (13)$$

where $q^* = Q_d/q_h$.

Several other processes are included in the model, namely, phytoplanktonic grazing by micro- and meso-zooplankton, detritus, remineralization and seabed processes. As the model is essentially a tool for water column microbiology, it does not explicitly represent the benthos, the animals and microorganisms living on and in the seabed, nor does it describe the dynamics of the larger planktonic animals (mesozooplankton and others) [10].

In order to account for the fate of the microplankton and detritus, as they sink to the seabed and accumulate there, as well as for the exchanges between the water column and the seabed, a loosely-packed layer named as "fluff" layer was defined in the microplankton and detritus compartments and in the sediment model. It is an unconsolidated layer, of few millimetres thick, situated at the bases of the water column, immediately above the sediment surface, which can be rapidly eroded during periods of strong currents and turbulent mixing. In the model the ''fluff'' layer has a very simplified representation and is included as a compartment of negligible physical thickness, as the material is considered to sink into this layer as a flux rather than a concentration [10]. On the other hand, zooplankton is not modelled but mesozooplankton grazing pressure is imposed. The ecosystem described by the model is therefore "open" at the second trophic level. The zooplankton nitrogen variable accumulates potential losses at this level. The chlorophyll concentration is not directly simulated, but is derived algebraically from the microplankton carbon and the nitrogen concentrations. The sediment model determines the time evolution and the transport of inorganic particulate material.

Having quantified all the source and sink terms related to each dissolved, suspended and particulate substances, the concentrations are updated in time by solving the transport equation for each state variable, which computes the physical transport by advection and diffusion.

The water column concentration of each state variables Ψ (Ψ represents biological state variables, sediments or contaminant concentrations) is, therefore, determined by solving a transport equation summarized as:

$$(I + A_h + A_v + A_s + D_v + D_h = P(\psi) - S(\psi) \tag{14}$$

where the left side of the equation represents the time derivative I, the advective terms A_h and A_v, the vertical sinking/upwelling term A_s, the diffusive terms, D_v and D_h. This equation includes all the contributions of the source terms $P(\Psi)$ and the sink terms $S(\Psi)$ for the biological variables, the sediments or the contaminants.

3. Results

The results presented in this section concern the physical and the biological responses of the model to typical upwelling and downwelling wind situations along the western Iberian coast. Figure 3 shows a temporal series of wind speed and direction for the Aveiro coastal zone, representing a typical upwelling situation along the coast.

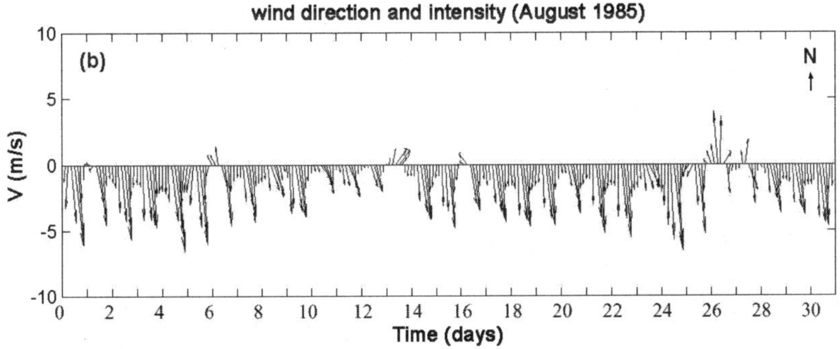

Figure 3. Temporal series of wind speed and direction for a typical summer situation at the Aveiro coastal zone.

Figure 4(a) shows the computed surface, middle and bottom maps of the horizontal temperature distributions, whereas Figure 5(a) shows maps of cross shelf vertical sections. They evidence a typical upwelling situation characterized by the setup of a layer of cold water along the coastal side and a warm mixed layer offshore.

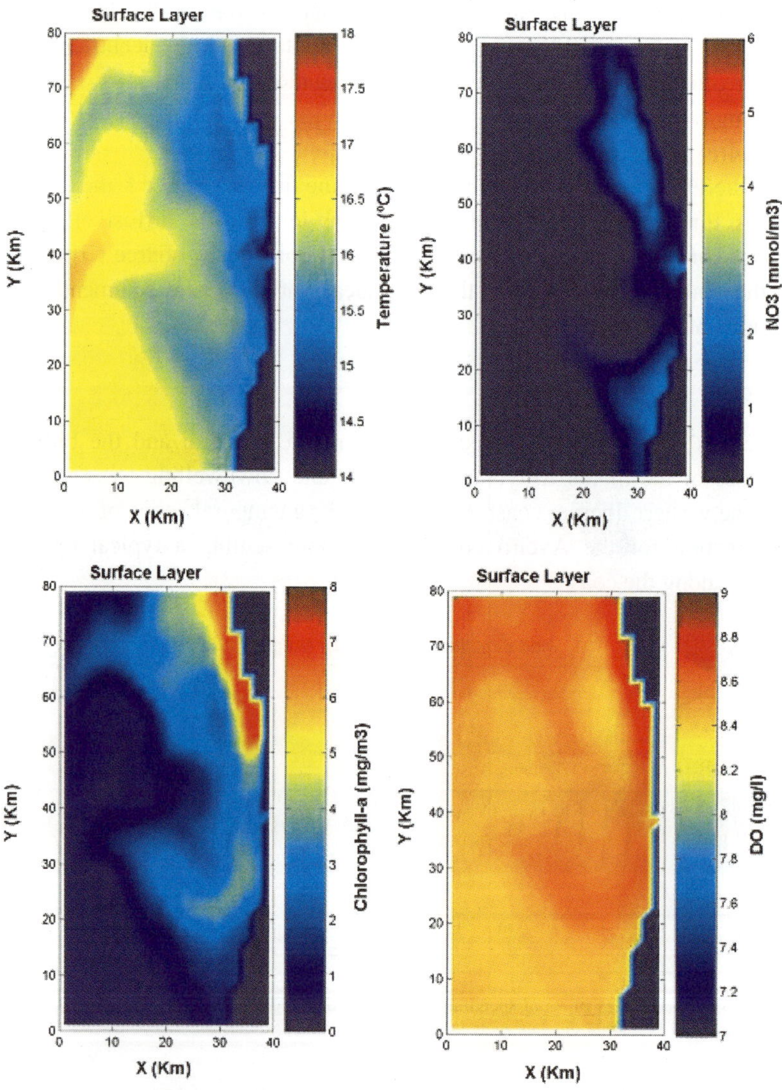

Figure 4. Summer surface distribution of: (a) Temperature, (b) NO$_3$ concentration (mmol m^{-3}), (c) Chl-a concentration (mg m^{-3}) and (d) DO concentration (g m^{-3}).

It can be observed the setup of a thermal stratification pattern characterized by a stratified warm mixed layer, located within the upper 30 m, superimposed to a homogeneous cold deep water layer. The surface temperature distribution is characterized by low inshore values, 15-16°C, and high offshore values, 17-19°C. The warm water is the result of the strong divergences associated to the Ekman transport, driven by the north-westerly winds. It is also evident the increase of the declivity of the isotherms close to the coast, resulting from the upwelling of the deep water into the surface layer.

Figure 4(c) shows the computed surface maps of the horizontal temperature, nitrates (NO_3), chlorophyll-a (Chl-a) and dissolved oxygen (DO) distributions, whereas Figure 5(c) shows maps of cross shelf vertical sections. The maps evidence a typical situation of summer phytoplankton bloom. The upwelling of rich in nutrients deep waters along the coastal side induces the shoaling of the nutricline, resulting in a situation of increasing nutrient availability near the surface layer.

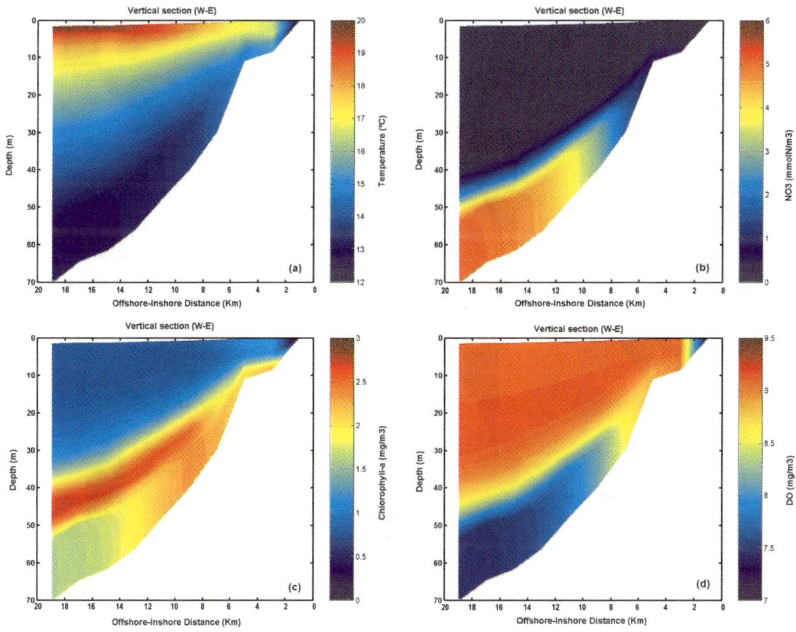

Figure 5. Summer cross shelf vertical section at FF of: (a) Temperature (°C), (b) NO_3 concentration (mmol m^{-3}), (c) Chl-a concentration (mg m^{-3}) and (d) DO concentration (g m^{-3}).

The vertical section of NO$_3$ (Figure 5(b)) evidences the shoaling of the nutricline resulting from the upwelling of deep water. The vertical *Chl-a* distribution (Figure 5(c)) shows the setup of a layer of subsurface chlorophyll-a maximum, mainly located within the euphotic zone. Its depth is about 10 m bellow the surface near the coast, deepening to values close to 40 m offshore (Figure 4(c)). The low NO$_3$ concentration values, near the coast, through the entire water column reflect, therefore, its depletion by the phytoplankton. As during summer there is no light limitation situation in the Aveiro coast, the upwelling event, is, therefore, a favourable situation for the phytoplankton growth as well as for its bloom occurrence.

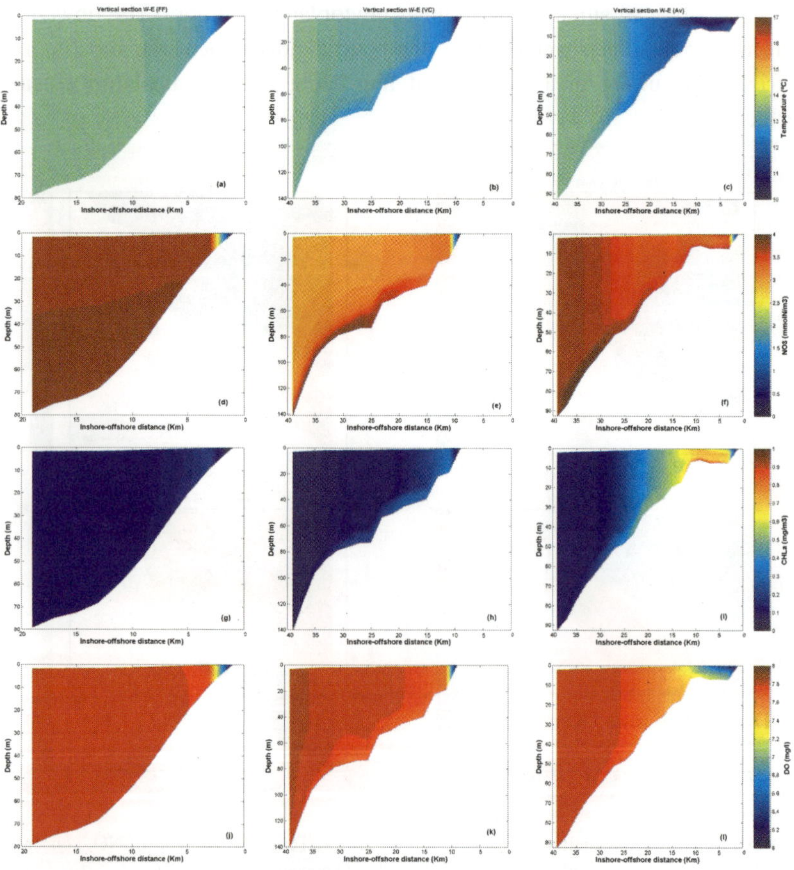

Figure 6. FF, Av and VC cross shelf vertical sections for winter of: Temperature (a-c), NO$_3$ concentration (mmol m^{-3}) (d-f), Chl-a concentration (mg m^{-3}) (g-i) and DO concentration (g m^{-3}) (j-l).

Figure 6 shows a winter situation, for which winds are predominantly southerly (not shown) along the Aveiro coast, representing a typical downwelling situation. The vertical distributions show that temperature is nearly homogeneous throughout the water column. The surface temperature is low, when compared to the summer situation, and varies between 14 ∘C (offshore) and 12 ∘C (inshore). In this situation the thermocline is almost inexistent. The surface NO3 concentration is homogeneous, with values of the order of 3.5–4 mmol m^{-3} (Figure 6). Contrarily to the summer scenario, the winter scenario shows high NO$_3$ surface concentrations, since the nutrients consumption by phytoplankton is lower. The surface Chl-a concentration stretches in a layer close the coast, where a maximum concentration of about 1mg/m3 is observed (Figure 7).

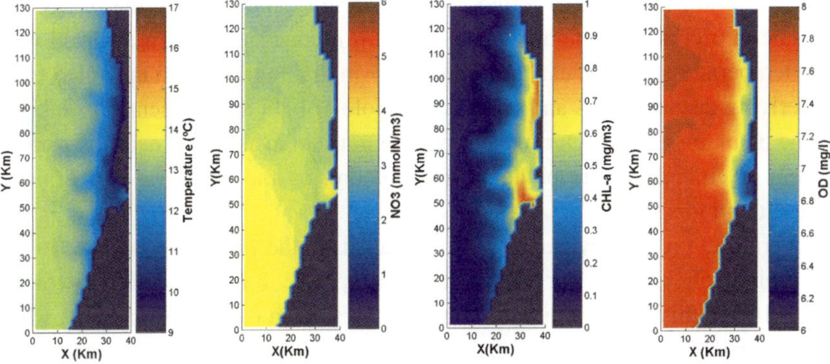

Figure 7. Winter surface distribution for: (a) Temperature (ºC), (b) NO$_3$ concentration (mmol N. m^{-3}), (c) Chl-*a* concentration (mg m^{-3}) and (d) DO concentration (g m^{-3}).

The maximum Chl-a concentration deepens, reflecting the deepening of the nutricline, as in a typical downwelling situation the phytoplankton must sink in order to get nutrients. Offshore the Chl-a concentration is homogeneous and low with values lower than 1 g m^{-3}.

4. Discussion and Conclusions

The main goal of this study is to apply a 3-D coupled biological–physical model in order to simulate the temperature and the phytoplankton distributions in the Aveiro coastal zone. The results show that the model it is capable of simulating the main pattern observed for the Aveiro coast. It has realistically simulated the response of the system to a summer upwelling situation as well as to a winter downwelling situation. The results evidence that, during an upwelling event the surface temperature distribution is characterized by isothermal patterns nearly parallel to the coast, and an increasing of their declivity toward the coast, which generates an offshore thermal stratification and an inshore well mixed structure. The model results provide a value for thermocline depth varying between 20 to 30 m below the sea surface and allow to estimate the depth of the euphotic zone, between 20 to 40m, values which are of the same order as those observed by Moita [12]. Nevertheless, it underestimates the surface mixed layer depth, which can be attributed to the Mellor-Yamada closure scheme, as pointed out by some authors [3, 26]. The model predicts satisfactorily the observed values of the maximum *Chl-a* concentration, the depth of the inshore subsurface chlorophyll maximum (25 m), even though underestimating its offshore value (40 m instead 50 m). It is evident the setup of a layer of high phytoplankton concentration near the coast and a chlorophyll-*a* maxima extending offshore, along the picnocline and the nutricline. The high values of the *Chl-a* concentrations, associated with low temperature values near the coast, confirm that nutrients are upwelled from underlying deeper waters and evidence the role of light and nutrients availability in the plankton growth. During summer the nutrient availability near the surface waters is crucial for the phytoplankton bloom, as the light is not a limiting condition in this season.

In overall, the results evidence the role played by the physical processes in the phytoplankton blooms, for which the atmospheric forcing is the key factor control- ling this event, which occurs after a situation of northerly winds, when both light and nutrients become available. Although light availability is necessary for the phytoplankton productivity, nutrients availability within the mixing layer is, therefore, essentially to the phytoplankton blooms occurrence.

Acknowledgements

This work was supported by the *Fundação para a Ciência e Tecnologia, Portugal* (FCT), through the project SIMCLAVE (POCI/MAR/56296/2004).

List of symbols

Symbol	Definition	Dimensions or Units
α	Photosynthetic efficiency for the light controlled growth rate	mmol C (mg Chl)$^{-1}$day^{-1} (W/m^2)$^{-1}$
b_a	Slope of the respiration-growth relationship for autotrophs	-
b_h	Slope of the respiration-growth relationship for heterotrophs	-
B	Microplankton carbon biomass	mmol C m^{-3}
Δ_{opt}	Thickness of the optical layer for the PAR decay	m
ε^X	Phytoplankton attenuation cross-section	m^2 mg^{-1} Chl^{-1}
Φ	Photosynthetic quantum yield	nmol C μE^{-1}
I_p	Photosynthetically active radiation (PAR)	W/m^2
k_1	Optical attenuation coefficient for infrared radiation	m^{-1}
k_2	Diffuse attenuation coefficient for monochromatic light	m^{-1}
k_3	Diffuse attenuation coefficient for PAR.	m^{-1}
μ	Micro-plankton growth rate	day^{-1}
η	Ratio of heterotroph to microplankton biomass	-
q_a	Autotroph nitrogen to carbon quota	mmol N (mmol C)$^{-1}$
q_h	Heterotroph nitrogen to carbon quota	mmol N. (mmol C)$^{-1}$
Q	Nitrogen quota	mmol N. (mmol C)$^{-1}$
Q_a	Autotrophic cell quota	mmol N (mmol C)$^{-1}$
Q_{rad}	the incident radiation at the sea surface	W/m^2
r_a	Autotrophic respiration losses	day^{-1}
r_b	Heterotrophic respiration losses	day^{-1}
r_{0a}	basal respiration rate for autotrophs	day^{-1}
r_{0b}	basal respiration rate for heterotrophs	day^{-1}
R_1	Infrared fraction of the solar radiation	-
R_2	Rapid attenuation of PAR in the near surface layer	-
χ_a	Chlorophyll to nitrogen ratio for autrotrophs	(mg Chl) (mmol C)$^{-1}$

References

1. T.V. Blanc, *J. Phys. Ocean.* **15**, 650 (1985).
2. A.F. Blumberg and G.L. Mellor, *Amer. Geophys. Un.* **4**, 1 (1987).
3. H. Burchard, K. Bolding, W. Kühn, A. Meister, T. Neumann, L. Umlauf, *J. Mar. Syst.* **61**, 180 (2006).
4. H. Charnock , *Quart. J. Royal Meteo. Soc.* **81**, 639 (1955).
5. E. Deleersnijder, J.M. Beckers, J.M. Campin, M. El Mohajir, T. Fichefet, P. Luyten, Some mathematical problems associated with the development and use of marine models, J.I. Diaz Ed., Springer Verlag, 39-86 (1987).
6. M.R. Droop, *Bot. Mar.* **26**, 99 (1983).
7. G.L. Geernaert, Bulk parameterizations for the wind stress and heat fluxes, Geernaert G.L. and W.J. Plant Ed., Kluwer Academic Publishers, 91-172 (1990).
8. A.E. Gill, Atmosphere-Ocean Dynamics. *Intern. Geophys. Ser.*, Vol. 30., Academic Press, (1982).
9. G.W. Hedstrom, *J. Comp. Phys.* **30**, 222 (1979).
10. P.J. Luyten, COHERENS – Dissemination and exploitation of a coupled hydrodynamical-ecological model for regional and shelf seas, MAS3-CT97-0088. Final Report. MUMM Internal Report, Management Unit of the Mathematical Models, p. 76 (1999).
11. G.L. Mellor and T. Yamada, *Rev. Geophys. Space Phys.* **20**, 851(1982).
12. M.T. Moita, Estrutura, Variabilidade e Dinâmica do Fitoplâncton na Costa de Portugal Continental. PhD Thesis. Faculdade de Ciências da Universidade de Lisboa, p. 272 (2001).
13. A. Moll and G. Radach, *Progr. Ocean.* **57**, 175 (2003).
14. Á. Peliz, T. Rosa, L., A.P. Santos, Miguel, J. L. Pissarra, *J. Mar. Syst.* **35**, 61 (2002).
15. R. K Reed, *J. Phys. Ocean.* **77**, 482 (1977).
16. R.M.M. Reis and M. Z. Gonçalves, O clima de Portugal, WLI. Instituto Nacional de Meteorologia e Geofísica, Lisboa, p. 160 (1988).
17. L. P. Røed and C. K. Cooper, A study of various open boundary conditions for wind-forced barotropic numerical ocean models, J.C.J. Nihoul and B.M. Jamart Ed., Elsevier, p. 305–335 (1987)
18. A. Rosati and K. Miyakoda, *J. Phys. Ocean.* **18**, 1601 (1988).
19. K.G. Ruddick, Modelling of coastal processes influenced by the freshwater discharge of the Rhine. Ph.D. Thesis, Univ. de Liège, Belgium, (1995).
20. W.C. Skamarock and J. B. Klemp, *J. Comp. Phys.* **227**, 3465 (2008).
21. S.D. Smith and E.G. Banke, *Quart. J. Meteo. Soc.* **101**, 665 (1975).
22. I. Stevens, M. Hamann, J. A. Johnson, A. F. G. Fiuza, *J. Mar. Syst.* **26**, 53 (2000).

23. P. Tett, Parameterising a microplankton model. Report, Napier University, Edinburgh (1998).
24. P. Tett and C. Grenz, *Vie et Milieu* **44**, 39 (1994).
25. P. Tett, and A. Walne, *Ophelia* **42**, 371 (1995).
26. R. Tian and C. Chen, Influence of model geometrical fitting and turbulence parameterization on phytoplankton simulation in the Gulf of Maine. *Deep-Sea Res.* **II (23–24)**, 2808 (2006).

Chapter 15

EXPERIMENTAL CALIBRATION OF A SIMPLIFIED METHOD TO EVALUATE ABSOLUTE ROUGHNESS OF VEGETATED CHANNELS

SERGIO DE FELICE, PAOLA GUALTIERI and GUELFO PULCI DORIA

Hydraulic, Geotechnical and Environmental Engineering Department, University of Naples Federico II, Naples, Italy

This chapter deals with the calibration of a new simplified experimental method to evaluate absolute roughness of vegetated channels. The method is based on boundary layer measurements in a short channel rather than on uniform flow measurements, as usual. The proposed method can be applied to any kind of rough bed, but it is particularly useful in vegetated beds where long channels are difficult to prepare. In this paper a calibration coefficient is experimentally obtained. In order to perform suitable comparisons with literature data relationships between ε absolute roughness and Manning's n coefficient are deepened. The results are successfully compared with literature experimental data with a very good fit. Finally, a particular dependence of ε values on the vegetation density are explained through further experiences. In conclusion it is possible to state that the proposed method, once calibrated, can provide reliable prediction of absolute roughness in vegetated channels.

Keywords: absolute roughness, boundary layer, calibration, Coriolis coefficient, cylinder, flow resistance, friction factor, LDA, Manning roughness factor, uniform flow, vegetated bed

1. Introduction

In the past, vegetation on river beds was considered an unwanted source of flow resistance. For this reason, vegetation was commonly removed to improve the water conveyance.

Nowadays, vegetation is regarded as a means to provide stabilization for banks and channels, habitat and food for animals, and pleasing landscapes for recreational use. Therefore, the preservation of vegetation plays a very important role in the water systems ecology.

For this reason, the study of the effects of vegetation on the river hydrodynamics represents one of the most important topics hydraulic engineers studies today. The characteristics of flows over vegetated beds have been deepened with either experimental and numerical methods. Numerous studies have been carried out to examine the streams flow resistance on a vegetated bed, and the main hydrodynamic characteristics of these streams, such as the mean flow, the turbulent structures and the sediment transport [1] [2] [4] [5] [9] [10]

[11] [12] [13] [14] [15] [17] [18] [19] [20] [21] [22] [23] [24] [26] [27] [28] [29] [30] [31] [32] [33] [34] [35] [36] [37] [38] [40] [41] [42].

2. Generalities and State of the Art about Resistance Formulas

In a review of Yen [39] it is observed that each resistance coefficient can be considered either as cross section coefficient or as reach one, and that the usual resistance formulas can be considered as reach formulas to be applied to uniform flows. It is possible to apply them also not to uniform flows, although to a very short reach, practically to a single cross section: therefore, in this case, the resistance coefficient can vary in the subsequent cross sections of the flow.

Referring to the reach formulas, the most frequently used, relating open-channel flow velocity V to flow resistance coefficients, are the Darcy-Weisbach, Manning and Chézy ones:

$$V = \sqrt{\frac{8g}{f}}\sqrt{RS} \; ; \;\; V = \frac{1}{n} R^{2/3} S^{1/2} \; ; \;\; V = C\sqrt{RS} \tag{1}$$

where f, n, and C are the Weisbach, Manning and Chézy flow resistance coefficients, and R = hydraulic radius, S = slope, g = gravitational acceleration.

In case of cross section formulas, the slope S must be substituted by the head slope J. Comparing these formulas, it is possible to obtain the following expressions:

$$\sqrt{\frac{f}{8}} = \frac{n\sqrt{g}}{R^{1/6}} = \frac{\sqrt{g}}{C} \tag{2}$$

Among these flow resistance formulas, the authors choose the Darcy-Weisbach approach, as it is the most suitable for an exact evaluation of the flow resistances. In this approach, the friction factor f is related to the absolute roughness, here called ε, through the Colebrook-White formula:

$$\frac{1}{\sqrt{f}} = -K\,log\left(\frac{\varepsilon/4R}{a} + \frac{b}{Re\sqrt{f}}\right) \tag{3}$$

with Re Reynolds number defined as $Re = 4VR/\nu$, and ν kinematic viscosity of the fluid.

The values of the constants a, b and K have been the object of many experimental surveys. In particular, in [39] values for a, b and K obtained by other researchers either for open channels with different ratios aspect and wide open channels, are suggested. In Moody-type diagram, relative to circular full pipes, the values $K = 2$; $a = 3.71$; $b = 2.51$ are used.

According to [25], it is always possible to use the values of Moody-type diagram relative to circular full pipes also for open channels flows. This requires adoption of a shape parameter ψ depending on the aspect ratio, so that the Colebrook-White formula becomes:

$$\frac{1}{\sqrt{f}} = -2\,log\,\frac{1}{\psi}\left(\frac{\varepsilon/4R}{3.71} + \frac{2.51}{Re\sqrt{f}}\right) \tag{4}$$

In particular, ψ assumes the value 0.83 for wide rectangular channels. The values of the so obtained friction factor f are very similar to those suggested by Yen for wide channels.

Referring to the previous considerations, in a water current, also the vegetation may be regarded as a kind of bed roughness that produces a resistance against the flow, and many papers dealt with the problem of attaining the corresponding resistance coefficients and/or the absolute roughness [1] [17] [18] [19] [20] [22] [23] [24] [26] [36].

3. Aim and Expected Results of the Paper

Usually the resistance coefficient is evaluated in uniform flow. In [7], a methodology to determine the flow resistance coefficient by investigation of a boundary layer developing in a short channel over a vegetated bed, was proposed and implemented.

As the channel length required to produce a boundary layer is shorter, therefore, shorter laboratory channels are needed. Therefore this methodology would simplify the determination of the flow resistance coefficient.

In this chapter, the methodology is upgraded in the following way: i) evaluation of a calibration factor through measurements on a uniform flow developing in a long channel with varying slope over a vegetated bed; ii) comparison of the calibrated absolute roughness values in seven different bed cases are compared with literature data.

Finally, some further considerations about the obtained values of the friction factor are reported.

4. Synthesis of the Methodology to Evaluate the Absolute Roughness Through Boundary Layer Measurements

The methodology [7], here synthetically described, was based on experimental data performed in a short channel, on an equilibrium boundary layer flow either with a smooth or a vegetated bed [6] [16].

In particular, equilibrium boundary layer is characterized by horizontal free surface. Therefore, it was necessary to suitably incline the channel to balance the head losses generated by the vegetation. The boundary layer developed along the first 50cm of the channel, where it attained the depth of the current.

The vegetation was modelled by means of 4mm diameter brass cylinders with three different heights (5mm, 10mm, 15mm) placed according to two different regular geometries (respectively, a rectangular mesh $5*2.5\text{cm}^2$ and a square mesh $2.5*2.5\text{cm}^2$), hereafter respectively called single density (s.d.) and double density (d.d.). Consequently, the projected vegetation area per water unit volume in the flow direction [38] were, respectively, 3.2m^{-1} and 6.4m^{-1}. Combinations of three different heights and two different densities produced six different vegetated beds.

In all the seven considered flow conditions (smooth bed and six different vegetated beds), the velocity of the free-stream was at 1.424m/s. The Reynolds number, evaluated as in eq. (3) and with $R=h$ was equal to 263.000. The channel slopes' values are reported in Table 1.

The test sections were set at 20, 30, 40 and 50cm from the channel inlet. In each test section, two measurement verticals were considered, differently positioned with respect to the cylinders. The first one was set at the centre of either a rectangular or a square mesh. The second one was set along a cylinder row and at the centre of the lateral side of the same rectangular or square mesh. It is clear that, in case of a smooth bed, there was no need of such a second measurement location.

Either in the case of smooth or vegetated bed, in each test section, along the chosen verticals, instantaneous velocities were measured in 20÷30 experimental points, through a suitable LDA system, and, in each point, the local mean velocity was obtained.

All these measurements were elaborated computing in each test section, the Coriolis α coefficient, and, afterwards, interpolating the subsequent α values through a parabolic function. Moreover, it was shown that in case of equilibrium boundary layer, the generic section friction factor f could locally be obtained through the following equation:

$$f = 8\alpha S - 4h\frac{d\alpha}{ds} \tag{5}$$

where S is obviously the slope of the channel, and s and h are, respectively, the distance from the inlet of the channel and the flow depth, of the generic section.

Therefore, on the ground of the parabolic interpolation, a particular reference section could be identified where the condition $d\alpha/ds = 0$ held and the simplified equation, $f = 8\alpha S$, was valid. As flow in this reference section most closely resembled uniform flow (where α=const), the corresponding friction factor f was assumed to be a representative reach-value of a uniform flow, having the same flow rate and water depth.

As final results, the friction factors were reported in Table 1, together with the slopes of the channel, the depths of the reference sections, and the absolute roughness, obtained through eq. (4).

Table 1. Values of bed's slope S, reference section depth h_0, friction factor f, and ε_{bl} absolute roughness.

		S	h_0(m)	f	ε_{bl}(m)
Smooth bottom		0.0025	0.00481	0.02038	0.000139
Single Density	5mm	0.0092	0.05083	0.07783	0.010102
	10mm	0.0160	0.05507	0.14202	0.031961
	15mm	0.0227	0.05716	0.21518	0.058847
Double Density	5mm	0.0115	0.05050	0.09626	0.015215
	10mm	0.0205	0.05428	0.18411	0.045692
	15mm	0.0295	0.05727	0.28642	0.082069

5. Experimental Calibration of the Methodology

In order to test and to calibrate the methodology synthetically described, measurements were performed in a uniform flow developing in a long channel with varying slope, endowed with the single density (s.d.) 15mm high vegetation type (Figure 1).

Figure 1. Channel used for the experimentation.

The channel had plexiglas walls and bed and was 40cm wide, 8m long and its walls are 40cm high. Moreover, it was fed by a tank connected to the hydraulic circuit of the laboratory. At the exit of the channel, there was a draining tank from which the water flows through a system of free surface channels straight into the tanks of the laboratory. In particular, the water was taken (through a system of lifting pumps with nominal capacity from 0.025 to 0.075m³/s) from one of the backwater tanks of the laboratory, and it was sent (passing through a device to measure the flow-rate) to the feeding tank, hold at a constant water level.

As the channel was sufficiently long, it was always possible to obtain a uniform flow condition. In case of supercritical flow, the current attained the uniform flow condition naturally. In case of subcritical flow, the use of a sluice-gate at the end allowed it to attain the uniform flow. The test section was located in the reach of uniform flow.

It was decided to measure the absolute roughness in six different flow conditions: three in a subcritical flow and three in a supercritical flow, with different flow-rates and currents depths. The absolute roughness was calculated through eq. 4, experimentally measuring the other terms (slope and mean section velocity as ratio between flow-rate and section area).

The flow conditions, and the results of measurements and calculations are reported in Table 2.

Table 2. Results of the performed experimentations.

Test n.	Flow conditions	S	Flow-rate [m³/s]	h_u [m]	V_u [m/s]	ε_{unif} [m]
1	subcritical	1%	0.0226	0.0785	0.719	0.035
2	subcritical	1%	0.0330	0.0970	0.850	0.035
3	subcritical	1%	0.0450	0.1155	0.974	0.035
4	supercritical	2%	0.0226	0.0644	0.876	0.033
5	supercritical	2%	0.0330	0.0803	1.027	0.035
6	supercritical	2%	0.0450	0.0954	1.178	0.035

As the ε_{unif} values coincide one another, within the experimental accidental errors, their mean value (0.0347m) was considered as ε true value.

This true value was compared with the correspondent one previously obtained through the boundary layer measurements relative to the same vegetation type (0.058847m in Table 1). The result of this comparison was that the value relative to uniform flow (ε_{unif}) appeared to be a little smaller than the correspondent one relative to boundary layer (ε_{bl}) probably due to the circumstance that the α value computed in the boundary layer was slightly higher than the one relative to uniform flow. It is noteworthy that this difference, at a first sight not at all slight, is strongly mitigates by the presence of the logarithm in the Colebrook-White equation.

In any case, at least at this stage, the ratio (0.590) between ε_{unif} and ε_{bl} was assumed as a calibration factor for the methodology.

Using this calibration factor, ε_{cal} values were calculated either in case of smooth bed or in the six models of vegetated bed (Table 3).

Table 3. Absolute roughness for the seven flow conditions.

		ε_{bl}(m)	ε_{unif}(m)	$\varepsilon_{unif}/\varepsilon_{bl}$	ε_{cal}(m)
Smooth bottom		0.000139	-	-	0.000082
Single Density	5	0.010102	-	-	0.005957
	10	0.031961	-	-	0.018846
	15	0.058847	0.0347	0.590	0.034700
Double Density	5	0.015215	-	-	0.008972
	10	0.045692	-	-	0.026943
	15	0.082069	-	-	0.048393

Finally, in Figure 2, ε_{cal} values of each vegetated bed are reported as a function of the cylinders height and density, including also the case of smooth bed.

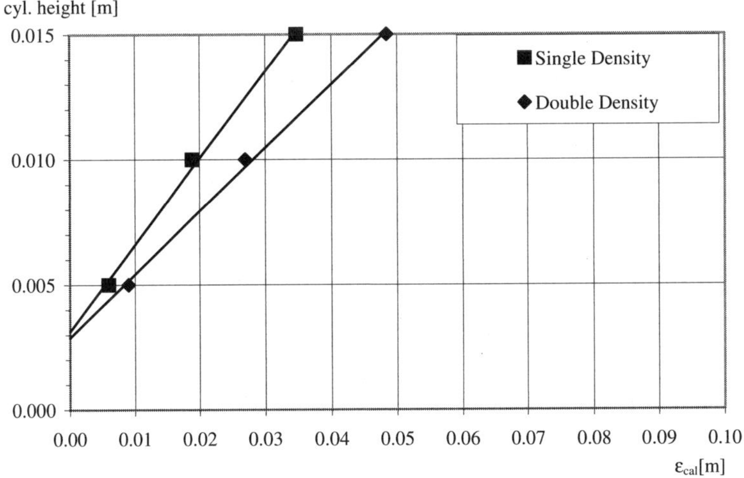

Figure 2. Absolute roughness values and interpolating functions.

The experimental points are very well aligned along straight lines which join themselves about in a single point on the abscissa axis. Probably the ε_{cal} value is almost zero if the cylinders height is no more than 3mm, because of the presence of the laminar sublayer. The two straight lines show an increasing trend with the cylinders height, more accentuated in case of double density with respect to single density. Over the height of 3mm cylinders, ε_{cal} values can be computed, in relation to the two vegetation densities, through following relationships:

$$\varepsilon_{s.d.} = 2.874\ h - 0.0089 \tag{6}$$

$$\varepsilon_{d.d.} = 3.942\ h - 0.0113 \tag{7}$$

where $\varepsilon_{s.d.}$ and $\varepsilon_{s.d.}$ represent the absolute roughness values in (s.d.) and (d.d.) cases.

6. Relation between Roughness Coefficients n and ε

The obtained absolute roughness will be compared with Manning's n literature data, therefore, it is interesting to deepen here the relations between ε and n.

It is known that the Manning's n depends not only on the bed roughness but also moderately on the hydraulic radius of the flow.

Ven Te Chow [3] already proposed some considerations about this matter:

"On the contrary, in planning problems, it was pointed out that n depends on the depth of the current flow, on the shape of the section and on the channel slope, which would make impossible to supply a univocal solution of the measuring problem.

However, it is also clear that n is linked to the product of the current mean velocity by the hydraulic radius.

In particular, this relation is specific for a certain type of vegetation, and it seems to be independent of the shape of the section and the channel slope.

Practically speaking, it was noticed that the curves n-VR relative to channels characterized by different types of vegetation and different shapes of the section and slopes, group themselves in reference curves n-VR, each one relative to some entity of flow resistance (low, mean, high…) classified according to the type of vegetation and its height and density."

It is to say that Chow expressed the variability of n for a fixed type of vegetation as a function of the VR product (practically a Reynolds number).

At present, the ε absolute roughness is more and more used instead of Manning's n to define the roughness. A fundamental hypothesis linked to the use of the ε absolute roughness is that it really depends only on the bed characteristics.

The hypothesis of the independence of ε of the hydraulic radius can explain the variability of n. Indeed, the joint use of equations (2) and (4) leads to the following equation:

$$n = \frac{R^{1/6}}{2\sqrt{8g}\ log\ \dfrac{14.84\psi R}{\varepsilon}} \tag{8}$$

Eq. (8) is wholly valid only if the term $2.51/Re\sqrt{f}$ is negligible with respect to the term $(\varepsilon/4R)/3.71$ within eq. (4), circumstance that happens practically always with vegetated beds. In case of almost smooth bed, eq. (8) can be yet considered only approximately valid. This equation clearly shows that, if in a current with a given bed type but with different R values, ε can be considered constant, consequently n must depend on R.

Moreover, an analytical study of the function $n(R)$, performed through the study of the sign of its derivative, shows that this sign depends on the value of the ratio ε/R in respect of the particular value $14.84\psi/e^6$. More precisely, if ε/R is less than $14.84\psi/e^6$ then n increases with increasing R, whereas if ε/R is larger than $14.84\psi/e^6$ then n decreases as R increases. In case of a large rectangular channel, this particular value is 0.03. Consequently, in case of ordinary roughness or also of very sparse and/or very low vegetation ($\varepsilon/R<0.03$) the first situation holds, whereas in case of high and/or dense vegetation ($\varepsilon/R>0.03$) the second situation holds.

As for the statement of Chow about the dependence on the VR product, it must be stressed that, with constant ε, increases of R cause increases of velocity too. Consequently it can be supposed that some little confusion between a dependence only on R and a dependence on the product VR could be happened.

As for the dependence of n on the R, starting from eq. (8), it is possible to state that the ratio between the n values relative to the same roughness but to different current's hydraulic radius R_1 and R_2 is the following one:

$$\frac{n_1}{n_2} = \left(\frac{R_1}{R_2}\right)^{1/6} \Bigg/ \frac{\log(14.84\,\psi\,R_1/\varepsilon)}{\log(14.84\,\psi\,R_2/\varepsilon)} \tag{9}$$

This equation can be considered almost exact also when eq. (8) could be considered only approximately valid.

At this point, it is obvious that, in possible comparisons between two different experimental works, whose results are reported in terms of ε against n, it is necessary to use both eqs. (8) and (9) in order to correctly perform the same comparisons.

     S. de Felice, P. Gualtieri and G.P. Doria

7. Comparison with Manning's *n* Literature Data

As it has been already stressed, the very little value of roughness coefficients, obtained in the case of smooth bed (Table 3), represents a first confirmation of the validity of the methodology.

But, in any case, another comparison was performed with Manning's *n* experimental data represented in a graph (Figure 3) by Lopez and Garcia [22], relative to experiments carried out with rigid submerged vegetation, realized through vertical cylinders, always 10cm height but with different densities and with currents depths of about 0.50m.

In this graph, in the abscissa, the projected area of vegetation per unit volume of water in the flow direction (vegetation density) and in the ordinate the Manning's *n* values are reported. For every value of vegetation density a single experimental point is plotted as a black round point. These points appear to be aligned along a horizontal line up to a value of about 0.025 for the vegetation density, and afterwards along an inclined line which is represented on the graph.

In the cases of authors experimental data, the projected area of vegetation per unit volume of water in the flow direction was worth 3.2m^{-1}, in the case of single density, and 6.4m^{-1} in the case of double density. Two dotted vertical lines, representing these values, were reported in the graph by Lopez and Garcia. Starting from ε_{cal} values from Table 3, through eq. (8) seven *n* values were obtained (third column of Table 4). Through eq. (9) it was possible to evaluate the ratio between *n* values relative to Lopez and Garcia current and *n* values relative to authors' current (fourth column of Table 4). Finally, *n* values relative to Lopez and Garcia current was calculated (last column of Table 4).

The six obtained values relative to vegetated beds are reported in the Lopez and Garcia graph in the shape of empty triangles, circles and squares placed along the verticals relative to their densities.

Table 4. Mannings' n values for the flow conditions relative to Lopez and Garcia current depth.

		ε_{cal} (m)	$R = h_0$(m)	n_{auth} (s/m$^{1/3}$)	$n_{L\&G}/n_{auth}$ (s/m$^{1/3}$)	$n_{L\&G}$ (s/m$^{1/3}$)
Smooth bottom		0.000082	0.00481	0.009300	1.169161	0.01087
Single Density	5	0.005957	0.05083	0.016992	0.981663	0.01668
	10	0.018846	0.05507	0.022371	0.893990	0.02000
	15	0.034700	0.05716	0.026796	0.834310	0.02236
Double Density	5	0.008972	0.05050	0.018640	0.951005	0.01773
	10	0.026943	0.05428	0.024901	0.855992	0.02132
	15	0.048393	0.05727	0.030113	0.793369	0.02389

Moreover, the two interpolating laws (6) and (7) were used to extrapolate the absolute roughness values to the case of 10cm high cylinders, obtaining 0.205m for the single density and 0.383m for the double density. It is clearly possible, starting from these values, and using eq. (8) with Lopez and Garcia current depth, to obtain the correspondent Manning's *n* values, namely 0.034 for the single density and 0.040 for the double density, represented as red circles in Figure 3.

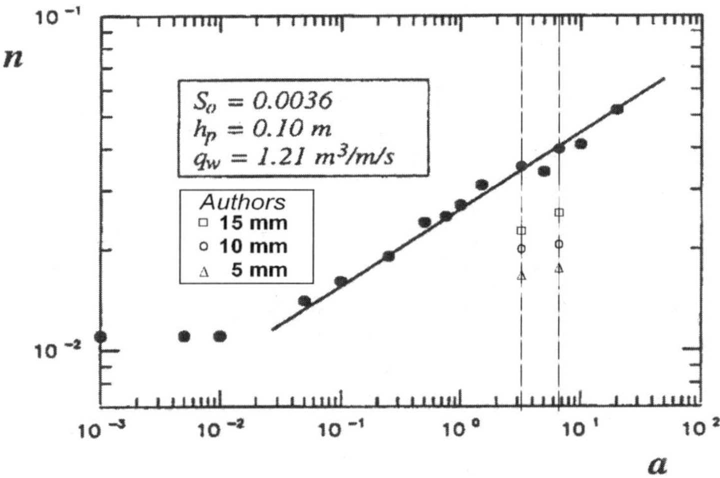

Figure 3. Comparison with Lopez and Garcia *n* values.

As final comparison, it is possible to stress the following observations.

The *n* value of the smooth bed is wholly correspondent to such a physical condition. The six *n* values of the vegetated beds are still in agreement with the Lopez and Garcia data, in the sense that they increase with increasing height and density of the cylinders, but always remain less than the Lopez and Garcia *n* values because these last ones are relative to the corresponding cylinder densities but to cylinder height 10cm. Finally, the two extrapolated *n* values fit very well with Lopez and Garcia *n* values.

8. Comparison Between *f* and *ε* Values in Single and Double Density

In this paragraph a supplementary observation about the obtained results data will be stressed.

The vegetated beds, that were taken into consideration, had three different heights, and two different densities. If we compare either the two friction factors

(Table 1) or the two absolute roughness (Table 3) relative to the two densities with the same cylinders height, we can observe that, in both cases, one is not double than the other (Table 5).

In particular, the doubling of density should double the friction factor in currents with the same depth and the same mean velocity. In the cases considered, the two currents compared each time have almost the same depth and almost the same mean velocity, but the friction factors differ from each other by a factor lesser than two. Clearly the absolute roughness ratios differ from the friction factors ratios due to the shape of eq. (4).

Table 5. Ratios between f and ε_{cal} values.

Cyl. height	$f_{s.d.}$	$f_{d.d.}$	$f_{s.d.}/f_{d.d.}$	$\varepsilon_{s.d.}$ (m)	$\varepsilon_{d.d.}$ (m)	$\varepsilon_{s.d.}/\varepsilon_{s.d.}$
5mm	0.07783	0.09626	1.2368	0.005957	0.008972	1.5061
10mm	0.14202	0.18411	1.2964	0.018846	0.026943	1.4296
15mm	0.21518	0.28642	1.3311	0.034700	0.048393	1.3946

In a recent paper [8] the authors supposed that this behaviour would depend on the mutual influence of the wakes of the single cylinder, and experimentally investigated the length of its influence zone.

In particular, they placed a single cylinder 15mm high in a uniform motion current with about the same depth and the same mean velocity than the observed boundary layer current, and measured, always through LDA system, the instantaneous velocities along verticals placed just behind that cylinder at distances of 1.7cm, 2.2cm, 3.2cm, 5.2cm, 10.2cm and 15.2cm.

For each vertical, the distributions of local mean velocity and variance, skewness and kurtosis of the fluctuating velocities (disturbed distributions) were obtained.

The distributions of the same statistical quantities were obtained along the same verticals but without the cylinder (undisturbed distributions).

The difference from disturbed and undisturbed distributions was greatest for the verticals nearer to the cylinder, and much smaller for the verticals farther from the cylinder.

In particular, for local mean velocities and for variances of fluctuating velocities, it was defined an index $\Delta A/A_{no\ stick}$ ($A_{no\ stick}$ is the area subtended to the mean velocities or variances distributions without the cylinder; ΔA respectively its reduction or increase of the area subtended to the statistical distribution, due to the cylinder) that represented the discrepancy between the disturbed and undisturbed distribution.

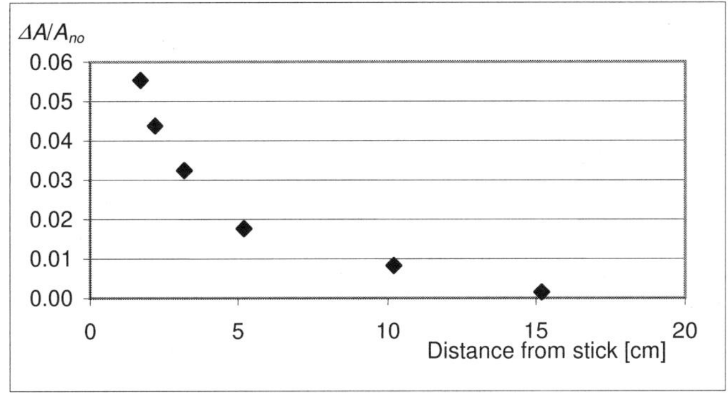

Figure 4. Discrepancy index for mean velocity versus the distance from the cylinder.

The trend of this index versus the distance from the cylinder is shown in Figures 4 and 5 where, according to [8], the cylinders were called "sticks".

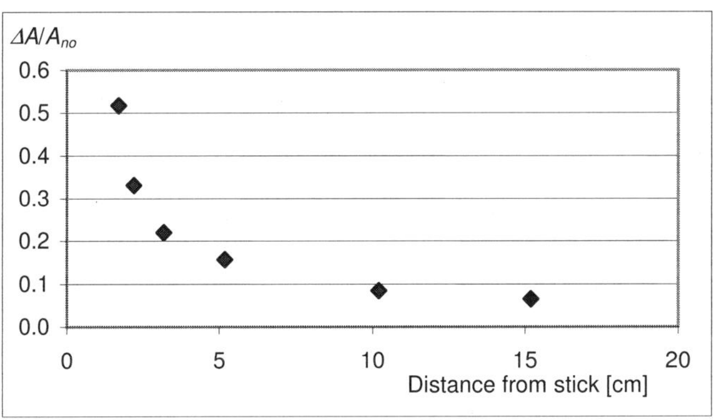

Figure 5. Discrepancy index for fluctuation velocities variances versus the distance from the cylinder.

Therefore, the results demonstrate that the wake of a single cylinder attains about 15cm after the same cylinder so that it affects at least three cylinders in single density, and even six cylinders in double density.

9. Conclusions

In this paper the calibration of a simplified methodology of evaluating the resistance coefficients of water currents, through experimental measurements carried out on a boundary layer stream instead of a uniform one, was presented.

                  S. de Felice, P. Gualtieri and G.P. Doria

The calibration consisted in a coefficient that must be applied to the absolute roughness values directly evaluated through the simplified methodology.

The calibrated absolute roughness values well fit literature data relative to vegetated beds, and some of their features can be interpreted also in the light of physical considerations about the length of the influence zone of a single element of vegetation.

Consequently, instead of requiring a very long experimental channel, the whole methodology lets evaluate the absolute roughness of a vegetated bed within the boundary layer of the current, therefore, in a shorter experimental channel.

APPENDIX - List of symbols

List of Symbols		
Symbol	**Definition**	**Dimensions or Units**
A	area subtended to statistical distributions with the cylinder	$[L^2]$
$A_{no\ stick}$	area subtended to statistical distributions without the cylinder	$[L^2]$
C	Chézy flow resistance coefficient	$[L^{1/2}T^{-1}]$
J	Head slope	
K	Parameter in Colebrook equation	
R	Hydraulic radius	$[L]$
Re	Reynolds number	
S	Slope	
V	Longitudinal velocity	$[LT^{-1}]$
V_0	Longitudinal velocity in reference section of boundary layer	$[LT^{-1}]$
V_u	Longitudinal velocity in uniform flow	$[LT^{-1}]$
a	Parameter in Colebrook equation	
b	Parameter in Colebrook equation	
f	Friction factor	
$f_{d.d.}$	Friction factor (double density vegetation)	
$f_{s.d.}$	Friction factor (single density vegetation)	
g	Gravitational acceleration	$[LT^{-2}]$

h	Flow depth	[L]
h_0	Flow depth in reference section of boundary layer	[L]
h_u	Flow depth in uniform flow	[L]
n	Manning roughness factor	$[L^{-1/3}T]$
n_{auth}	Manning roughness factor relative to authorss current	$[L^{-1/3}T]$
$n_{L\&G}$	Manning roughness factor relative to L. & G. current	$[L^{-1/3}T]$
s	Abscissa along the channel	[L]
ΔA	Reduction or increase of area subtended to statistical distribution due to the cylinder	$[L^2]$
α	α Coriolis coefficient	
ε	Absolute roughness	[L]
ε_{bl}	Absolute roughness (boundary layer experiments)	[L]
ε_{cal}	Calibrated absolute roughness	[L]
$\varepsilon_{d.d.}$	Absolute roughness (double density vegetation)	[L]
$\varepsilon_{s.d.}$	Absolute roughness (single density vegetation)	[L]
ε_{unif}	Absolute roughness (uniform flow)	[L]
ψ	Shape parameter in Colebrook equation	

References

1. Baptist M.J., V. Babovic, J. Rodriguez Uthurburu, M Keijzer, R.E. Uittenbogaard, A Mynett, and A. Vevey, On inducing equations for vegetation resistance, Journal of Hydraulic Research, **45** (4) pp.435-450, 2007

2. Ben Meftah M., De Serio F., Malcangio D., Petrillo A. and Mossa M., Experimental study of flexible and rigid vegetation in an open channel, *Proceedings of River Flow 2006*, Lisbona, Portugal, 2006

3. Chow Ven Te, Open-Channel Hydraulics, McGraw-Hill Classic Textbook Reissue Series, 1950

4. Cui J., Large-eddy simulation of turbulent flow over rough surfaces, PhD dissertation Mechanical Engineering, University of Iowa, Iowa City, Iowa, USA, 2000

5. Cui J. and Neary V.S., Large eddy simulation (LES) of fully developed flow through vegetation, *Proceedings of the 5th International Conference on Hydroinformatics*, Cardiff, UK, 2002

6. De Felice, S. and Gualtieri P., Interactions between turbulent Boundary Layer and rigid vegetated beds: Comparison among hydrodynamic characteristics in different points around the cylinders, *I.A.H.R. Congress, Seoul (Korea)*, pp.2434-2445, 2005

7. De Felice, S., Gualtieri P., Pulci Doria G., A simplified experimental method to evaluate equivalent roughness of vegetated river beds, Int. Congr. on Environ. Model. and Soft. 4th Biennial Meeting of iEMSS, published on CD-ROM 11 pages, July 2008

8. De Felice, S., Gualtieri P., Pulci Doria G., Influence field of a single rigid submerged stick in a liquid current and its relations with a sticks array, *Proceedings of International Symposium on EchoHydraulics*, Concepcion (Chile), January 2009

9. Defina A. and Bixio A.C., Mean Flow and turbulence in vegetated open channel flow, *Water Resources Research*, **41**(7), 2005

10. Finnigan j., Turbulence in plant canopies, *Annual Review of Fluids Mechanics*, **32**(1), pp.519-571, 2000

11. Fischer-Antze T., T. Stoasser, P. Bates & N.R.B. Olson, 3D numerical modeling of open-channel flow with submerged vegetation, *Journal of Hydraulic Research* **39**(3) pp.303-310, 2001

12. Freeman G.E., R.E. Copeland, W. Rahmeyer, D.L. Derrick, Field determination of Manning's n value for shrubs and woody vegetation, *Engineering Approaches to Ecosystem Restoration, Proc. Wetlands Engrg. And River restoration, Conf. ASCE*, New York, USA, 1998

13. Ghisalberti M. and Nepf H., The structure of shear layer flow over rigid and flexible canopy, *Environmental Fluid Mechanics*, **6**, pp.277-301, 2006

14. Gioia G., and F.A. Bombardelli, Scaling and similarity in rough channel flows, *Physical Review Letters*, **88**(1), pp.14501-14504, 2002

15. Green J.C., Modelling flow resistance in vegetated streams: review and development of a new theory, *Hydrological Processes*, **19**, pp. 1245-1259, 2005

16. Gualtieri, P. and Pulci Doria G., Boundary layer development over a bed covered with rigid submerged vegetation, *Fluid Mechanics of Environmental Interfaces*, Taylor&Francis, Leiden, The Netherlands pag 241-298, 2008

17. Huthoff F., Augustijn D.C.M. and Hulscher S.J.M.H., Depth-averaged flow in presence of submerged cylindrical elements, *Proceedings of River Flow 2006*, Lisbona, Portugal, 2006

18. Huthoff F., Augustijn D.C.M. and Hulscher S.J.M.H., Analytical solution for the depth-averaged flow velocity in case of submerged rigid cylindrical vegetation, *Water Resources Research*, **43**, W06413, 2007

19. Keijzer M., Baptist M., Babovic V. and Uthurburu J.R., Determining equation for vegetation induced resistance using genetic programming, *Proceedings of GECCO'05*, Washington, DC, USA, 2005

20. Klopstra D., Barneveld H.J., Van Noortwijk J.M. and Van Velzen E.H., Analytical model for hydraulic roughness of submerged vegetation, *Proceedings of the 27th IAHR Congress*, San Francisco, USA, pp.775-780, 1997

21. Kutija V. and Hong H.T.M., A numerical model for assessing the additional resistance to flow introduced by flexible vegetation, *Journal of Hydraulic Research*, **34**(5), pp.99-114, 1996

22. Lopez F. and Garcia M., Open-Channel Flow Through Simulated Vegetation: Turbulence Modeling and Sediment Transport, *US Army of Engineers Waterway Experiment Station Wetlands Research Program Technical Report WRP-CP-10*, 1997

23. Lopez F. and Garcia M., Open channel flow through simulated vegetation: Suspended sediment transport modeling, *Water Resources Research*, **34**(9), pp.2341-2352, 1998

24. Lopez F. and Garcia M., Mean Flow and Turbulence Structure of Open-Channel Flow Through Non-Emergent Vegetation, *Journal of Hydraulic Engineering*, **127**(5), pp.392-402, 2001

25. Marchi, E., Il moto uniforme delle correnti liquide nei condotti chiusi ed aperti, *L'Energia Elettrica*, **23** (4)(5), 1961

26. Mejier D.G. and Van Velzen E.H., Prototype-scale flume experiments on hydraulic roughness of submerged vegetation, *Proceedings of the 28th IAHR Congress*, Graz, Austria, 1998

27. Neary V.S., Numerical model for open-channel flow with vegetative resistance, *IAHR's 4th International Conference on Hydroinformatics*, july 23-27, Cedar Rapids, Iowa, USA, 2000

28. Neary V.S., Numerical solution of fully developed flow with vegetative resistance, *Journal of Engineering Mechanics*, **129**(5), pp.558-563, 2003

29. Nepf H.M., and E.R. Vivoni, Drag, turbulence and diffusion in flow through emergent vegetation, *Journal of Geophysical Research*, 105(2), pp.479-489 , 1999

30. Nezu I., Sanijou M. and Okamoto T., Turbulent structure and dispersive properties in vegetated canopy, *Proceedings of River Flow 2006*, Lisbona, Portugal, 2006

31. Nezu I., Sanijou M., Turbulent structure and coherent motion in vegetated canopy open-channel flows, *Journal of Hydro-environment Research, 2*, pp.62-90, 2008

32. Patel V.C., and Yoon J.Y., Application of turbulence models to separated flows over rough surfaces, *Journal of Fluids Engineering* 117(2), pp. 234-241, 1995

33. Poggi D., Porporato A. and Ridolfi L., Albertson J.D., Katul G.G., The effect of vegetation on canopy sub-layer turbulence, *Boundary-Layer Meteorology*, **111**, pp.565-587, 2004
34. Raupach M.R. and Shaw R.H., Averaging procedures for flow within vegetation canopies, *Boundary Layer Meteorology*, **22**, pp.79-90, 1982
35. Raupach M.R., and Thom A.S., Turbulence in and above plant canopies, *Annual Review of Fluids Mechanics*, **13**, pp.97-129, 1981
36. Stone B.M. and Tao Shen H., Hydraulic Resistance of Flow in Channels with Cylindrical Roughness, *Journal of Hydraulic Engineering*, **128**(5), pp.500-506, 2002
37. Tsujimoto T., and Kitamura T., Velocity profile of flow in vegetated-bed channels, *KHL Progressive Report, Hydraulic Laboratory, Kanazawa University,* 1990
38. Tsuijimoto, T., Shimizu, Y., Kitamura, T. and Okada, T., Turbulent open-channel flow over bed covered by rigid vegetation, *Journal of Hydroscience and Hydraulic Engineering*, **10**(2), pp.13-25, 1992
39. Yen B.C., Open channel flow resistance, *Journal of Hydraulic Engineering*, **128**(1), pp. 20-39, 2002
40. Wilkerson G., Flow through trapezoidal and rectangular channels with rigid cylinders, Journal of Hydraulics Engineering, 133 (5) pp.521-533, 2007
41. Wilson N.R. and Shaw R.H., A higher closure model for canopy, *Journal of Applied Meteorology,* **16**, pp.1197-1205, 1977
42. Zhang J., Numerical model for flow motion over vegetation, Journal of Hydrodynamics, **20** (2), pp.172-178, 2008

Index